Advances in Experimental Medicine and Biology

Volume 1011

More information about this series at http://www.springer.com/series/5584

Bin Li • Fan Pan
Editors

Immune Metabolism in Health and Tumor

Springer

Editors
Bin Li
Shanghai Institute of Immunology, Department of Immunology and Microbiology
Shanghai Jiao Tong University
Shanghai, China

School of Medicine
Shanghai Jiao Tong University
Shanghai, China

CAS Center for Excellence in Molecular Cell Science, CAS Key Laboratory of Molecular Virology and Immunology, Unit of Molecular Immunology, Institut Pasteur of Shanghai, Shanghai Institutes for Biological Sciences
University of the Chinese Academy of Sciences, Chinese Academy of Sciences
Shanghai, China

Fan Pan
Immunology and Hematopoiesis Division, Department of Oncology, Sidney Kimmel Comprehensive Cancer Center
Johns Hopkins University School of Medicine
Baltimore, MD, USA

ISSN 0065-2598 ISSN 2214-8019 (electronic)
Advances in Experimental Medicine and Biology
ISBN 978-94-024-1168-3 ISBN 978-94-024-1170-6 (eBook)
DOI 10.1007/978-94-024-1170-6

Library of Congress Control Number: 2017951122

Printed on acid-free paper

This Springer imprint is published by Springer Nature
The registered company is Springer Science+Business Media B.V.
The registered company address is: Van Godewijckstraat 30, 3311 GX Dordrecht, The Netherlands

Contents

Chapter 1
Metabolism in Immune Cell Differentiation and Function

Nicole M. Chapman, Sharad Shrestha, and Hongbo Chi

Abstract The immune system is a central determinant of organismal health. Functional immune responses require quiescent immune cells to rapidly grow, proliferate, and acquire effector functions when they sense infectious agents or other insults. Specialized metabolic programs are critical regulators of immune responses, and alterations in immune metabolism can cause immunological disorders. There has thus been growing interest in understanding how metabolic processes control immune cell functions under normal and pathophysiological conditions. In this chapter, we summarize how metabolic programs are tuned and what the physiological consequences of metabolic reprogramming are as they relate to immune cell homeostasis, differentiation, and function.

Keywords Metabolism • T cells • Treg cells • NK cells • B cells • mTOR • AMPK

1.1 Introduction

The immune system is comprised of the innate and adaptive immune cells, which develop from bone marrow-derived progenitors cells. The response of the innate immune cells is more rapid than adaptive immune cells. Innate immune cells are activated by germ line-encoded receptors, including various pattern recognition receptors and cytokine receptors. The engagement of these and other receptors allows innate immune cells to engulf pathogens and other foreign antigens, to produce antimicrobial products, and to secrete cytokines and chemokines. Innate immune cells, especially dendritic cells (DCs), can serve as antigen-presenting cells (APCs), which process and present acquired antigens in the context of short peptides on major histocompatibility complex (MHC) molecules. Trafficking of these cells to secondary lymphoid tissues (e.g., spleen, lymph nodes) allows them to

N.M. Chapman • S. Shrestha • H. Chi (✉)
Department of Immunology, St. Jude Children's Research Hospital,
262 Danny Thomas Place, Memphis, TN 38105, USA
e-mail: Hongbo.Chi@stjude.org

B. Li, F. Pan (eds.), *Immune Metabolism in Health and Tumor*, Advances in Experimental Medicine and Biology 1011, DOI 10.1007/978-94-024-1170-6_1

interact with adaptive immune cells. While the innate immune response is rapid, adaptive immunity takes days to weeks to form. T and B cells make up the adaptive immune system, and both of these cell populations express antigen receptors, called the T cell antigen receptor (TCR) and the B cell antigen receptor (BCR). The TCR must bind antigens expressed in the context of MHC molecules, whereas BCR engagement is MHC independent. These antigen receptors ensure that the adaptive immune response is selective and specific. Additionally, both T and B cells require co-stimulatory receptor engagement for maximum activation, and cytokines further shape their specific functions. The coordinated actions of the innate and adaptive immune responses promote pathogen clearance and the formation of long-term immunological memory, which allows for rapid immune responses following reinfection. However, dysregulated immune cell responses are connected to many diseases, ranging from tumorigenesis to autoimmunity. Therefore, it is of interest to understand how immune responses are regulated at the steady state and in various disease states.

Metabolism is the net result of both the breaking down (catabolic) and de novo synthesis (anabolic) of nutrients. In addition to providing the cell with energy, metabolic by-products from one pathway can tune other metabolic programs. Metabolites also serve as important regulators of gene transcription and protein translation, localization, activity, and expression. Metabolism plays a crucial role in shaping immune cell differentiation and function. In this chapter, we provide an overview of metabolic processes used by immune cells and how metabolic reprogramming is regulated at the molecular level (Figs. 1.1 and 1.2). We also summarize our current understanding of how metabolism shapes immune cell differentiation and function, with a particular emphasis being placed upon macrophages and DCs, natural killer (NK) cells, conventional T cells, and regulatory T (Treg) cells (Figs. 1.3, 1.4, and 1.5). We conclude with a brief discussion on how metabolism regulates B cell responses and future challenges facing the immunometabolism field.

1.2 Mechanistic Regulation of Cellular Metabolism

1.2.1 Overview of Catabolic Metabolism

Immune cells utilize adenosine triphosphate (ATP) derived from diverse nutrients as an energy source to support their differentiation and specialized functions. Glucose is one major source of ATP for both resting and activated immune cells. After it has been taken up into the cell, glucose is converted into glucose-6-phosphate (G6P) by hexokinase (HK), of which there are four isoforms: HK1, HK2, HK3, and glucokinase (GCK; also called HK4). During glycolysis, several important intermediaries are produced, including 3-phosphoglycerate (3PG), phosphoenolpyruvate (PEP), and nicotinamide adenine dinucleotide (NADH) (Fig. 1.1). Glycolysis culminates in the production of two molecules of ATP and pyruvate, a metabolite that can be

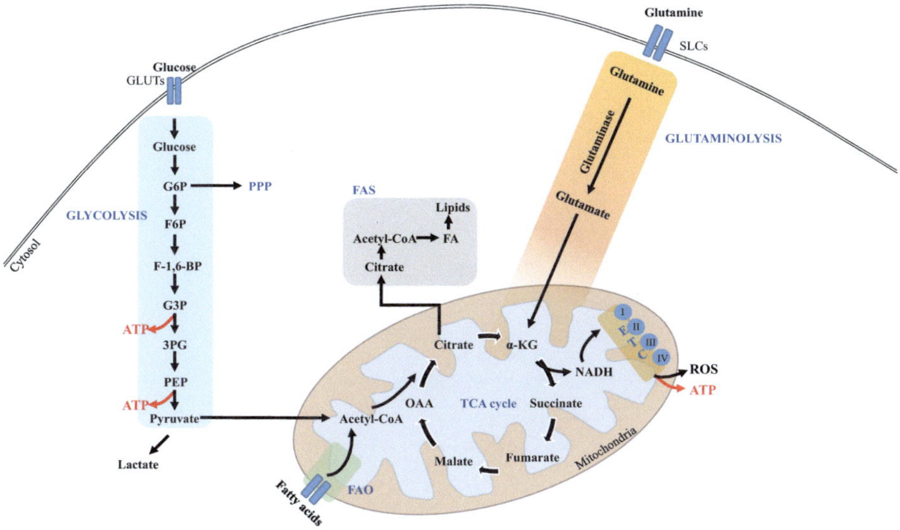

Fig. 1.1 Overview of cellular metabolism
The major metabolic pathways of immune cells are glycolysis, glutaminolysis, fatty acid synthesis (*FAS*), fatty acid oxidation (*FAO*), and oxidative phosphorylation (*OXPHOS*) fueled by the tricarboxylic acid (*TCA*) cycle. The pentose phosphate pathway (*PPP*) also controls select immune cell functions. Glycolysis, FAS, and glutaminolysis reactions occur in the cytosol, while the TCA and OXPHOS occur in the mitochondrial matrix. These pathways generate metabolites critical for multiple cellular functions and also produce cellular energy in the form of adenosine triphosphate (*ATP*). The reader is encouraged to visit Sect. 1.1 for a more detailed discussion of these pathways. *Abbreviations*: *G6P* glucose-6-phosphate, *F6P* fructose-6-phosphate, *F-1,6-BP* fructose-1, 6-bisphosphate, *G3P* glyceraldehyde-3-phosphate, *3-PG* 3-phosphoglycerate, *PEP* phosphoenolpyruvate, *FA* fatty acid, *ROS* reactive oxygen species, *SLCs* solute carrier family of amino acid transporters, *GLUTs* glucose transporters

further processed into lactate via lactate dehydrogenase (LDH). Glycolysis is favored under conditions where oxygen is limiting. However, seminal studies by Otto Warburg demonstrated that highly proliferative cells convert glucose into lactate in the presence of oxygen, a phenomenon termed the Warburg effect or aerobic glycolysis. Roles for aerobic glycolysis for energy production and for providing metabolic by-products that support immune cell functions have been recently uncovered, which we discuss throughout this chapter.

Pyruvate derived from aerobic glycolysis can also be converted to ATP energy via the tricarboxylic citric acid (TCA) cycle (also known as the citric acid cycle or the Krebs cycle) and mitochondrial oxidative phosphorylation (OXPHOS) (Fig. 1.1). To enter into the TCA cycle, pyruvate is transported into the mitochondria and oxidized to generate acetyl coenzyme A (acetyl-CoA). Citrate synthase then combines acetyl-CoA with oxaloacetate to generate citrate. During this multistep process, NADH and flavin adenine dinucleotide ($FADH_2$) are produced. These products are critical electron donors for the electron transport chain (ETC) that generates ATP via OXPHOS. The ETC is comprised of five protein complexes that shuttle electrons

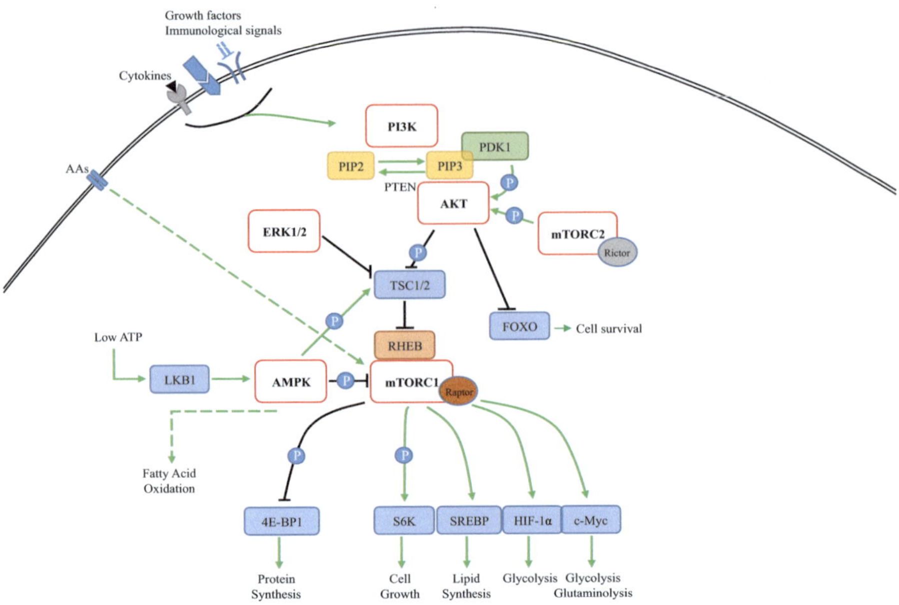

Fig. 1.2 Kinases and transcription factors cooperatively regulate immune cell metabolism
Immune cells receive immunological and environmental cues (growth factors, nutrients) that tune metabolic pathways. The alterations in cellular metabolism are shaped by intracellular serine/threonine kinases, including ERK1/2, AKT, mTORC1, mTORC2, LKB1, and AMPK. Additionally, the lipid kinase PI3K is a crucial regulator of immune responses, because it can modulate AKT, mTORC1, and mTORC2 activities, among many others. The mTORC1 pathway is a major determinant of metabolic fitness in immune cells. Upon its activation, mTORC1 phosphorylates 4E-BP1 and S6K to influence protein translation. Additionally, it induces expression of key metabolic enzymes, including c-MYC, HIF-1α, and SREBPs. The reader should refer to Sect. 1.2 for more details about how these pathways are tuned

through the inner mitochondrial membrane. NADH donates an electron to complex I; $FADH_2$ donates an electron to complex II and another electron to succinate to generate fumarate to feed back into the TCA cycle. Electrons transferred to complexes I and II are shuttled to complex III via coenzyme Q (also known as ubiquinone). Cytochrome c then transmits the electrons to complex IV. Electron movement across complexes I, III, and IV is coupled to proton pumping from the inner mitochondrial membrane matrix into the intermembrane space and thus creates a gradient that enables complex V (also known as ATP synthase) to produce ATP [417]. How OXPHOS controls immune responses is discussed below.

Reactive oxygen species (ROS) are produced during the process of mitochondrial OXPHOS. Superoxide is generated when electrons are not efficiently passed along the ETC and transferred to oxygen. Superoxide generated from complex I or complex II is released into the mitochondrial matrix, while superoxide generated from complex III can be present in either the matrix or intermembrane space. Superoxide dismutase (SOD) localized in either the matrix (SOD2) or intermembrane space (SOD1) can convert superoxide to hydrogen peroxide, which readily crosses the inner and outer mitochondrial membranes and can participate in antimicrobial responses or peroxisome metabolism. Alternatively, superoxide found in the intermembrane space can be exported into the cytosol via a voltage-dependent anion exchange channel [417]. Metabolic and cellular stresses that accompany active immune responses increase superoxide production, the biological consequences of which are discussed below.

Pyruvate is also generated from fatty acids via the process of fatty acid β-oxidation (FAO). This catabolic process occurs within the mitochondria. To begin FAO, free fatty acids are modified by fatty acyl-CoA synthetase (FACS; also called fatty acyl-CoA synthase) to form a fatty acyl-CoA. This product is not permeable to the mitochondrial membrane, so carnitine palmitoyltransferase 1 (CPT1) adds a carnitine moiety onto the fatty acyl-CoA to generate fatty acylcarnitine. The carnitine translocase (CAT) allows the fatty acylcarnitine to enter into the inner mitochondrial membrane via carnitine exchange. Within the inner mitochondrial membrane, CPT2 converts the fatty acylcarnitine back into fatty acyl-CoA. FAO is complete when the fatty acyl-CoA is converted into acetyl-CoA, which enters into the TCA cycle. NADH and $FADH_2$ are also produced during this process to feed the ETC-OXPHOS pathway and generate ATP (Fig. 1.1).

Dietary proteins can also be used as an energy source. Proteins are first hydrolyzed into individual amino acids, which can feed into various parts of the TCA cycle to fuel ATP production. Alanine, glycine, threonine, cysteine, serine, and tryptophan can be converted directly into pyruvate. Additionally, asparagine and aspartate can be used to produce oxaloacetate, which is converted into PEP by the mitochondrial enzyme phosphoenolpyruvate carboxykinase 2 (PCK2). PEP can then be subsequently converted back into pyruvate to generate acetyl-CoA. Catabolism of additional amino acids, including arginine, glutamine, and glutamate, also produces TCA cycle intermediates. The breakdown of glutamine via glutaminolysis is crucial for immune cell biology. During this process, glutamine is converted into α-ketoglutarate (α-KG) via a two-step process requiring the activities of glutaminase (GLS), which converts glutamine to glutamate, and glutamate dehydrogenase (GDH), which converts glutamate to α-KG [128, 404]. This TCA cycle intermediate can then fuel OXPHOS for cellular energy. Because glutaminolysis can produce TCA cycle intermediates, it can also promote anabolic processes by serving as a carbon or nitrogen donor to support cell growth and proliferation. We discuss how amino acids contribute to biosynthesis in more detail below.

1.2.2 Overview of Anabolic Metabolism

Anabolic processes are also important for immune cell fitness. Aside from entering into glycolysis, G6P can also be shuttled into the cytosolic pentose phosphate pathway (PPP) after its conversion into 6-phosphogluconolactone by G6P dehydrogenase (G6PD). NAD^+ and NADPH are two by-products of the PPP. NAD^+ is an important electron acceptor for catabolic processes driving OXPHOS, while NADPH is utilized in reducing reactions important for the production of pentose sugars, nuclear hormone receptor ligands, and fatty acids. Derivatives from the PPP can also integrate back into glycolysis to feed ATP production and are also important for generating nucleotides. We will discuss how the PPP influences immune cell functions in this chapter. The glycolytic by-products 3PG and pyruvate are important for serine, cysteine, glycine, and alanine biosynthesis. Moreover, oxaloacetate and α-KG generated via the TCA cycle are important for aspartate, asparagine, proline, and arginine synthesis. Thus, glycolysis and TCA cycle intermediates support cell growth and functions through both catabolic and anabolic processes.

Fatty acid and cholesterol synthesis are crucial anabolic processes supporting cellular functions. To initiate fatty acid synthesis (FAS) in the cytosol, acetyl-CoA is shuttled out of the mitochondria by citrate and subsequently converted into malonyl-CoA via acetyl-CoA carboxylase (ACC) in a manner that requires biotin (also called vitamin B_7). Malonyl-CoA is then reduced by fatty acid synthase (FASN) and its cofactor NADPH to generate palmitate, a 16-carbon fatty acid. Among several functions, palmitate serves as the backbone for other long-chain fatty acids and phospholipids and can also modify proteins, a process termed palmitoylation. Additionally, accumulation of palmitate perturbs FAS via the feedback inhibition of ACC. Acetyl-CoA can also serve as a substrate for cholesterol biosynthesis. The sequential activities of thiolase, 3-hydroxy-3-methylglutaryl-CoA (HMG-CoA) synthase (HMGCS), and HMG-CoA reductase (HMGCR) convert acetyl-CoA into mevalonate, the metabolic precursor required for cholesterol synthesis. Cholesterol biosynthesis is important for maintaining the plasma membrane and for producing sterol hormones, vitamin D, and various oxysterols [e.g., 25-hydroxycholesterol (25-HC)]. Mevalonate-derived isoprenoids produced during cholesterol biogenesis also serve important cellular functions. For instance, isoprenoids modify proteins in a process termed prenylation [85]. Statins are effective inhibitors of HMGCR function and thus block cholesterol and isoprenoids-dependent functions. How FAS and cholesterol biosynthesis control immune cell fate is discussed throughout this chapter.

Serine metabolism is also crucial for anabolic processes supporting nucleotide and lipid biosynthesis. Glucose-derived 3PG serves as a precursor for serine biosynthesis, whose rate-limiting step is catalyzed by phosphoglycerate dehydrogenase (PHGDH). In the presence of glutamine-derived glutamate, 3-phosphohydroxypyruvate is catalyzed into 3-phosphoserine via the enzymatic activity of phosphoserine aminotransferase (PSAT). This reaction also generates α-KG that can be used for other cellular purposes. Dephosphorylation of 3-phosphoserine generates serine, which is important for one-carbon metabolism [306]. One-carbon metabolism is a process whereby a carbon unit derived from serine or glycine is cycled through the folate and methionine cycles to fuel biosynthetic pathways. This pathway aids in lipid, nucleotide, and protein biosynthesis and the generation of products important for redox reactions (i.e., those that neutralize ROS species) and methylation reactions (e.g., for epigenetic modifications like DNA methylation) [232, 306]. Dietary folate (also called vitamin B_9) is essential for one-carbon metabolism. After entering the cell, folate is reduced to tetrahydrofolate (THF). Then, a carbon unit from serine or glycine is donated to THF via the enzymatic activity of serine hydroxymethyl transferase (SHMT), forming methylene-THF (me-THF) and glycine. In addition to folate, vitamins B_2, B_6, and B_{12} serve as important cofactors for these reactions. How serine and one-carbon metabolism regulate cancer and immune cell biology is under active investigation [232, 306].

1.2.3 Nutrient Transporters Regulate Catabolic Metabolism

To initiate catabolic or anabolic processes, nutrients must be made available to the cells. Extracellular nutrients are delivered to cells via nutrient transporters. Glucose transporters (GLUTs) mediate glucose uptake into many cell types. GLUT1 is the preferential transporter expressed by immune cells, but other GLUTs can play redundant or compensatory in the absence of GLUT1. Glucose diffusion via GLUTs is a passive process, so the expression of GLUTs in the plasma membrane is tightly controlled to regulate glucose import. In activated T cells, for example, strong TCR or suboptimal TCR signals combined with CD28 co-stimulatory signals upregulate GLUT1 expression [121, 173]. Then, GLUT1 translocates from the cytosol to the membrane via mechanisms requiring CD28 co-stimulation and phosphatidylinositol-3-kinase (PI3K)-AKT activation [418]. The physiological consequences of altering glucose transport into immune cells are discussed in this chapter.

Extracellular fatty acids must enter into the cytoplasm for transport into the mitochondria for FAO. Free fatty acids can readily diffuse across the plasma membrane into the cytosol. However, most fatty acids require transport facilitated by protein surface receptors. Fatty acid transport proteins (FATPs) are transmembrane proteins involved in fatty acid uptake, but these proteins do not appear to be expressed at high levels in immune cells [99]. Instead, fatty acid translocase (FAT or CD36) facilitates long-chain fatty acid and oxidized fatty acid transport across the plasma membrane [354]. Additionally, G protein-coupled receptors can recognize fatty acids of different lengths and promote their transport into immune cells. GPR40 [also known as free fatty acid receptor 1 (FFA1)] and GPR120 bind long-chain fatty acids; GPR84 recognizes medium-chain fatty acids; and short-chain fatty acids (SCFAs) are ligands for GPR43 (also known as FFA2) and GPR41 (also known as FFA3) [87]. The roles of these transporters as they relate to immunity are discussed in later sections of this chapter.

Essential amino acids cannot be synthesized via intrinsic metabolic programs and must therefore be obtained from dietary sources. There are nine essential amino acids: histidine, isoleucine, leucine, lysine, methionine, phenylalanine, threonine, tryptophan, and valine. Under conditions of cellular stress, glutamine and other amino acid levels become limiting, so cells also utilize extracellular sources of these amino acids for energy and biosynthetic processes. The system L amino acid transporters allow neutral amino acids to enter into immune cells. LAT1 (also known as Slc7a5) is an anti-porter that transports leucine into the cell and glutamine out of the cell and is a crucial regulator of immune cell responses as discussed below [312]. ASCT2 (also known as Slc1a5) transports glutamine into cells and also has reported roles in immune responses [265]. Despite the importance of extracellular amino acids, the breakdown of amino acids by various enzymes, including cytosolic branched-chain aminotransferase (BCATc), indoleamine 2,3-dioxygenase (IDO), and arginase (Arg), also plays crucial roles in controlling metabolic programs and immune cell fates (Fig. 1.2).

1.2.4 Intracellular Kinases Regulate Metabolic Programs

Four major kinase pathways cooperate to control metabolic reprogramming in immune cells: PI3K, mechanistic target of rapamycin (mTOR), AMP-activated protein kinase (AMPK), and mitogen-activated protein kinases [MAPKs; including the extracellular-related kinases (ERKs), the p38 kinases, and the c-Jun N-terminal kinases (JNKs)]. Below, we review how these pathways are coupled to metabolic reprogramming (Fig. 1.2).

1.2.4.1 PI3K

Phospholipids are important second messengers that control protein localization, expression, and functions. The PI3K pathway is a major regulator of phospholipid turnover. The class I PI3Ks proteins are a heterodimer containing a regulatory subunit (p85α, p85β, and p55) and a catalytic subunit (p110α, p110β, p110δ, or p110γ). To promote PI3K activity, the Src homology 2 (SH2) domain of the PI3K regulatory subunit binds membrane-associated proteins containing a pYXXM motif, where pY is a phosphotyrosine, M is a methionine, and X is any amino acid. Alternatively, PI3K activity is triggered downstream of the membrane-anchored RAS GTPases (HRAS, NRAS, and KRAS), which directly bind to the RAS-binding domain (RBD) within the catalytic p110 subunit [61]. The p85 regulatory subunit also associates with GAB, which can bind growth factor receptor bound 2 (Grb2) and be indirectly recruited to membrane proteins containing pYXN motifs [61]. After binding the plasma membrane, the catalytic PI3K subunit converts phosphatidylinositol-(4,5)-bisphosphate (PIP_2) into phosphatidylinositol-(3,4,5)-trisphosphate (PIP_3). Among other functions, PIP_3 allows proteins containing pleckstrin homology (PH) domains to bind the plasma membrane, which modulates their enzymatic activity. Phosphatase and tensin homolog (PTEN) and SH2 domain-containing inositol 5′-phosphatase (SHIP) antagonize PI3K signaling by converting PIP_3 to PIP_2 and PI-(3,4)-P_2, respectively. These phospholipids and their by-products are also important signaling molecules for immune cells [338, 408].

AGC kinases are important regulators of cell proliferation, cell growth, cytoskeletal rearrangements, survival, and cellular metabolism. The activation of these kinases requires phosphorylation of both a residue in the activation segment of the kinase domain and a hydrophobic motif residue. PI3K-dependent mechanisms activate 3-phosphoinositide-dependent kinase 1 (PDK1). PDK1 phosphorylates other AGC kinases, including AKT, ribosomal S6 kinase (S6K, also called $p70^{S6K}$), serum- and glucocorticoid-regulated kinase (SGK), and protein kinase C (PKC). PDK1 is constitutively active, because it *trans*-autophosphorylates its own activation loop residue. PDK1 is expressed throughout the cell and is made assessable to its substrates via multiple mechanisms. The PH domain of PDK1 binds PIP_3 and, to a lesser extent, PIP_2, which helps localize PDK1 in proximity with its substrate AKT (isoforms AKT1, AKT2, and AKT3). Further, PDK1 binds soluble inositol phosphates in the cytosol. This association, coupled with substrate interactions mediated by its PIF (PDK1-interacting fragment) pocket, promotes the PDK1-dependent phosphorylation of proteins like S6K and SGK. PDK1-dependent functions are linked to metabolic reprogramming and immune cell functions as discussed below.

Like PDK1, AKT is recruited to the plasma membrane via PH domain-PIP_3 interactions. Binding of AKT to PIP_3 promotes a conformational change in AKT, which enables the PDK1-dependent phosphorylation of AKT threonine 308. The phosphorylation of AKT serine 473 by mTOR complex 2 (mTORC2) promotes maximal AKT activity by allowing the PDK1-PIF pocket to bind AKT S473, driving the phosphorylation of AKT T308. Of note, the activity of mTORC2 is also regulated by PIP_3 [127], although the receptor-specific activation mechanisms of mTORC2 activation remain unclear. This model for AKT activation is, however, context specific, as excess PIP_3 levels can drive AKT activation when mTORC2 activity is inhibited [166]. Glycogen synthase kinase 3 (GSK3), tuberous sclerosis 2 (TSC2), and forkhead box O (FOXO) proteins are substrates of AKT. Further, AKT activity influences glucose metabolism, in part by regulating the expression of GLUT1 and mTORC1 activity as discussed more below.

S6K and SGK do not have PH domains, but the phosphorylation of their activation loop and hydrophobic motif residues is linked to PI3K activity. The S6Ks are phosphorylated on their hydrophobic motif residues by mTOR complex 1 (mTORC1). Phosphorylation of the hydrophobic domain residue of S6K allows the PIF domain of PDK1 to bind S6K and phosphorylate its activation segment residue. Similarly, the mTORC2-dependent phosphorylation of the hydrophobic motif residue of SGK promotes PDK1 binding to and subsequent activation of SGK. The S6K signaling network influences mRNA processing, translation initiation and elongation, protein folding, and cell growth. The mTORC1-S6K axis also influences cellular metabolism by modulating the expression of sterol regulation element-binding proteins (SREBPs). SGK can influence cellular metabolism by phosphorylating FOXO transcription factors [43]. We discuss how mTORC1, mTORC2, SREBPs, and FOXOs govern metabolic processes below.

1.2.4.2 mTOR

The conserved serine/threonine kinase mTOR is another crucial regulator of immunometabolism. mTOR-induced signaling alters protein synthesis, lipid and energy metabolism, and autophagy, which ultimately help shape the lineage fate and function of immune cells. In mammalian cells, the mTOR protein associates with different proteins to form two multi-protein complexes, mTORC1 and mTORC2, defined by the obligate adaptor proteins, regulatory-associated protein of mTOR (RAPTOR) and rapamycin-insensitive companion of mTOR (RICTOR), respectively. The PI3K-AKT axis is among the best-studied nodes by which mTORC1 is activated, which is achieved through at least three known mechanisms: (1) AKT phosphorylates and subsequently impairs TSC complex activity. TSC, a heterodimer consisting of TSC1 and TSC2, is a key negative regulator of mTORC1 and functions as a GTPase-activating protein for Ras homolog enriched in brain (RHEB). GTP-bound RHEB strongly activates mTORC1 kinase activity; (2) AKT phosphorylates PRAS40, a negative regulator of mTORC1, to promote its degradation; (3) AKT directly phosphorylates RAPTOR, which stabilizes its association with mTOR. Additionally, PI3K-dependent and AKT-independent mechanisms involving PDK1-dependent mTORC1 activation have been reported [117, 414]. Because PTEN antagonizes PI3K signaling, it can inhibit mTOR activation. The LKB1-AMPK axis also suppresses mTOR activation, which we detail more below. In addition to the PIP_3-dependent regulation of mTORC2 discussed above, ribosome synthesis and assembly induce mTORC2 activity [452]. The activities of mTORC1 and mTORC2 themselves appear to be antagonistic, where loss of mTORC1 activity enhances mTORC2 functions and vice versa. This observation is important to note, because such regulation complicates interpretations as the relative roles mTORC1 and mTORC2 play in different cellular systems.

Amino acids also influence mTORC1 activation. In resting cells, mTORC1 is localized to the cytosol, where it is inactive. Different amino acids activate mTORC1 via discrete mechanisms reviewed extensively elsewhere [177, 349, 350]. Briefly, the vacuolar H^+-adenosine triphosphatase (v-ATPase), which associates with the RAGULATOR complex (comprised of LAMPTOR1-5) and the RAG heterodimers, senses amino acids in the lysosome lumen. Leucine binds and activates the RAG heterodimers, which are comprised of RAGA or RAGB bound to RAGC or RAGD. In these complexes, RAGA/B is loaded with guanosine triphosphate (GTP), and RAGC/D is bound with guanosine diphosphate (GDP). The RAGULATOR complex is the guanine nucleotide exchange factor (GEF) for RAGA/B [334, 349, 350], while the GTPase-activating protein (GAP) for RAGC/D is the folliculin-folliculin-interacting protein (FLCN-FNIP) complex [349, 350, 389]. The sestrin 2 (SESN2)-GATOR2-GATOR1 axis controls the GDP-bound status of RAGA/B [20, 302, 349, 350]. GATOR1 is the RAGA/B GAP, which is antagonized by GATOR2. In the absence of leucine, SESN2 inhibits GATOR2, thus favoring the GDP-bound state of RAGA/B. After the active RAG heterodimer binds mTORC1, it is recruited to GTP-loaded RHEB binds and activates mTORC1 [177, 335, 349, 350]. Of note, RAPTOR associates with the RAG complexes regardless of their guanine nucleotide binding status, albeit with different efficiencies [291]. In contrast to leucine, glutamine-induced mTORC1 activation occurs independently of the RAG or RAGULATOR complex but requires the v-ATPase and ADP-ribosylation factor 1 (ARF1), a Golgi-associated GEF [177, 350]. Arginine activates mTORC1 activity via a mechanism that appears to involve the arginine transporter SLC38A9, which associates with the v-ATPase [350, 406]. Interestingly, amino acids also activate mTORC2 via a mechanism that involves the PI3K-AKT pathway [382]. These data may mechanistically explain, in part, how amino acids can directly promote mTORC1 activation in the absence of growth factor receptor-mediated TSC complex inactivation [308].

The canonical substrates of mTORC1 are S6K and the eIF4E-binding proteins (4EBPs), which regulate protein translation and synthesis. mTORC1 coordinates cellular metabolism by activating key transcription factors, such as hypoxia-inducible factor-1α (HIF-1α) and c-MYC, which primarily regulate glycolytic enzymes and transporters [403, 414, 434]. Further, mTORC1 promotes the synthesis of fatty acids, cholesterol, and triglycerides by controlling the functions of the SREBPs. mTORC1 also regulates lipid homeostasis through peroxisome proliferator-activated receptor γ (PPARγ)-dependent mechanisms [215]. Inactivation of mTORC1 leads to inhibition of lipogenesis. The catabolic process of autophagy, a process wherein cytosolic contents (e.g., organelles, proteins) are sequestered and degraded for nutrient use, is sensitive to mTOR-dependent regulation [130, 195]. The role of autophagy in the immune system is discussed in a later chapter of this book.

Compared to mTORC1, relatively less is known about mTORC2, which is only sensitive to high-dose or long-term treatments of rapamycin [69, 215]. By driving AKT activity, mTORC2 inhibits the transcriptional activity of the FOXO transcription factors, which are essential for the expression of genes involved in cell survival, proliferation, and metabolism [54]. mTORC2 also influences cell cycle progression via AKT, which phosphorylates the cell cycle regulatory protein p27 [351], and by indirectly inhibiting the tumor suppressor p53 [400]. mTORC2-mediated control of AKT signaling can also modulate glucose uptake by regulating GLUT surface expression [29, 39, 418]. Additionally, mTORC2 influences cell survival, proliferation, growth, differentiation, and trafficking by modulating cytoskeletal dynamics or ion transport through the activation of PKC isoforms and SGK, respectively [69]. The roles of mTOR signaling in immunometabolism are discussed throughout this chapter.

1.2.4.3 AMPK

AMPK is one critical sensor of nutrient and intracellular energy levels in eukaryotic cells. The AMPK complex is comprised of one catalytic subunit (AMPKα) and two regulatory subunits (AMPKβ and AMPKγ). When intracellular ATP levels are low, AMP or ADP binds AMPKγ, which subsequently drives a conformation change that allows liver kinase B1 (LKB1; also known as STK11) to phosphorylate threonine 172 in the activation loop of the AMPK kinase domain. AMPK T172 phosphorylation is also triggered by calcium-dependent activation calcium/calmodulin-dependent protein kinase 2 (CaMK2; also known as CaMKβ) and by transforming growth factor beta-activated kinase 1 (TAK1; also known as MAP 3K7) in a manner that may also require LKB1. AMPK is activated under conditions of hypoxia and glucose deprivation and in response to inflammatory-inducing signals, such as antigen receptor or toll-like receptor (TLR) stimulation, in immune cells [32, 203, 212, 262, 370, 379]. Additionally, pharmacological agents like the glycolysis inhibitor 2-deoxyglucose (2-DG) and the AMP-mimic 5-aminoimidazole-4-carboxamide-1-b-d-ribofuranoside (AICAR) activate AMPK, while metformin inhibits AMPK function. Substrates of AMPK regulate pathways linked to transcription, cell polarity, cell growth, autophagy, and lipid or glucose metabolism. The processes are discussed below and in other resources [254].

Because it is activated in response to cellular stress and nutrient deprivation, catabolic metabolism is favored when AMPK activity is high. Fatty acid and cholesterol biosynthesis are halted by the AMPK-dependent inhibitory phosphorylation of ACC and HMGCR [123, 155]. Further, AMPK suppresses FAS by reducing the transcription of ACC and FASN via the phosphorylation of the transcription factor SREBP [225]. AMPK can also modulate metabolic gene expression at the epigenetic level of which readers are encouraged to explore in more detail elsewhere [254]. In other cell lineages, AMPK also promotes GLUT1 and GLUT4 surface expression to augment glycolysis needed to support energy production [254, 425], but the role of AMPK in glucose transport for immune cells is poorly defined. The phosphorylation of TSC2 by AMPK stabilizes TSC2 expression, which inactivates mTORC1 [194]. AMPK activity also increases nutrient availability by inducing autophagy [194]. We will describe how AMPK controls immunity throughout this chapter.

1.2.4.4 MAPKs

The MAPKs are also regulators of cellular metabolism. There are four subfamilies of MAPKs: ERK1/2, ERK5 (also known as BMK1 or MAPK7), the JNKs (JNK1, JNK2, and JNK3), and the p38 kinases (p38α, p38β, p38γ, and p38δ). MAPKs are activated downstream of many receptors in immune cells, including antigen receptors, TLRs, and growth factor receptors. The activation of these kinases requires the phosphorylation of the threonine and tyrosine residue within the TXY motif, where T is threonine, Y is tyrosine, and X is glutamate, proline, or glycine [260]. MAPKs are activated via a signaling cascade: a MAPK kinase kinase (MAPKKK or MAPK3) phosphorylates a MAPK kinase (MAPKK or MAPK2), which then phosphorylates the MAPK itself. For example, after RAS is activated, the ERK1/2 kinases are activated by RAF (the MAPKKK) and MEK (the MAPKK).

The MAPKs play crucial roles in immune cell functions and have been demonstrated to influence metabolism under conditions of cellular stress [131]. The mechanisms linking MAPK activation and metabolic rewiring have not been extensively studied in immune cells. In nonimmune cells, the MAPK-dependent metabolic rewiring is largely linked to changes in gene and protein expression. ERK1/2 can also influence fatty acid uptake and FAO by controlling CD36 expression and mediating the inhibitory phosphorylation of lipid synthesis-related enzymes, including ACC [390]. The p38 kinases can regulate glucose uptake via poorly understood mechanisms. By modulating the activity of FOXO proteins, JNK can control cell-intrinsic metabolic responses. JNK signaling in immune cells also has implications in controlling systemic metabolism. Of note, many of the mechanisms by which the MAPKs serve to promote metabolic reprogramming have also been attributed to mTOR-related signaling [131], so it is likely that there are cell and context-dependent requirements for these two pathways in metabolism. We discuss how MAPKs and metabolic programs intersect to control immune cell biology in sections below.

1.2.5 Transcription Factors Control Metabolic Programs

1.2.5.1 c-MYC

c-MYC (also known as MYC) is a basic helix-loop-helix leucine zipper (bHLP-LZ) transcription factor. c-MYC-mediated transcription occurs when it heterodimerizes with another bHLP-LZ transcription factor Myc-associated factor X (MAX), which results in gene expression changes that tune cell proliferation, cell growth, apoptosis, metabolism, and biosynthesis. Aberrant function of c-MYC or its family members (N-MYC and L-MYC) is strongly linked to tumorigenesis. Therefore, the expression of c-MYC is tightly controlled, wherein quiescent or nonproliferating cells express low levels of c-MYC and proliferating cells upregulate c-MYC expression. In immune cells, c-MYC expression is upregulated following engagement of activating receptors, such as antigen receptors [403, 434]. The increase in expression is controlled at both the transcriptional and posttranscriptional levels. *MYC* transcription is upregulated in response to signaling downstream of Janus kinase (JAK), which subsequently activates signal transducer and activator of transcription (STAT) transcription factors [204]. mTORC1- and MAPK-dependent signaling enhance *MYC* translation. c-MYC expression is also regulated at the posttranslational level: the PI3K-mTORC2-AKT and RAS-MAPK pathways increase the protein stability of c-MYC [204].

As noted above, c-MYC is a crucial regulator of metabolic reprogramming. c-MYC regulates the expression of genes important for catabolism. For instance, the genes encoding HK2 and LDH are upregulated by c-MYC-dependent transcriptional programs. LAT1, ASCT2, and GLS expression are also induced by c-MYC, which can increase glutaminolysis [403]. c-MYC transcriptional programs can also influence OXPHOS by increasing mitochondrial biogenesis. Importantly, c-MYC-dependent programs can feedback enhance mTOR-dependent signaling, such as via the suppression of TSC2 expression [323]. Anabolic processes, including FAS, are also influenced by c-MYC function [204]. We will discuss the role of c-MYC-related metabolic reprogramming in immune cell differentiation and function throughout this chapter.

1.2.5.2 HIF-1

Hypoxia-induced factor-1 (HIF-1) is a heterodimeric transcription factor consisting of the constitutively expressed HIF-1β subunit and the inducible HIF-α (isoforms HIF-1α, HIF-2α, or HIF-3α) subunit. HIF-1 transcriptional function is controlled at the protein expression level and by two transactivation domains, which are bound by p300/CBP. The protein expression of HIF-α is controlled by prolyl hydroxylation via proly-4-hydroxylase (PHD) and other members of the Fe(II) and 2-oxoglutarate (2-OG)-dependent dioxygenase superfamily, which require molecular oxygen as a cofactor. When oxygen levels are sufficient, PHD phosphorylates HIF-α, leading to its degradation. During this process, succinate is produced and can feed into the TCA cycle. By contrast, insufficient oxygen levels or high levels of ROS inactivate PHD and subsequently stabilize HIF-α expression. HIF-α is also subjected to asparaginyl hydroxylation, which is mediated by factor inhibiting HIF (FIH). This process suppresses the transcriptional activity of HIF-1, in part by blocking the recruitment of the coactivators, p300/CBP. Other metabolites have also modulate HIF-α expression or transcriptional function at the level of PHD as discussed elsewhere [248].

In addition to oxygen-related regulation, other mechanisms modulate HIF-1 functions. The mTOR pathway induces expression of HIF-α by modulating its protein translation. The mTORC1-4E-BP-eIF4E and mTORC2-AKT pathways cooperate to control HIF-1α expression, while HIF-2α expression appears to be dependent upon mTORC2-AKT2 signaling [386]. Interestingly, by modulating the expression of regulated in development and DNA damage responses 1 (REDD1), an activator of the TSC complex, HIF-1 feedback suppresses mTORC1 activation. These results suggest that AKT enhances HIF-α translation via mTORC1-dependent and mTORC1-independent mechanisms. The MAPKs have also been implicated in controlling HIF-1 activity. For instance, MAPK-dependent phosphorylation of p300 was shown to stabilize p300/CBP-HIF-1 interactions to promote HIF-1 transactivation and transcriptional activity [336]. Lysine acetylation by p300/CBP potentiates HIF-1-dependent functions, whereas sirtuin (SIRT)-1/3/6-mediated deacetylation antagonizes HIF-1 functions.

HIF-1 is a critical regulator of metabolism [248]. Briefly, HIF-1-dependent transcription drives expression of glycolysis-related genes, including the GLUTs, HK, and the monocarboxylic acid transporters (MCT) that drive lactate efflux [248]. HIF-1-dependent programs also inhibit the TCA cycle and OXPHOS. For example, HIF-1 inhibits the pyruvate dehydrogenase (PDH)-dependent conversion of pyruvate to acetyl-CoA by increasing the expression of pyruvate dehydrogenase kinase (PDHK) [197]. Additionally, HIF-1 suppresses the expression of acyl-CoA dehydrogenases to inhibit FAO [164]. It is important to note, however, that HIF-1-dependent programs are subjected to feedback regulation, such as by ROS produced due to inefficient OXPHOS. Moreover, hypoxia-driven HIF-1 activity is linked to the upregulation of SREBP function and the subsequent upregulation of FAS expression. We discuss these transcription factors in the next section.

1.2.5.3 SREBPs

The SREBP transcription factors (SREBP1a, SREBP1c, and SREBP2) are major regulators of lipid and cholesterol biosynthesis. These transcription factors are anchored to the endoplasmic reticulum (ER) membrane. The N-terminus of SREBP contains basic helix-loop-helix motifs found in the basic helix-loop-helix leucine zipper family of transcription factors. To initiate the transcriptional activity of SREBPs, the N-terminal domain must be cleaved in a process termed regulated intramembrane proteolysis (RIP). SREBP cleavage-activating protein (SCAP) senses changes in intracellular sterols, including cholesterol and its derivatives, to regulate SREBP activity. SCAP is bound to the ER membrane in cholesterol-replete cells, where it associates with insulin-induced gene (INSIG) and SREBP. The interaction between INSIG and the SCAP-SREBP complex sequesters SREBP in the ER, where it is inactive. Upon cholesterol depletion, the SCAP-INSIG interaction is disrupted. The SCAP-SREBP complex then associates with COPII-coated vesicles, which transport the SCAP-SREBP complex from the ER to the Golgi. Here, Site-1 protease (S1P) and Site-2 protease (S2P) sequentially cleave the N-terminal cytosolic portion of SREBP, allowing it to translocate to the nucleus. The transcriptional activity of SREBP drives expression of genes related to cholesterol biosynthesis. The transcriptional activity of SREBP-1 (specifically SREBP-1c in mice and SREBP-1a and SREBP-1c in humans) is also induced by oxysterols and unsaturated fatty acids, known agonists of the nuclear hormone receptor liver X receptor (LXR) that bind LXR response elements present in the *Srebf1* promoter region [346, 440]. Thus, SREBP-dependent functions can regulate both cholesterol and fatty acid synthesis.

SREBP activity is also controlled by posttranscriptional mechanisms. The activity of mTORC1 is necessary and sufficient to stimulate SREBP-1 activity in multiple cell types, including immune cells. Mechanistically, mTORC1 phosphorylates lipin-1, a phosphatidic acid phosphatase that converts phosphatidic acid into diacylglycerol. The phosphorylation of lipin-1 blocks its nuclear localization and activates SREBP. The enzymatic activity of lipin-1 is required to suppress SREBP function, suggesting that lipin-1 could modulate triacylglyceride or phospholipid concentrations in the nucleus to control SREBP functions. The PI3K-AKT axis also induces SREBP transcriptional activity by inactivating GSK3, which phosphorylates SREBP and drives its subsequent proteasomal degradation. AMPK also directly phosphorylates SREBP to limit lipid biosynthesis when nutrients are limiting. Finally, the lysine acetyltransferases p300/CBP drive SREBP acetylation, which stabilizes SREBP expression [133], whereas SREBPs are deacetylated by SIRT1 and subsequently degraded. It is noteworthy that lysine acetyltransferases use acetyl-CoA as a cofactor to acetylate proteins, and SIRT1 requires NAD^+ as a cofactor for its enzymatic function. Thus, other metabolites aside from cholesterol and lipids influence SREBP expression and functions by regulating posttranslational modifications. How metabolic pathways shape protein modifications is discussed in other chapters of this book. We will discuss this topic briefly as it relates to macrophage and T cell functions below.

1.2.5.4 PPAR

The PPARs are ligand-inducible transcription factors belonging to the nuclear receptor superfamily. Three isoforms [PPARα, PPARδ (also called PPARβ or PPARδ/β), and PPARγ] with overlapping functions exist in mammalian cells. PPARs associate with their obligate heterodimer retinoid X receptor (RXR) and bind PPAR-responsive elements in their responsive genes. The PPARs are receptors for dietary fats, select lipid metabolites, and oxidize phospholipids. Not surprisingly, PPAR activity is high in mice fed with high-fat diets (HFD) and is strongly implicated in disease pathologies associated with human obesity [2]. The activation of the PPARs plays important roles in systemic and cell-specific metabolic responses. PPARγ activation is linked to adipogenesis, which contributes to the systemic availability of free fatty acids. In skeletal muscle and the liver, PPARα and PPARδ activities are linked to FAO. Similarly, PPARγ upregulates the transcription of genes associated with fatty acid uptake and FAO, including CD36 and acetyl-CoA synthetase (ACS). PPARγ can also control glucose homeostasis by upregulating GLUT4 expression downstream of the insulin receptor and has thus been linked to insulin sensitivity. Consistent with this idea, thiazolidinediones (TZDs) are a class of drugs that activate the PPARγ and induce insulin sensitivity [2, 401]. We discuss how PPAR activation controls immune responses below.

As with other transcription factors, PPAR functions are tuned by posttranslational modifications. The phosphorylation of PPAR by various serine/threonine kinases modulates its activity. For instance, MAPK-dependent phosphorylation inhibits PPAR activation [2]. Additionally, acetylation and sumoylation events appear to positively and negatively regulate the transcriptional activity of PPAR, respectively. mTORC1 can also regulate PPAR activity through several proposed mechanisms: by increasing its expression, by inhibiting the activity of its coactivator lipin-1, and by altering the availability of its lipid ligands via the activation of the SREBPs [214, 215]. Although inflammatory mediators that drive immune responses, such as TLRs, modulate PPAR activity, the activation of PPAR in immune cells has predominantly been linked to anti-inflammatory effects [2, 401]. This topic will be discussed throughout this chapter.

1.2.5.5 FOXO

The FOXO family of transcription factors regulates cell survival, proliferation, differentiation, and migration but is also implicated in controlling metabolism [106, 156]. There are four FOXO family proteins, characterized by a distinct forkhead DNA-binding domain: FOXO1 (also called FKHR), FOXO3 (also known as FKHRL1), FOXO4 (also termed AFX), and FOXO6. The FOXOs are shuttled out of the nucleus to suppress their transcriptional activity, which is induced upon the AKT-dependent phosphorylation of three consensus sites (threonine 24, serine 253, and serine 316) [106]. The acetylation of FOXO appears to be mediated by the reciprocal actions of p300/CBP and SIRT1, which inactivate and activate FOXO transcriptional functions, respectively. The FOXOs are also subjected to ubiquitination, which controls FOXO function by incompletely understood mechanisms [106]. Arginine methylation of the FOXO transcription factors by protein arginine methyltransferase 1 (PRMT1) prevents the AKT-mediated phosphorylation of FOXO to enhance its transcriptional activity [106, 430]. Thus, multiple posttranscriptional modifications control FOXO-dependent gene expression.

The ability of FOXOs to reprogram metabolism has been studied extensively in insulin-sensitive tissues. In the liver, for instance, FOXO1-dependent functions are linked to gluconeogenesis (i.e., the de novo production of glucose) [142]. By contrast, the FOXOs appear to antagonize glycolytic programs, an idea supported by the observations that AKT-dependent programs favor glycolysis but inactivate FOXO functions. Interestingly, mTORC2 activation promotes FOXO1 and FOXO3 acetylation via the inactivation of class IIa histone deacetylases. This acetylation inactivates FOXO1/3, which disrupts the transcription of the c-MYC targeting miR-34c microRNA network. Thus, mTORC2 induces c-MYC activity and glycolytic upregulation via an AKT-independent mechanism [249]. FOXO-dependent programs are linked to OXPHOS. Peroxisome proliferator-activated receptor γ coactivator-1 (PGC-1; isoforms PCG-1α and PGC-1β) is a transcriptional coactivator whose functions favor mitochondrial biogenesis and respiration [114]. The promoter region of the PGC-1 contains FOXO1 binding sites, and PGC-1α transcriptional activity is inhibited by AKT signaling [114]. Thus, FOXO-dependent programs are also likely linked to OXPHOS by modulating the expression of mitochondria-related proteins. We discuss the implications of FOXO-dependent programs in immunity below.

1.3 Roles of Metabolism in Myeloid Cells

1.3.1 Overview of Myeloid Cells

The innate immune system is the first line of defense for host immunity. Innate immune cells residing at barrier surfaces such as the skin, lung, and intestines play crucial roles in antimicrobial responses. There are several classifications of innate immune cells, including granulocytes, mast cells, and neutrophils. Among the innate immune cells, the metabolic requirements driving DC and macrophage activation and functions have been the best studied, so we will focus upon these cell types in this section. However, others have more comprehensively reviewed how innate immune cell metabolism is tuned [128, 305].

Macrophages and DCs are found in both lymphoid tissues and nonlymphoid tissues. Macrophages engulf and degrade damaged cells and pathogens, whereas quiescent DCs have low phagocytic capacity. The sensing of these pathogen-associated molecules via TLRs or danger signals via multiple intracellular senses causes both macrophages and DCs to undergo morphological changes and upregulate their capacity to migrate into distal tissues. Activated macrophages and DCs also produce chemokines and cytokines to drive the recruitment and differentiation of other immune cells, including T cells. Additionally, these activated cells increase their capacity to capture, process, and present protein-derived peptides on MHC molecules. The expression of various co-stimulatory ligands, including CD80 and CD86 (also called B7-1 and B7-2, respectively), is also increased. In vivo, there are several specialized DC subsets that are distinguished based upon their expression of surface molecules, anatomical location, and functions: plasmacytoid DCs, conventional DCs (cDCs; including $CD8\alpha^+$ cDCs, $CD103^+$ cDCs, and $CD11b^+$ cDCs), Langerhans cells, and monocyte-derived DCs [129]. Macrophages and DCs can also be generated in vitro from bone marrow cells in response to various cytokines and are called bone marrow-derived macrophages (BMDMs) or bone marrow-derived DCs (BMDCs), respectively. It is important to note that much of our knowledge of metabolic reprogramming comes from work performed using BMDMs stimulated under M1 or M2 conditions (see below) or BMDCs stimulated with lipopolysaccharide (LPS), a TLR4 agonist (Fig. 1.3). Further work will elucidate how metabolic programs equip different DC subsets with their specialized phenotypes and functions.

1.3.2 Glycolysis Supports Pro-inflammatory Macrophage and DC Functions

For nearly 50 years, it has been appreciated that macrophages undergo metabolic reprogramming upon activation: glucose and glutamine consumption increases, while oxygen utilization decreases [152, 269]. This metabolic shift facilitates the upregulation of aerobic glycolysis and downregulation of OXPHOS [270]. Increased glucose consumption and upregulated expression of HK2 and G6PD enhance aerobic glycolysis in macrophages [269]. The precise molecular mechanisms driving metabolic reprogramming in different macrophage subsets, both in the context of inflammation and immune suppression, are under active investigation. The upregulation of glycolysis in macrophages is coupled to signaling induced by TLR ligation and IFN-γ sensing and the upregulation of GLUT1 expression and 6-phosphofructo-2-kinase (PFK2) activity [122, 327]. Similarly, TLR signaling will increase glycolysis in DCs to support their functional maturation [108, 109]. However, the induction

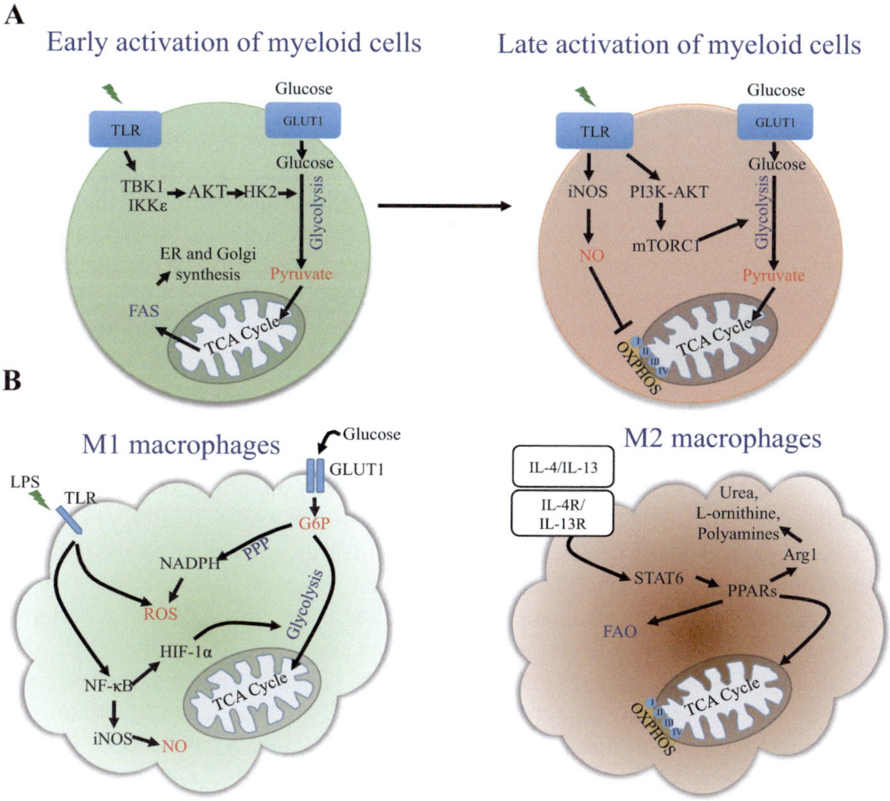

Fig. 1.3 Metabolic control of dendritic cell and macrophages
(**a**) During DC activation, aerobic glycolysis is regulated early on by TLR and AKT signaling which helps drive additional biosynthetic intermediates necessary for fatty acid, ER, and Golgi synthesis, essential for DC maturation. Following activation, glycolytic program supported by PI3K-AKT-mTOR signaling pathway plays a major role in metabolic reprogramming. OXPHOS fueled by mitochondrial activity is diminished by regulation of iNOS and NO. (**b**) M1 macrophages rely on the induction of glycolysis for energy, which also helps in lactate production and intermediates necessary for TCA cycle. M1 macrophages have increased levels of iNOS, NO, HIF-1α and ROS. Signals inducing M2 macrophages increase OXPHOS and lipid oxidation orchestrated by STAT6 and PPARs. M2 macrophages express arginase (*Arg1*) necessary for production of urea, polyamines and ornithine upon arginine degradation. The reader should refer to Sect. 1.3 for more details regarding macrophage metabolism

of glycolysis is temporally regulated by distinct signaling mechanisms in DCs. Early after TLR stimulation, AKT activity is upregulated in a PI3K-independent and TBK1- and IKKε-dependent manner. The early burst of PI3K-independent glycolysis increases the levels of metabolites that can shuttle into the PPP and drive NADPH production or enter into the TCA cycle to produce citrate necessary for FAS. The de novo production of these fatty acids supports endoplasmic reticulum and Golgi apparatus synthesis to facilitate cell growth and biomass accumulation [108]. The role of the PPP in macrophage and DC activation is discussed more below.

Pro-inflammatory signals driving PI3K activity continue to shape macrophage and DC functions and metabolism after more prolonged stimulation [203]. Glycolysis becomes the dominant program of inflammatory macrophages (termed M1 macrophages) and DCs, with OXPHOS being downregulated [109, 175, 203]. This metabolic balance is important for M1 macrophage differentiation, as treatment with rotenone and antimycin A (i.e., ETC inhibition) favors M1 macrophage-like functions [327, 396]. Similarly, 2-DG treatment suppresses DC maturation [109]. One hallmark of M1 macrophages and activated DCs is the expression of inducible nitric oxide synthase (iNOS), which generates nitric oxide (NO). The production of NO inhibits OXPHOS by competing with oxygen for binding to cytochrome c, the final electron donor of the ETC. In turn, this inhibition prevents oxidation of NADH to NAD^+, requiring the upregulation of aerobic glycolysis to drive NADH accumulation when pyruvate is converted into lactate [109, 279]. Indeed, iNOS expression is required for the full maturation of DCs and supports the glycolytic program of these cells [109]. The attenuation of ETC activity is also crucial to support M1 macrophage functions upon TLR stimulation, as the generation of mitochondria-derived ROS plays an important role in bacterial killing [417].

Excessive NO and/or ROS production could also contribute to the HIF-1α-dependent regulation of glycolysis and M1 macrophage-related functions [36, 81, 197, 297, 310, 311, 377, 381]. LPS, IFN-γ, and hypoxia have all been reported to increase HIF-1α or HIF-2α expression in macrophages to facilitate the pro-inflammatory phenotypes and functions [49, 171, 377, 378]. Further, the increased activity of HIF-2α is positively correlated with tumor growth, in part by minimizing macrophage recruitment into tumors [171, 187, 221, 250, 378]. ATP, ROS, and NO production by macrophages appear to be independent of HIF-2α expression [171]. Thus, HIF-α isoforms appear to facilitate metabolic adaptions required for macrophage functions at sites of inflammation.

Pro-glycolytic transcriptional programs also modulate DC activation. Stimulation of multiple TLRs induces HIF-1α expression in DCs [175, 296, 365], and HIF-1α-deficient DCs express lower levels of CD80 and CD86, resulting in diminished T cell activation [31, 175]. Although a role for MYC proteins in metabolic reprogramming of DCs has not been reported, mature DCs selectively express the gene encoding L-MYC (*Mycl1*). L-MYC is important for the proliferation and survival of several DC subsets, and it supports $CD8\alpha^+$ DC-induced T cell priming [188]. Future work will continue to elucidate how upstream signals regulate the balance of metabolic reprogramming in different DC subsets.

1.3.3 OXPHOS Supports Anti-inflammatory Functions of Macrophages and DCs

Macrophages can also adopt an alternative fate, termed alternatively activated or M2 macrophages. M2 macrophages promote tissue repair and wound healing and are activated by type 2 cytokines, including IL-4 and IL-13; thus, these cells have important roles in antiparasitic immune reactions. M2 macrophages also express high levels of scavenger, mannose, and galactose receptors, express high levels of Arg1 and its downstream metabolites, and produce high levels of IL-10 [126]. The metabolic program of M2 macrophages is shifted in favor of OXPHOS over glycolysis [396]. In contrast to pro-inflammatory DCs that activate T cells, tolerogenic DCs promote T cell tolerance. These cells are phenotypically distinct from pro-inflammatory DCs, expressing lower levels of MHC class II molecules and co-stimulatory ligands. They also express anti-inflammatory mediators, including IDO, IL-10, and TGF-β [224]. These DCs also have different metabolic signatures than pro-inflammatory DCs. It was demonstrated that human tolerogenic DCs have high mitochondrial function and produce more cellular ROS and mitochondria-derived superoxide than inflammatory DCs. Further, human tolerogenic DCs have elevated levels of OXPHOS fueled by mitochondrial FAO and, to a lesser extent, glycolysis [243]. Of note, our understanding of tolerogenic DCs is limited, but these cells are of great therapeutic interest [355], so future studies will continue to explore the molecular mechanisms driving DC tolerance.

The AMPK axis is a crucial regulator of OXPHOS and the acquisition of anti-inflammatory functions of macrophages and DCs. AMPK activity also regulates metabolic programs and fate decisions of macrophages and DCs. AMPK activity suppresses the upregulation of glycolysis and supports mitochondrial biogenesis and FAO in TLR-primed DCs [60, 203]. AMPK signals positively regulate the M2 macrophage program, and loss of AMPK activity reduces M2 macrophage function and enhances M1 macrophage-related functions, including pro-inflammatory cytokine production [261, 332]. Upstream signals such as LPS and TNF-α reduce AMPK activity to limit fatty acid metabolism and upregulate glycolysis [125, 368, 439]. It is noteworthy, however, that while many studies have demonstrated that AMPK agonists can limit macrophage-mediated inflammation [125, 178, 455], some of these effects are independent of AMPK activity [332]. To promote FAO and OXPHOS, AMPK phosphorylates ACC, and mice bearing a knock-in mutation where the AMPK phosphorylation site of ACC is mutated have reduced rates of FAO and increased FAS [123]. Similarly, AMPK-deficient macrophages have a correlated decrease of both ACC1 phosphorylation and FAO [125]. Of note, although macrophages preferentially express ACC1, ACC1 and ACC2 play redundant functions for FAO in macrophages [261]. AMPK also reduces the levels of mitochondria available for OXPHOS by inducing mitochondrial autophagy, also called mitophagy [105]. Consistent with this role, defective mitochondria with reduced OXPHOS capacity accumulate upon AMPK inactivation [277].

PGC-1α and PGC-1β are crucial regulators of M2 macrophage and tolerogenic DC differentiation downstream of AMPK. AMPK-mediated inhibition of glycolysis and upregulation of mitochondrial respiration is mediated by PGC-1α activity [60, 203]. Indeed, the upregulation of PGC-1α antagonizes DC functions [201, 304]. Similarly, PGC-1β is required for IL-4R-STAT6-induced fatty acid uptake and oxidative metabolism that increases M2 macrophage differentiation [396]. Mechanistically, STAT6 and PGC-1β can directly bind to the promoters of M2 macrophage-associated genes, including *Arg1* [396]. However IFN-γ-induced STAT1 activation can induce ROS production via estrogen-related receptor-α

(ERRα), which is also positively regulated by PGC-1β [362]. Thus, PGC-1β coactivates different downstream transcriptional programs to direct macrophage fates.

PGC-1α and PGC-1β are important coactivators of PPARγ, which controls M2 macrophage differentiation in selected contexts, such as in adipose tissue [38, 259, 283, 453]. Moreover, the PPARs are complex regulators of DC biology. PPARγ agonists suppress IL-6, IL-12, IL-15, and TNF-α production, reduce CD80 expression, and induce CD86 expression in DCs [112, 139, 268, 375]. The migration of select DC subsets is also impaired in the presence of PPARγ agonists [7, 9, 268]. These results indicate that PPARγ plays an anti-inflammatory role in DCs, although it is controversial as to the effects of PPARγ activation in the DC-mediated priming of T cells [201, 268, 375]. PPARγ activation blocks DC activation via mechanisms involving MAPK and nuclear factor κB (NF-κB) transcription factor activation [9]. Of note, a recent report suggests that the upregulation of PPARα-dependent FAO supports pDC but not cDC functions as discussed more fully below [423]. Thus, there are context-dependent requirements for FAO as it relates to the differentiation of anti-inflammatory versus pro-inflammatory DC responses.

1.3.4 The AKT-TSC Axis Is a Crucial Regulator of Macrophages and DCs

The PI3K-AKT pathway is one major determinant of M1 versus M2 macrophage polarization. Both IL-4R and TLR4 engagement activate PI3K-AKT-mTOR signaling but via distinct mechanisms [52, 388]. The activation of AKT appears to promote M2 macrophage differentiation [52, 326]. However, AKT1 appears to be the dominant isoform for inducing M2 macrophage skewing, because loss of AKT1 enhances the differentiation of M1 macrophages, but AKT2 deletion drives increased M2 macrophage skewing [12]. AKT may mediate its effects on macrophage differentiation via control of the transcription factors, FOXO1, C/EBPβ, or NF-κB [12, 80, 110]. Future studies are needed to further explore how the precise activation of AKT isoforms differentially impacts macrophage polarization, and if these differences impact macrophage functions in the context of different diseases.

Because TSC2 is a point of convergence for both AMPK and AKT signals, several groups have investigated the role of the TSC complex in macrophage and DC fate decisions. Although rapamycin studies have suggested an important role of mTORC1 in DC function and survival [5, 37, 55, 150, 167, 287], the hyperactivation of mTORC1 in TSC1-deficient DCs impairs their development and functions [410]. The TSC complex is under complex control in macrophages. The expression of TSC1 is upregulated in macrophages upon IL-4 stimulation, but not LPS stimulation. However, TSC complex activity is downregulated via the phosphorylation of TSC2 upon both LPS and IL-4 stimulation, which in turn upregulates mTORC1 activity [451]. It has been demonstrated that TSC1-deficient BMDMs are hyperresponsive to M1 macrophage polarizing stimuli, leading to increased production of inflammatory cytokines and iNOS expression [111, 451]. By contrast, IL-4-induced M2 differentiation is impaired in the absence of TSC1 [52, 451]. TSC1 deficiency influences the activation of different downstream signaling pathways in M1 and M2 BMDMs. In the former, the RAS-RAF1-MEK-ERK cascade is hyperactivated in an mTORC1-independent manner. Additionally, TSC1 deficiency augments JNK activation. In M2 macrophages, mTORC1 activity is upregulated, which appears to be independent of increased AKT activity [451]. In fact, mTORC1 hyperactivation activity appears to inhibit AKT activation downstream of IL-4R stimulation [52]. Thus, in the context of in vitro M2 polarization, excess mTORC1 signaling is detrimental to the PI3K-AKT-driven polarization of these cells. However, a recent study suggested that TSC1-deficient peritoneal macrophages acquire both M1 and M2 signatures [111].

What could account for the apparent discrepancies in studies analyzing how TSC1 affects macrophage differentiation? One potential answer lies in the differences between BMDMs and peritoneal macrophages. Another possibility is that TSC1-deficient macrophages have increased mTORC1 activity at the expense of mTORC2 activity [410, 433]. The specific deletion of *Rptor* in macrophages diminishes inflammation in murine models of obesity and atherosclerosis [3, 179]. Consistent with these studies, HIF-1α-deficient macrophages also have impaired M1-related macrophage functions [377]. By contrast, *Rictor*-deficient macrophages have exaggerated M1 phenotypes, and its deletion in macrophages renders mice more susceptible to sepsis [42, 115]. Mechanistically, mTORC2 appears to inhibit the TLR4-induced activation of FOXO1 [42]. A loss of mTORC2 activity in TSC1-deficient BMDMs could also account for the reduction of AKT activation essential for M2 macrophage polarization [52]. Future studies will explore how the timing and strength of mTOR signaling influences site or tissue-specific macrophage differentiation and function.

Given that it influences mTOR, MAPK, and AKT activation, it is not surprising that TSC1 deficiency also affects metabolism. TSC1-deficient DCs have augmented levels of glycolysis and OXPHOS [410]. It was demonstrated that IL-4-induced FAO is reduced in TSC1-deficient BMDMs [52]. Of note, TSC1-deficient peritoneal macrophages have increased levels of ROS [111], but whether this is derived from the mitochondria is unknown. Cell growth and the expression of nutrient transporters are increased in an mTORC1-dependent manner in TSC1-deficient peritoneal macrophages, while cell survival is impaired [111]. Intriguingly, however, the loss of TSC1 did not affect the gene expression of *Hk2*, *Ldha*, *Tpi1*, or *Hif1a* in peritoneal macrophages [111]. While these results seem incongruent with an increase of M1 macrophage differentiation in the absence of TSC1, glycolysis could still be upregulated by stabilizing HIF-1α expression. In this regard, the increased expression of amino acid transporters could support glutaminolysis, which in turn would produce α-KG. The conversion of α-KG into succinate could then inhibit PHD activity and stabilize HIF-1α expression [381]. PHD activity may also be inhibited by excessive ROS production in TSC1-deficient macrophages [111, 272]. Future studies will continue to investigate how alterations in the TSC1-AKT-mTOR axis influence macrophage and DC fates and functions in different diseases and tissues.

1.3.5 The PPP and TCA Cycle Intermediates in Macrophage and DC Functions

Glucose shuttling can increase PPP activation, which is important for purine and pyrimidine synthesis supporting cell growth [279]. Additionally, upregulation of PPP activity generates NADPH for NADPH oxidase (NOX) activation, a protein complex whose function is linked to the generation of superoxide species important for antimicrobial responses [295]. NADPH generation also maintains glutathione in a reduced state, which limits ROS-induced cell damage in M1 macrophages [279]. M1 macrophages have thus adopted several strategies to upregulate NADPH. For instance, they express high levels of the PPP enzymes, G6PD, and 6-phosphogluconate dehydrogenase [271, 279]. Relative to M2 macrophages, M1 macrophages express lower levels of carbohydrate kinase-like protein (CARKL) [154], an enzyme that catalyzes the formation of sedoheptulose-7-phosphate to suppress PPP cycling [279]. Indeed, overexpression of CARKL limits and promotes M1 and M2 macrophage functions, respectively, while repressing its gene expression has the opposite effect [154]. Finally, the expression of glutamate dehydrogenase is upregulated in LPS-stimulated macrophages, which can increase glutamine metabolism to support NADPH production [271, 279].

The upregulation of glycolysis in M1 macrophages is associated with reduced TCA cycle activity [203, 381]. While this acts as an effective means to generate mitochondrial-derived ROS that can contribute to microbial killing, certain TCA cycle intermediates are crucial for macrophage functions. For instance, citrate can be broken down into acetyl-CoA via the activity of ATP-citrate lyase (ACL), which then provides the cell with a source of cytoplasmic acetyl-CoA that can be used for lipid and phospholipid synthesis [279]. The generation of acetyl-CoA via ACL activity is also linked to histone acetylation reactions that enhance the expression of genes associated with glycolysis, including *Hk2* [416]. Additionally, oxaloacetate generated during the ACL-mediated breakdown of citrate can subsequently be converted into malate, pyruvate, and lactate via the activities of malate dehydrogenase, the malic enzyme, and LDH, respectively. Each of these chemical reactions also converts $NADP^+$ to NADPH and regulates the balance between glycolysis and OXPHOS [279]. How do macrophages increase cytoplasmic citrate levels to promote these processes? Upon LPS stimulation, the expression of mitochondrial-associated solute carrier family 25 member 1 (Slc25a1; also called mitochondrial citrate carrier CIC) is upregulated, which in turn promotes citrate transport from the mitochondria into the cytosol [172]. Inhibition of Slc25a1 expression or function is correlated with reduced LPS-triggered inflammatory responses, including the production of phospholipid-derived prostaglandins, ROS, and NO [172]. Thus, mitochondrial-derived citrate tunes LPS-induced macrophage responses.

Succinate levels are also increased in LPS-stimulated macrophages [381]. The intracellular concentration of succinate increases due to glutaminolysis, where α-KG can serve as an upstream intermediate for succinate production [294, 327]. Glutamine can also enter into the γ-aminobutyric acid (GABA) shunt to form succinate [381]. Interestingly, an inhibitor of the GABA shunt enzyme GABA transaminase is protective in an LPS-induced model of septic shock [381], suggesting that this pathway contributes to pro-inflammatory functions of innate immune cells. Succinate derived from these pathways can then be catalyzed into malate and subsequently help augment NADPH levels necessary for M1 macrophage functions. Succinate also regulates pro-inflammatory macrophage functions by inhibiting PHD enzyme activity, which stabilizes HIF-1α expression [381]. Extracellular succinate has also been demonstrated to augment TLR3/7-induced TNF-α production by DCs [330], but this role has not been explored in macrophages. Finally, succinyl groups can be added to lysine residues of certain proteins, including many enzymes related to glycolysis [381]; however, the functions of these succinate-dependent posttranslational modifications are unknown.

1.3.6 Cholesterol and Lipid Metabolism in Macrophages and DCs

Cholesterol and lipid metabolism also control macrophage and DC responses. Upon M-CSF stimulation, SREBP-1c-dependent and SREBP-2-dependent gene expression are upregulated and downregulated in human monocytes, respectively. These changes are associated with increased FAS gene expression and decreased cholesterol synthesis gene expression [104]. Indeed, cholesterol levels diminish during M-CSF-induced differentiation, while fatty acid and phospholipid synthesis increase. When FAS-related pathways are inhibited during monocyte to macrophage differentiation, the resulting macrophages are smaller, have fewer organelles, and are less phagocytic [104]. Thus, cholesterol and lipid biosynthesis support organelle and membrane biogenesis, hence promoting cell growth and macrophage maturation.

Lipid or cholesterol homeostasis also affects mature macrophage responses. Bensinger and colleagues demonstrated that cholesterol and long-chain fatty acid pools are diminished upon TLR3 or type I interferon (IFN-I) stimulation of BMDMs [442]. These effects are mediated by reduced expression of *SREBP2*, *SQLE*, and *FASN* [34, 442]. However, the total lipid and cholesterol pool is enhanced upon TLR3 or IFN-I signaling, because CD36 and macrophage scavenger receptor (MSR)-dependent lipid uptake increases [442]. Limiting cholesterol flux via suppression of the SCAP-SREBP2-inducible mevalonate pathway engages IFN-I signaling, which promotes macrophage resistance to viral infection. It is important to note that while SCAP-deficient macrophages have no apparent changes in the total cellular cholesterol or lipid pools, culturing SCAP or SREBP2-deficient macrophages with excess, cell-permeable cholesterol restores their function [442]. Mechanistically, changes in cholesterol synthesis in the ER are linked to the stimulator of cyclic GMP-AMP (cGAS)-stimulator of interferon genes (STING)-TBK1-interferon regulatory factor 3 (IRF3) pathway. STING localizes to the ER membrane and is activated by cyclic dinucleotides that are generated via cGAS activity. Activated

STING promotes TBK1 phosphorylation and allows IRF3 to enter into the nucleus, where it drives IFN-I and IFN-I-responsive gene expression. SCAP-deficient macrophages are more sensitive to dicyclic nucleotide-induced STING activation, which is mitigated when free cholesterol is available to the cells. Thus, ER-associated STING senses changes in ER cholesterol synthesis to shape the IFN-I response of macrophages.

Cholesterol degradation also modulates macrophage functions. IFN-I treatment drives production and secretion of the cholesterol derivative, 25-HC, via the STAT1-dependent induction of *Ch25h*; in turn, 25-HC has potent antiviral effects by limiting viral replication and viral fusion [33, 231]. *Ch25h*-deficient mice are more susceptible to viral infections, LPS-induced septic shock, and experimental autoimmune encephalomyelitis (EAE; an autoimmune model of multiple sclerosis) but display enhanced clearance of bacterial infections [231, 325]. These phenotypes are linked, in part, to hyperproduction of IL-1β by *Ch25h*-deficient macrophages, but altered T cell and neutrophil responses also contribute to these disease alterations. The CH25H-dependent production of 25-HC suppresses SREBP2, but not LXR, functions [33, 325]. Of note, 25-HC is catalyzed into 25-HC-3-sulfate (25-HC-3-S), which antagonizes the 25-HC-induced upregulation of LXR activity [238]. These data suggest that 25-HC conversion into 25-HC-3-S may account for the LXR-independent functions of 25-HC in antiviral responses of macrophages [33]. 25-HC-3-S antagonizes LXR and SREBP-1-dependent lipid synthesis gene expression, such as ACC and FAS, which increases and decreases cell proliferation and apoptosis, respectively [238]. However, the role of SREBP-1 in macrophages is complex: SREBP-1a is more highly expressed in BMDMs than SREBP-1c, and SREBP-1a deficiency compromises the production of IL-1β upon LPS stimulation [170]. Future studies will continue to uncover the context-specific roles of lipid and cholesterol biosynthesis in macrophage function.

In addition to controlling cholesterol biosynthesis, the mevalonate pathway produces isoprenoids, such as farnesyl pyrophosphate (FPP) and geranylgeranyl pyrophosphate (GGPP), which serve as protein posttranslational modifications. Statin drugs, including lovastatin, inhibit HMGCR activity and thus block not only cholesterol synthesis but also mevalonate-dependent protein prenylation. It was demonstrated that lovastatin treatment impairs macrophage viability due to reduced protein prenylation [226]. These defects were linked to defective activation of the GTPases, Cdc42, and Rac, as well as defective JNK-c-Jun activity [226]. The production of 25-HC by M1 macrophages impairs protein prenylation, which potentiates antiviral effects by preventing viral entry into the cells [33]. Recent reports also demonstrate that mevalonate-dependent protein prenylation is linked to IFN-I responses in macrophages. In the absence of protein geranylgeranyl transferase-1β (GGTase I; encoded by the gene *Pggt1b*), macrophages hyperproduce inflammatory cytokines downstream of TLR stimulation [4]. Of note, this hyperactivation appears to require myeloid differentiation factor D88 (MyD88) adaptor protein-dependent signaling, as MyD88-independent TLR3 stimulation does not augment pro-inflammatory cytokine production from *Pggt1b*-deficient macrophages [4, 278]. It was demonstrated that the GGTase I-dependent protein geranylgeranylation of KRAS mediates its interaction with the p110δ catalytic subunit of PI3K to promote PI3K activity and suppress IFN-I signaling. Consistent with this idea, *Pggt1b*- or *p110δ*-deficient macrophages produce excessive amounts of pro-inflammatory cytokines upon TLR activation due to the spontaneous activation of the pyrin inflammasome [4]. Pyrin inflammasome activation-dependent IFN-I signaling is also inhibited downstream of the RhoA, whose activation is also shaped by protein geranylgeranylation or farnesylation [301]. Thus, the mevalonate pathway regulates protein prenylation to ensure proper macrophage function.

As we discussed in detail above, the FAS program supports DC maturation, whereas FAO inhibits their pro-inflammatory functions. However, a recent study demonstrated that pDC cells favor FAO over FAS. TLR7 and TLR9 agonists activate the pDC subset, which causes them to produce pro-inflammatory cytokines like IFN-α and IFN-β [423]. Wu et al. demonstrated that TLR9-induced IFN-I acts in autocrine manner to signal via the IFN-α receptor (IFNAR) to enhance FAO in pDCs. In contrast to cDCs, the upregulation of these metabolic programs support pDC activation and energy production. Mechanistically, IFN-I signaling in pDCs induces pyruvate uptake into the mitochondria, which can subsequently serve as a substrate for FAS. Then, these de novo fatty acids can serve as a substrate for FAO. The pyruvate used to fuel FAS and subsequent FAO is derived from glycolysis. The divergent metabolic programs of pDCs and cDCs appear to be influenced by the PPARα network: pDCs express PPARα, while cDCs express low levels of this transcription factor. The induction of FAO and OXPHOS in pDCs is at least partially dependent upon PPARα [423]. Further exploration is required to dissect how TLR7/9-induced autocrine IFNAR signaling alters the metabolic programs of pDCs; however, it is reasonable to hypothesize that the histidine solute carrier SLC15A4 plays a role in this process. This protein is an important regulator of pDC cell activation, promotes endolysosomal acidification, and drives the mTOR-dependent upregulation of IFN-I production downstream of TLR7/9 [35, 202]. Thus, by regulating mTORC1 activation at the lysosome, SLC15A4 may induce the metabolic programs necessary to drive pDC activation [304, 306, 454].

1.4 Roles of Metabolism in Innate Lymphoid Cells

Innate lymphoid cells (ILCs) are immune cells enriched at mucosal surfaces, including the lung, skin, and intestines. ILCs express the common gamma chain (γc, CD132), IL-7Rα (CD127), IL-2Rα (CD25), and Thy1 (CD90) and are morphologically similar to lymphocytes, but they lack expression of molecules associated with T cells, B cells, myeloid cells, or granulocytes. Similar to innate immune cells, ILCs produce pro- and anti-inflammatory cytokines in response to other cytokines and microbial-derived signals. The ILC family is divided into two separate lineages: the cytotoxic ILCs and the noncytotoxic ILCs. NK cells are the only known cytotoxic ILC, whereas the noncytotoxic ILCs are further divided into group 1 (ILC1), group 2 (ILC2), or group 3 (ILC3) ILCs based on differences in transcription factor expression, cytokine profiles, and effector functions as discussed briefly below [13, 361]. Of these populations, the roles of cellular and systemic metabolism in noncytotoxic ILC cells is only just beginning to be understood [40, 153, 258]. Therefore, we limit our discussion below to NK cells.

Cytotoxic NK cells play roles in antiviral and antitumor responses by producing perforin, granzymes, and cytokines (e.g., IFN-γ) that can influence the activity of other immune cells. The expression of various activating and inhibitory receptors and cytokines promote degranulation and cytokine production by NK cells. Further, NK cells can lyse cells via CD16 binding to the Fc portions of antibody-bound targeted cells, a process called antibody-dependent cellular cytotoxicity (ADCC). Recent studies have investigated how metabolic programs contribute to NK cell functions. The cells are activated in response to surface receptors, such as NK1.1, NKG2D, Ly49D, and TLRs, and cytokines, including IL-15, IL-12, and IL-18. The resting metabolic activity of quiescent NK cells is low [190, 245], and they appear to preferentially utilize OXPHOS over aerobic glycolysis [190]. FAO does not appear to fuel OXPHOS in resting NK cells, because treating NK cells with etomoxir (CPT1 inhibitor) does not reduce ATP production [190]. The basal metabolic profile of NK cells is not altered upon short-term stimulation with various cytokines (IL-12 + IL-18; IL-12 + IL-15; IL-15 alone) or with agonistic antibodies (anti-NK1.1; anti-Ly49D) for activating receptors. However, the basal levels of OXPHOS and glycolysis are both required for NK1.1-induced IFN-γ production in vitro, whereas cytokine-dependent upregulation of IFN-γ is transcriptionally regulated and does require either OXPHOS or glycolysis when analyzed early (<6 h) after stimulation [190]. Thus, the resting metabolic state of NK cells is relatively low and can transiently support limited, stimuli-induced effector functions.

After extended activation, NK cells must undergo metabolic reprogramming to support their effector functions. Marcais et al. demonstrated that NK cells activated with IL-15 in vitro, with poly(I:C) (TLR3 and RIG-I ligand) in vivo, or with influenza A virus in vivo displayed increased cell size, granularity, expression of transferrin receptor (CD71) and CD98 (a component of the LAT1 amino acid transporter), and glucose uptake. Moreover, activated NK cells express higher levels of genes associated with protein and lipid biosynthesis and glucose catabolism [245]. Consequently, activated NK cells upregulate both glycolysis and OXPHOS after extended activation, with the balance shifted in favor of glycolytic metabolism [100, 245]. Inhibition of glycolysis reduces activation stimuli-induced upregulation of CD69, granzyme B, and IFN-γ. Mechanistically, mTORC1 activity increases glycolysis to support NK cell effector functions [100, 245]. NK cell functions may also be regulated by mTORC2-dependent inhibition of FOXO1 activity [95].

NK cells have been reported to have memory-like responses under certain conditions, displaying enhanced effector functions upon secondary antigen challenge [79, 276, 372]. Activated NK cells must balance proapoptotic and prosurvival pathways to become memory-like NK cells. Using a model of murine cytomegalovirus (MCMV) infection, O'Sullivan et al. found that activated NK cells accumulate defective mitochondria, characterized by lower membrane potentials and higher production of ROS. At the contraction-to-memory phase transition, these dysfunctional mitochondria are cleared via mitophagy [282]. NK cells require mitophagy to survive and form memory, as *Atg3*$^{-/-}$ NK cells undergo more extension contraction and generate fewer memory-like cells following MCMV challenge [282]. AMPK activity activates and mTORC1 activity suppresses autophagy induced by MCMV infection [282]. Moreover, FOXO1 colocalizes with Atg7 and is required for the induction of autophagy in NK cells [407]. These data suggest that metabolic reprogramming also supports NK cell survival and differentiation into memory-like NK cells.

It is reasonable to hypothesize that memory NK cells require FAO, because AMPK activation and mTORC1 inhibition augment both memory NK cell and memory $CD8^+$ T cell (T_M) differentiation [10, 282, 307, 321]. Indeed, AMPK-triggered autophagy could degrade intracellular lipid stores as an energy source for FAO [307, 358]. However, it is also feasible that memory NK cell and T_M cell differentiation require distinct metabolic programs. In support of this view, mitochondrial mass is lower, and mitochondrial membrane potential is higher in memory-like NK cells than effector or contraction-phase NK cells, whereas T_M cells have increased mitochondrial mass and function [393]. The microenvironment is a critical regulator of NK cell responses, so it will be important to determine how both cell-intrinsic and cell-extrinsic metabolites regulate NK cell responses in different pathological conditions. How NK cell responses are shaped in different microenvironments is a topic covered elsewhere [16].

Nutrient deprivation also influences NK cell functions. Tumor microenvironments and sites of inflammation are hypoxic. This state causes quiescent and activated NK cells to express lower levels of activating receptors, such as NKp46, NKp30, NKp44, and NKG2D [19]. NK cells cultured under hypoxic conditions are less cytotoxic, despite expressing normal levels of perforin and granzyme B; however, ADCC is retained in hypoxic states, consistent with the observation that CD16 expression is not inhibited by hypoxia [19, 337]. It has been reported that IL-2 treatment can restore NK cell function induced by hypoxia [337]. NK cells upregulate HIF-1α expression to survive under states of hypoxia [19]. Moreover, RNAi targeting of HIF-1α in the T-cell lymphoma tumor microenvironment promotes the NK cell-dependent regression of the tumor [373], suggesting that HIF-1α may impede NK cell functions in certain cancers. How local changes in metabolism influence site-specific NK cell responses is unknown.

1.5 Roles of Metabolism in Conventional T Cells

T cells originate from hematopoietic stem cell progenitors in the bone marrow and complete their development in the thymus. Upon maturation and egress from the thymus as naïve, single-positive $CD4^+$ or $CD8^+$ T cells, these T cells recirculate between secondary lymphoid organs via the blood and lymphatic systems. These naïve T cells are quiescent. Only upon receiving TCR and co-stimulatory activation signals from APC in the presence of cytokines will these cells undergo extensive proliferation and differentiate into effector T cells, known as $CD4^+$ T helper (T_H) and effector $CD8^+$ cytotoxic T lymphocytes [CTLs; also called effector $CD8^+$ T cells (T_E) in this chapter]. T cells play a vital role in cell-mediated immunity. Specifically, $CD4^+$ T cells are involved in serving other immune cells, such as B cells for antibody production and activation of macrophages to promote clearance of infectious agents. $CD8^+$ T cells directly kill infected host cells. A network of transcriptional and metabolic programs coordinates T cell quiescence, activation, and differentiation, which we discuss further below. For brevity, our discussion is primarily limited to mature T cells, but other references are available describing how metabolic reprogramming contributes to conventional T cell development [46, 242]. The reader should refer to Fig. 1.4 for a model summarizing the information below.

A

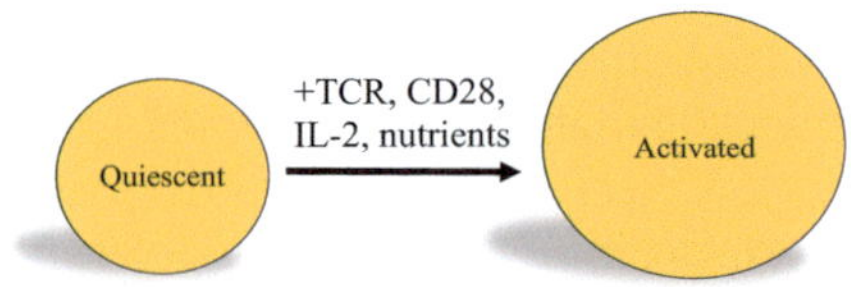

Metabolic profile	Low glycolysis; OXPHOS, & FAO	High glycolysis, glutaminolysis; OXPHOS, FAS, & cholesterol synthesis
Kinases involved	High AMPK, Low mTORC1	Low AMPK, High mTORC1
Transcription factors	FOXO1, Foxp1, KLF2	ERRα, c-MYC, SREBPs

B

Metabolic profile	High aerobic glycolysis Low OXPHOS (?) High glutamine uptake Low leucine uptake (?)	High aerobic glycolysis High OXPHOS (?) Low glutamine uptake (?) Low leucine uptake (?)	High aerobic glycolysis High FAO High glutamine uptake (?) High leucine uptake (?)	Aerobic glycolysis OXPHOS Lipid metabolism
Kinases involved	High mTORC1 High mTORC2	High mTORC1 High mTORC2 (?)	High mTORC1	mTORC1 mTORC2
Transcription factors	c-MYC, ERRα, SREBPs, T-bet	c-MYC, ERRα, SREBPs, GATA3	HIF-1α, ERRα, RORγt	Bcl6

Fig. 1.4 Metabolic changes are associated with T cell activation and CD4+ T cell differentiation The metabolic requirements of T cells vary depending upon their activation or differentiation state. This figure compares the metabolic and molecular signatures (kinases, transcription factors) of (**a**) quiescent, naïve T cells versus activated T cells or (**b**) various effector $CD4^+$ T cell populations. The reader is encouraged to visit Sect. 1.5 of this chapter for more information

1.5.1 Quiescent T Cells

Naïve T cells are actively maintained in a quiescent state. Quiescent T cells are characterized by smaller cell size and lower metabolic rates than activated T cells. Energy is required for naïve T cell survival, which is fueled by OXPHOS. Pyruvate feeding into this catabolic pathway can be generated from glucose, FAO, or amino acids. Tonic TCR (i.e., low-level signaling in response to self-peptide/MHC molecules) and IL-7-IL-7 receptor (IL-7R) signals control glucose uptake to mediate naïve T cell survival. Consistent with this idea, without TCR engagement, GLUT1 expression and glucose uptake are reduced, ultimately causing naïve T cell death and atrophy [322]. IL-7R signaling through STAT5 mediates AKT activation to promote glucose uptake and prevent cell death [420]. Tonic TCR signals also suppress the transcriptional activity of the FOXO family, Foxp1, and KLF2 to maintain T cell quiescence [57, 113, 144, 191, 207, 292, 413]. Foxp1 deletion disrupts quiescence by inhibiting IL-7Rα expression and negatively regulating ERK, thereby simultaneously repressing key pathways in both cellular metabolism and cell cycle progression [113]. FOXO- and KLF2-dependent programs also regulate IL-7Rα expression, as well as the expression of trafficking molecules to maintain the quiescent, peripheral T cell pool [57, 144, 191, 292]. KLF2 overexpression causes T cells to exit from the cell cycle, in part by decreasing *Myc* mRNA expression [47]. How metabolism links into the homeostatic properties of naïve T cells is unknown.

The proper balance of mTORC1 and mTORC2 signals also controls the metabolic demands of naïve T cells. In the absence of TSC1, there is an increase of mTORC1 activity and a concomitant decrease of mTORC2 activity. This imbalance of mTOR signaling, especially excessive mTORC1 signaling, drives metabolic dysregulation that promotes spontaneous cell growth, proliferation, loss of quiescence, and ultimately cell death [275, 313, 426, 433, 446]. Similarly, TSC2-deficient peripheral T cells display a hyperactivated phenotype indicative of defective quiescence, although the survival of TSC2-deficient T cells appears to be intact [313]. These differences appear to reflect changes in mTORC2 activity and TSC1 expression, since only TSC1 deficiency diminishes both [275, 313, 426, 433]. LKB1 is another upstream inhibitor of mTOR whose function is linked to T cell development and peripheral homeostasis. LKB1 deficiency limits thymocyte development [56, 241], and the peripheral T cells that arise in the absence of LKB1 have an activated phenotype [241]. Further, LKB1-deficient T cells have an increased glycolytic profile, displaying elevated glucose uptake and expression of GLUT1 and HK2 [241]. Of note, these phenotypes may be the consequence of lymphopenia-driven proliferation, but the metabolic alterations are similar in AMPKα-deficient T cells, which do not have developmental defects [241]. Because PI3K signaling induces mTOR activity and indirectly suppresses FOXO functions, one could predict that PTEN deficiency also impairs T cell quiescence. However, PTEN-deficient T cells undergo tumorigenesis but remain quiescent until tumor formation [118, 148]. It is possible that the mTORC2-AKT axis is able to suppress FOXO in the context of PTEN deficiency, thus maintaining quiescence.

1.5.2 T Cell Activation and Proliferation

1.5.2.1 Mitochondrial Respiration Is Essential for T Cell Activation

Upon activation, naïve T cells proliferate and differentiate into effector $CD4^+$ T cells or $CD8^+$ T cells, and this is associated with a shift in their metabolic programs. T cell activation is marked by changes in cell size, increased expression of activation-associated molecules (e.g., CD44, CD25, CD154), and proliferation. While naïve T cells are more catabolic, activated T cells adopt a more anabolic program to support cell growth and proliferation. The changes in metabolic programs are associated with shifts in intracellular metabolites that enable cell growth and proliferation, including amino acids, polyamines, lipids, and nucleotides [403]. OXPHOS programs increase in activated T cells [46, 64, 242]. OXPHOS supports T cell activation, as inhibition of the ETC diminishes proliferation and activation molecule expression [23, 64]. Moreover, in the absence of *Uqcrfs* (a gene encoding a subunit of complex III), T cells produce less IL-2 and are unable to proliferate in response to antigen stimulation, although lymphopenia-driven proliferation is retained [345]. The observation that FAO diminishes after T cell activation suggests that fatty acids do not fuel OXPHOS [403]. Interestingly, mitochondrial-derived ROS produced during OXPHOS activates nuclear factor of activated T cells (NFAT) to promote T cell functions [345]. Of note, exogenous ROS, especially those derived from phagocytic cells, have been demonstrated to negatively impact T cell survival and functions [24]. Deletion of Tfam, a protein that regulates mitochondrial DNA dynamics and hence OXPHOS, impairs peripheral T cell homeostasis and reduces activation-induced proliferation [17]. In this model, defects in mitochondrial respiration were also demonstrated to impair lysosome function, endolysosomal trafficking, and autophagy; these cells also accumulate NAD^+ which impairs their functions [17]. Thus, mitochondria respiration controls T cell activation on multiple levels.

1.5.2.2 Glycolysis and Glutaminolysis Are Essential for T Cell Activation

Glucose uptake and glycolysis are rapidly upregulated after TCR and CD28 stimulation, which supports T cell proliferation and survival during the early stages of activation [64, 121, 403]. Additionally, glutamine consumption increases in activated T cells [403]. Carbon-tracing experiments have demonstrated that glucose-derived carbons can enter into the TCA cycle to support OXPHOS [64, 403]. However, glucose-derived pyruvate is primarily catabolized into lactate [64, 403], with approximately 10% of glucose entering into the TCA cycle [280]. Interestingly, despite this low rate of glucose-derived pyruvate entering into the TCA cycle and the observation that FAO is diminished upon T cell activation [403], Roos et al. demonstrated that nearly 85% of ATP was derived from OXPHOS [329]. The conclusion that OXPHOS supports ATP production by T cells was also recently confirmed using apoptosis-inducing factor (AIF)-deficient T cells [255]. Therefore, other nutrients appear to contribute to ATP generation through OXPHOS. It has also been demonstrated that glutaminolysis increases after T cell activation due to the c-MYC-dependent upregulation of GLS2 and other glutaminolytic enzymes and transporters [403]. This process provides cells with α-KG that feeds back into the TCA cycle, which can fuel OXPHOS and also provide intermediates for protein, nucleotide, and membrane synthesis required to rapidly proliferate and grow. Of note, carbon-tracing experiments demonstrated that glucose and glutamine-derived carbons only account for approximately 25% of labeled carbons in activated primary mouse T cells [161], suggesting that other nutrient sources largely contribute to biomass accumulation upon T cell activation. Given that both activated T cells and cancer cell lines use Warburg metabolism [306], it is likely that the remaining carbon sources are from other amino acids, including valine and serine [161]. We discuss how amino acid metabolism influences T cell fates in more detail below.

To upregulate glycolysis and glutaminolysis, activated T cells increase GLUT1 expression, which is linked to effective T cell activation. Indeed, GLUT1 deficiency impairs T cell clonal expansion, growth, and survival during activation [240], while overexpression of GLUT1 increases cell size and activation molecule expression [173]. Initial studies suggested that TCR- and CD28-driven PI3K activation increases GLUT1 expression to enhance glucose uptake [121, 173, 420]. However, PI3K-deficient T cells have minor growth and proliferation defects as compared to c-MYC-deficient T cells [288, 289, 403]. Of note, those initial studies characterizing the role of PI3K used the inhibitor LY294002, which targets the activity of not only PI3K but also mTORC1 and the Pim family kinases [116]. The use of more selective pharmacological inhibitors does not impair glucose uptake [239]. Similarly, T cells expressing a mutant PDK1 that cannot effectively bind PIP_3 and activate AKT display normal proliferation and growth [411]. Finally, deletion of *Rictor* impairs T cell survival but not proliferation [434]. Thus, investigators have suggested that PI3K-AKT activation might not be critical for initiating T cell activation [116].

What other kinases could account for T cell activation-related metabolic reprogramming? The inhibition of mTORC1 diminishes TCR- and CD28-induced c-MYC upregulation [403, 434], suggesting that mTORC1 is a critical regulator of T cell responses. In support of this idea, metabolic reprogramming is compromised in RAPTOR-deficient T cells [434], and *Mtor* and *Rptor* deletion suppresses TCR- and CD28-induced proliferation and growth [92, 434]. Interestingly, there is a temporal requirement for mTORC1-dependent signals driving proliferation, where mTORC1 activity is more important during T cell priming (i.e., during initial TCR/CD28 stimulation) than 24–72 h after activation [434]. Additionally, glucose utilization by activated T cells is enhanced by ERK-dependent signals, where ERK helps initiate transcription of metabolic genes. ERK inhibition also impairs HK2 activity, which limits glycolysis [59]. The MAPKs also promote signaling that augments the expression of the sodium-dependent neutral amino acid transporters 1 and 2 (SNAT1 and SNAT2), which subsequently increases glutamine uptake [246]. However, these receptors may only play a role in T cell function in select contexts, as ASCT2 appears to be the dominant glutamine transporter for T cells [265]. How the mTORC1 and MAPK pathways intersect to control metabolic reprogramming and whether there is a temporal requirement for ERK signaling for engaging glycolysis and glutaminolysis during T cell activation remain unexplored.

The activation of c-MYC and mTORC1 may also induce metabolic reprogramming via ERRα [84]. It was demonstrated recently that ERRα-deficient T cells have reduced growth and proliferation [253]. Mechanistically, these defects are linked to impaired GLUT1 expression, limited glucose uptake, and reduced expression of many glycolytic genes. While this study largely linked the ERRα-deficient T cell defects to alterations in glycolysis, this transcription factor also induces gene expression related to mitochondrial biogenesis [256], so reduced glucose-derived pyruvate shuttling into the mitochondria may account for some of these defects. Indeed, addition of free fatty acids restores ATP production and proliferation upon ERRα suppression, which is likely mediated by increased expression of CPT1a; however, free fatty acids are not sufficient to restore maximum cytokine effector responses [253]. We will discuss how metabolic pathways influence cytokine production below.

In vivo, T cell activation occurs in the context of oxygen tension [251]. T cells activated under hypoxic conditions in vitro display reduced T cell proliferation and cytokine production; interestingly, the metabolic profiles of T cells activated under hypoxic conditions are more similar to those activated in vivo [14, 15, 53, 216, 251, 387]. TCR stimulation induces HIF-1α expression, which is further enhanced by hypoxia [53]. Because mTORC1 activity promotes HIF-1α expression in naïve T cells [434], it has been of interest to characterize how HIF-1α controls T cell responses. Although HIF-1α activity promotes glycolysis, HIF-1α is dispensable for early glycolytic reprogramming in T cells [403]. Instead, HIF-1α has reported negative roles for controlling T cell responses [25, 210, 235, 384]. As noted above, mitochondrial ROS supports T cell activation [345], but whether ROS-dependent control of HIF-1α expression contributes to this control is unknown. We further discuss the roles of HIF-1α in effector T cell differentiation in the next subsection.

1.5.2.3 Cholesterol and Lipid Metabolism Influence Early T Cell Responses

The biosynthetic demands of T cells are also met via the cholesterol and lipid biosynthesis programs. Lipid synthesis increases while oxidation decreases in activated T cells [403]. How is this metabolic shift achieved? The expression of CPT1a is rapidly downregulated upon T cell activation [403]. Additionally, glucose and glutamine sensing inhibits FAO by attenuating AMPK activation in favor of mTORC1 activation [32, 252, 328]. However, studies have suggested that AMPK activation and FAO support T cell functions when glycolysis is limiting [32, 253]. Moreover, AMPK-dependent FAO promotes effective antitumor immune responses and supports T cell effector functions during infection [32, 319, 370], scenarios when T cells are in competition for nutrients. The activation of AMPK in these contexts could be supported by TCR-driven calcium flux activating CaMK2 [379]. Additionally, LKB1-mediated upregulation of CD98 could also support the survival of proliferating T cells when nutrients become limiting [380], although it is not clear if LKB1 acts via AMPK in this context. Interestingly, aberrant upregulation of FAO and CPT1a expression is linked to excessive T cell responses in graft-versus-host disease and systemic lupus erythematosus (SLE) disease models and SLE patients [51, 441]. Moreover, despite playing inhibitory roles in T cell activation [75, 97, 159, 437], PPARγ is required for the development of graft-versus-host disease and for $CD4^+$ T cell-mediated, lymphopenia-driven autoimmunity [162]. Thus, FAO can play both beneficial and detrimental roles in T cell functions.

In cells using Warburg metabolism, glucose-derived citrate from the TCA cycle is exported into the mitochondria and converted back into acetyl-CoA via the activity of mitochondrial ACLY [22]. This acetyl-CoA can then serve as a precursor for lipid and cholesterol biosynthesis. How, then, does fatty acid or cholesterol synthesis affect T cell activation? The inhibition of ACC1 limits both homeostatic and activation-induced proliferation of T cells [30, 217]. However, it has been suggested that FAS is not required to induce but sustain T cell activation, as ACC1 deficiency only reduces the survival of proliferating T cells [217]. Of note, ACC2 is dispensable for $CD8^+$ T cell responses [218]. In contrast, cholesterol synthesis programs are important for enforcing cell growth and proliferation during the exit from quiescence. Indeed, TCR triggering is sufficient to upregulate HMGCS and HMGCR expression to enforce cholesterol synthesis, and statin drugs profoundly impair T cell activation [62, 141, 193, 385]. Statins primarily suppress isoprenoid generation in T cells [103, 141, 385], so more work is needed to ascertain how these products mechanistically contribute to T cell immunity.

The mTORC1-dependent upregulation of the SREBPs induces lipid and cholesterol synthesis to support $CD8^+$ T cell growth and proliferation [193, 434]. Additionally, SREBP acts independently of c-MYC, HIF-1α, or AMPK to increase glycolysis and glutaminolysis upon T cell activation [193]. The loss of SREBP activity in SCAP-deficient T cells is correlated with a reduction of intracellular lipid and cholesterol pools, and exogenous cholesterol restores the growth and proliferation of SCAP-deficient T cells [193]. Thus, cellular cholesterol homeostasis is an important determinant of T cell activation. Bensinger et al. demonstrated that the enzyme SULT2B1, which modifies oxysterols with sulfur groups, regulates intracellular cholesterol homeostasis. This SULT2B1-mediated cholesterol modification drives oxysterol export from the cell through the ABCC1 cholesterol transporter and also reduces the availability of oxysterols that can activate LXR activity, an antagonist of TCR-mediated cell cycle progression [28]. The depletion of the cholesterol pool can then drive the SCAP-dependent activation of SREBP1 and SREBP2 to promote cholesterol and lipid biosynthesis [28, 193]. Recent studies have also suggested that cholesterol and its derivatives modulate TCR signaling [374, 402, 435], but TCR signaling appears to be intact in SCAP-deficient T cells [193]. How SCAP deficiency precisely impacts T cell responses awaits further exploration.

1.5.2.4 Amino Acids Impact T Cell Activation

Within hours of activation, naïve T cells upregulate the expression of multiple amino acid transporters, including LAT1-CD98, ASCT2, SNAT1, and SNAT2, among others [59, 356, 403]. Therefore, the influence of amino acids on T cell activation has been explored. Arginine depletion blocks T cell proliferation and glycolytic activity, but not OXPHOS [119]. Localized arginine depletion impairs T cell activation in the context of infections and cancers and can be mediated by both pathogens and APCs [41, 77]. Mechanistically, arginine depletion may inhibit T cell activation by limiting mTORC1 function [350, 406]. Similarly to Arg1-mediated arginine depletion, the tryptophan-catabolizing enzyme IDO arrests T cell proliferation and promotes anergy if expressed at high levels by APCs or tumor cells [77, 263]. Mechanistically, this is linked to increased activity of nonderepressible 2 (GCN2) in responding T cells, resulting in translational repression [263]. TCR triggering promotes BCATc expression, which catabolizes branched-chain amino acids (e.g., leucine, isoleucine, and valine). This process inhibits mTORC1-dependent glycolytic reprogramming [6, 448].

Several groups have also investigated how amino acid transporters regulate T cell activation. As one would predict given the role of glutaminolysis for T cell activation, deletion of glutamine transporters impairs proliferation [59, 356]. It has been shown that deficiency in the leucine/glutamine anti-porter complex LAT1-CD98 impairs clonal expansion [209, 356]. Of note, T cells lacking the CD98 heavy chain have impaired proliferation, but this might be attributed to the ability of this molecule to associate with integrins [209]. Interestingly, however, ASCT2-deficient T cells have no defects in activation-induced expansion, survival, or IL-2 production [265]. Thus, different amino acid transporters appear to have different effects on T cell activation and function. Below, we discuss how metabolism further governs the differentiation of T cells into unique functional subsets.

Although there is a clear correlation between amino acid uptake and increased mTORC1 activation, the mechanisms governing this activation in T cells remain unclear. As we mentioned above, the TSC1/2-RHEB-RAGA/B axis is a conserved pathway driving mTORC1 activity in response to different amino acids [334, 349, 350], while the GAP for RAGC/D is the FNIP-FLCN complex [349, 350, 389]. Consistent with this idea, TSC1/2-deficient T cells have elevated mTORC1 activation [313, 434]. Further, RHEB-deficient T cells display reduced mTORC1 activation; however, this inactivation is incomplete and transient, whereas RAPTOR-deficient T cells have sustained mTORC1 activation [434]. Interestingly, while amino acids can upregulate mTORC1 activation independently of TCR engagement, mTORC1 activation is only sustained in the presence of TCR stimulation and amino acids [265, 356]. Because TCR engagement is required for T cell activation, it is conceivable that amino acid signals prime mTORC1 activation in T cells but that these signals are rapidly downregulated to prevent inappropriate T cell activation. If and when appropriate TCR engagement occurs, signaling activated downstream of the TCR and accessory receptors increase amino acid flux and/or cooperate with the RHEB-mediated activation of mTORC1 to sustain mTORC1 activation necessary to exit quiescence and acquire effector phenotypes as discussed below. Indeed, mTORC1 activation can be induced by multiple AKT-independent pathways in T cells [117, 136, 137, 151], and the CARMA1-BCL10-MALT1 complex promotes ASCT2 function to increase glutamine and leucine uptake, allowing for maximum mTORC1 activity [265]. Future studies will continue to explore how amino acids influence the molecular programs of T cells to ultimately shape their functions.

1.5.3 *Effector CD4⁺ T Cell Differentiation*

1.5.3.1 Overview of Effector CD4⁺ T Cells

After receiving appropriate TCR and co-stimulatory signals, activated T cells further differentiate into effector T cell lineages. The subset differentiation process for activated $CD4^+$ T cells is dictated by the coordinated actions of transcription factors, signature cytokine profiles, and immune signals. In the presence of appropriate cytokines, $CD4^+$ T cells differentiate into T helper 1 (T_H1), T_H2, T_H17, and T follicular helper (T_{FH}) cells. Functionally, these subsets play discrete roles in the clearance of infectious pathogens. T_H1 cells that express the transcription factor T-bet (encoded by *Tbx21*) and produce large amounts of IFN-γ aid in the eradication of intracellular pathogens. T_H2 cells (defined by the expression of GATA3) produce IL-4, IL-5, and IL-13 and promote clearance of parasites. RORγt⁺ T_H17 cells produce cytokines of the IL-17 family (predominantly IL-17A and IL-17F), IL-21, and IL-23 and are important for the clearance of extracellular pathogens, such as fungi. Finally, T_{FH} cells expressing the transcription factor Bcl6 are present in the lymphoid follicle, where they interact with B cells to promote germinal center reactions that facilitate high-affinity, class-switched antibody production. Dysregulated $CD4^+$ effector T cell responses are also linked to autoimmunity or inflammatory disease [185]. IL-9-producing T_H9 and IL-22-producing T_H22 subsets that play a role in allergic diseases in the lung and skin have also been described [186]. However, little is known about the metabolic programs driving T_H22 differentiation. We discuss how metabolic programs control $CD4^+$ effector T cell differentiation below.

1.5.3.2 T_H1 and T_H2 Cells

Glycolysis is the dominant metabolic program of T_H1 and T_H2 cells. Compared to naïve T cells, in vitro-differentiated T_H1 and T_H2 cells express high levels of GLUT1, with T_H2 cells expressing the highest levels of this glucose transporter [252]. Glycolysis inhibition via 2-DG treatment suppresses T_H1 proliferation and survival [132]. Further, T-bet expression is not efficiently upregulated under T_H1-skewing conditions, which diminishes their ability to produce IFN-γ [132, 448]. These data are similar to results showing that 2-DG treatment or culture with galactose diminishes IFN-γ production by $CD4^+$ T cells [64, 448]. The AKT-mTOR axis controls metabolic rewiring to support T_H1 and T_H2 differentiation. Loss of either RHEB or RAPTOR diminishes T_H1 differentiation [93, 324, 434], demonstrating that mTORC1 directs T_H1 differentiation, while the role of mTORC2 signaling remains controversial [93, 220]. T_H2 differentiation also requires mTORC1 and mTORC2 activity [93, 220, 434], but because mTORC1 promotes T cell activation, its inhibition more potently suppresses T_H2 differentiation. Moreover, mTORC1 appears to regulate at least two phases of T_H2 differentiation: exit from quiescence and commitment to the T_H2 effector program [434]. Precisely how mTORC1 promotes T_H2 subset commitment is unclear, but metabolic genes or metabolites induced by mTORC1 activation may influence T_H2 programs analogous to glycolysis-regulated IFN-γ production [64, 158]. While mTORC1-dependent signaling transcriptionally influences metabolic reprogramming when T cells are exiting from quiescence [434], it is also possible that mTORC1 via its control of metabolism and translation influences protein expression and function at the translational and posttranslational levels. Indeed, a recent report demonstrated that mTORC1 can influence metabolic gene expression in activated $CD8^+$ T cells [166]. How mTORC2 regulates T_H2 and/or T_H1 differentiation is also not clear. However, mTORC2-mediated control of AKT is also likely important for this process, since AKT supports glycolytic metabolism in T cells [121]. Further, mTORC2 may promote c-MYC and SREBP activity by increasing RAPTOR-mTORC1 activity via AKT [434]. Finally, the mTORC2-dependent regulation of FOXO1/3a could influence this process. Future studies will continue to extrapolate how mTOR signaling drives T_H1 and T_H2 differentiation.

Recent investigations have helped answer how glycolysis supports T_H1 effector functions. When in vitro-activated $CD4^+$ T cells or in vivo-polarized T_H1 cells receive secondary TCR signals under glucose-limiting conditions, they produce less IFN-γ [64, 158]. Pearce and colleagues demonstrated that aerobic glycolysis is linked to the translation of *Ifng* mRNA. Mechanistically, the glycolytic enzyme glyceraldehyde-3-phosphate dehydrogenase (GAPDH) binds *Ifng* mRNA when aerobic glycolysis is not permissive (e.g., when T cells are activated in the presence of galactose), blocking *Ifng* translation [64]. The reduction of *Ifng* mRNA translation could also be linked to reduce AKT and mTORC1 activities or increase AMPK activity [32, 158, 370], but this remains to be addressed. Glycolytic metabolites also impact TCR signaling via the Ca^{2+}-NFAT axis [158]. Specifically, the glycolytic metabolite PEP enables intracellular Ca^{2+} from ER stores to accumulate in the cytosol, drive extracellular Ca^{2+} flux, and activate NFAT transcriptional activity upon secondary TCR engagement. Interestingly, TCR signaling appears to be hardwired to facilitate PEP accumulation. PEP is converted into pyruvate via the activities of PKM, of which PKM1 is more active than PKM2. Because TCR signaling selectively upregulates PKM2 expression [403], PEP can accumulate more readily [394], thus promoting appropriate Ca^{2+}-NFAT activation.

Although glycolysis supports T_H1 and T_H2 differentiation, analysis of Tfam-deficient mice suggests that OXPHOS antagonizes the differentiation of T_H1 cells [17]. When Tfam-deficient T cells are differentiated in vitro toward the T_H1 lineage, they express increased levels of IFN-γ and T-bet. Interestingly, similar elevations in

IFN-γ and T-bet expression are observed when cells are polarized under T_H2 conditions, suggesting that the balance of OXPHOS in differentiating cells may tune T-bet expression and thus determine the lineage commitment of T_H1 versus T_H2 cells. Indeed, Tfam deficiency attenuates IL-4 production by in vitro-derived T_H2 cells. How does mitochondrial dysfunction exacerbate T_H1 differentiation? Due to reduced levels of NAD^+, Tfam-deficient T cells accumulate dysfunctional lysosomes, leading to increased levels of sphingomyelin and triacylglycerides. Indeed, naïve T cells activated in the presence of a lysosome inhibitor of sphingomyelin produce excessive levels of IFN-γ, and increasing the levels of NAD^+ available to Tfam-deficient T cells restores their T_H1 profile. Future work will uncover whether additional mitochondrion-regulated processes also contribute to aberrant T_H1 differentiation.

Amino acid uptake and metabolism also influence T_H1 and T_H2 differentiation. Naïve T cells cultured in glutamine-free media fail to become IFN-γ-producing T_H1 cells. The uptake of glutamine is partially mediated by ASCT2, which promotes T_H1 cell expansion [265]. By contrast, ASCT2-deficient T_H2 cells express higher levels of GATA3 and IL-4, suggesting that ASCT2-dependent glutamine uptake suppresses T_H2 differentiation in favor of T_H1 differentiation. ASCT2 facilitates maximum glutamine uptake by activated T cells and also contributes to glucose and leucine uptake. ASCT2-related nutrient uptake is linked to increased TCR- and CD28-induced mTORC1 activation, which enhances c-MYC expression to support aerobic glycolysis in activated T cells. However, OXPHOS activity is only temporarily impaired in ASCT2-deficient T cells, which could help explain why activation-induced proliferation and IL-2 production are not halted in its absence. It is noteworthy, however, that there appears to be a temporal or differentiation-related requirement for ASCT2 in driving mTORC1 activity in T_H1 cells. Despite having reduced frequencies of T_H1 cells in its absence, ASCT2-deficient T_H1 cells appear to have normal levels of phosphorylated S6 (indicative of mTORC1 signaling) and glutamine uptake. Moreover, only excess glutamine can rescue IFN-γ production when naïve T cells are cultured under T_H1-polarizing conditions, whereas excess leucine inhibits T_H1 differentiation in both ASCT2-deficient and ASCT2-sufficient T cells. The mechanisms underlying the leucine-mediated inhibition of T_H1 polarization require further investigation.

1.5.3.3 T_H17 Cells

In vitro-differentiated T_H17 cells are also highly glycolytic relative to naïve T cells, and IL-17-producing cells express high levels of GLUT1 during EAE [132, 252, 348]. Glycolysis is functionally important for T_H17 differentiation in vitro and in vivo, because it supports T_H17 cell proliferation, survival, and the upregulation of RORγt expression [132, 348]. Mechanistically, the differentiation of T_H17 from naïve T cells is regulated by mTORC1 [92, 93, 348]. In the absence of mTOR activity, IL-6-induced STAT3 phosphorylation is impaired, which in turn limits the upregulation of

STAT3-dependent genes encoding IL-21 and RORγt [92]. Further, mTOR-deficient T cells do not upregulate IL-23R expression, which stabilizes the phenotype of T_H17 cells. These effects appear to be attributed to mTORC1 activity, because T_H17 differentiation is normal in the absence of RICTOR [93, 220]. The upregulation of IL-23R enhances mTORC1 activation in T_H17 cells [66]. Additional signals like the cytokine IL-1 and short-chain fatty acids can also maintain high levels of mTORC1 activity in T_H17 cells [66, 145, 169, 208, 299]. Of note, mTORC2-mediated activation of SGK1 downstream of IL-23R could also regulate the pathogenicity of T_H17 cells [200, 422], but this remains to be formally tested.

If glycolysis and mTORC1 activity are important for T_H1, T_H2, and T_H17 differentiation, what molecular and metabolic signatures determine which fate $CD4^+$ T cell adopt? T_H17 cells express higher levels of HIF-1α than T_H1 or T_H2 cells. Moreover, HIF-1α-deficient T cells do not efficiently upregulate T_H17-related cytokines, IL-23R expression, and glycolysis-related genes under T_H17 conditions [86, 348], whereas T_H1 or T_H2 differentiation does not require HIF-1α [86]. HIF-1-dependent functions are also involved in maintaining the human T_H17 cell population [205]. Mechanistically, HIF-1α not only increases the expression of RORγt at the transcriptional level but also activates RORγt transcriptional activity by recruiting p300 to the RORγt-bound *Il17a* promoter [86]. In addition to increasing glycolysis, HIF-1α supports the T_H17 program by promoting the proteasomal degradation of forkhead box P3 (FoxP3), a transcription factor that promotes Treg cell differentiation [86, 120, 160, 192]. We discuss Treg cells in the next subsection. HIF-1α expression is markedly upregulated in an mTORC1-dependent fashion in T_H17 cells [348], but how is it so potently increased in T_H17 cells and not T_H1 or T_H2 cells? Activated STAT3 binds directly to the *Hif1a* gene promoter, suggesting that mTOR-dependent regulation of IL-6-induced STAT3 activation increases HIF-1α expression in T_H17 cells. Metabolic programs induced by mTORC1 activation could also influence HIF-1α stability through the actions of succinate or ROS-mediated inhibition of PHD [86]. In support of this view, T_H17 differentiation is enhanced under hypoxic conditions, where PHD activity is likely inhibited [86]. Pyruvate and lactate also influence HIF-1α activity in the absence of hypoxia [90, 234, 363]. Rathmell and colleagues demonstrated that T_H17 cells produce higher levels of both pyruvate and lactate than T_H1 cells, likely due to the increased expression of PDHK1 [132]. Indeed, PDHK1 activity is crucial for T_H17 but not T_H1 differentiation. Recent reports demonstrated that exogenous sodium lactate induces RORγt and IL-17A production by $CD4^+$ T cells and that T_H17 responses are positively correlated with increased serum lactate levels in humans [147, 347]. Whether T_H17 cell-derived lactate is important for their differentiation remains unexplored.

Although glycolysis is essential for their differentiation and function, T_H17 cells also upregulate FAO more efficiently than T_H1 or T_H2 cells [252]; however, exogenous free fatty acids suppress T_H17 differentiation in vitro by impairing their survival [252]. Further, AICAR-induced AMPK activation is correlated with enhanced fatty acid uptake and FAO, and this chemical agent suppresses T_H17 differentiation [143]. This effect may be mediated by AMPK-dependent inhibition of mTORC1 activity [32, 252]. However, the role of AMPK in T_H17 differentiation is complex.

AMPK signaling suppresses T_H17 responses and hence is protective in EAE [266, 267]. These alterations appear to be cell extrinsic, however, because AMPK-deficient $CD4^+$ T cells do not have impaired T_H17 differentiation [241]. Of note, deletion of LKB1, which activates AMPK, enhances T_H17 differentiation in vitro [241], although it is conceivable that these effects are AMPK independent. Despite the controversial role of AMPK in directing T_H17 differentiation and function, the induction of PPARγ downstream of AMPK does appear to inhibit T_H17 development. PPARγ activity is inversely correlated with the development of EAE, which is attributed to the inhibition of T_H17 differentiation through the antagonism of STAT3 transcriptional activity [201, 318, 438]. One question to address is, if FAO appears to antagonize T_H17 cell responses, why do they upregulate FAO programs? T_H17 cells may upregulate FAO as a consequence of increased fatty acid uptake. Indeed, T_H17 cells upregulate their expression of the epidermal fatty acid-binding protein (E-FABP) that can sequester fatty acids and hydrophobic molecules in the cytosol, and E-FABP-deficient T cells have defects in differentiating into T_H17 cells [223]. Functionally, E-FABP antagonizes PPARγ to increase T_H17 differentiation. By serving as inhibitors of HDACs, short-chain fatty acids augment T_H17 differentiation in vitro and in vivo [299]. Moreover, recent work suggests that short-chain and long-chain fatty acid sensing in the intestines shapes T_H17 responses in the peripheral tissues, including the central nervous system [149]. Future work will continue to dissect how fatty acid sensing impacts T_H17 differentiation and function.

As noted above, FAS and cholesterol biosynthesis are crucial regulators of naïve T cell activation. How do FAS and cholesterol biosynthesis-related molecular programs influence T_H17 differentiation? LXR increases SREBP-1 expression, which antagonizes T_H17 development [82]. Mechanistically, SREBP-1 binds to the *Il17a* promoter, blocking its transcriptional upregulation by the aryl hydrocarbon receptor (AHR) [82, 198, 397]. As noted above, studies have suggested that AMPK might antagonize T_H17 differentiation. The AMPK-dependent phosphorylation of ACC1 activity disrupts de novo FAS. AMPK can also inhibit ACC2 activity, whose conversion of mitochondrial acetyl-CoA into malonyl-CoA is important for the allosteric inhibition of CPT1a; thus, ACC2 can suppress FAO. It was found that naïve T cells cultured under T_H17 conditions upregulate the expression of ACC1/2 [30]. This upregulation is important for T_H17 differentiation, as ACC1 inhibition reduces T_H17 differentiation [30, 107]. Intriguingly, ACC1 inhibition impairs aerobic glycolysis and limits glucose shuttling into the TCA cycle for energy generation, and FAS not only suppresses FAO but also impairs the upregulation of aerobic glycolysis [30]. Of note, T_H1 and T_H2 differentiation are also impaired in the absence of ACC1 [30], suggesting that ACC1-related metabolic programs are important for effector $CD4^+$ T cell functions. Interestingly, a recent study demonstrated that ACC1 also controls T_H17 differentiation in the context of obesity by modulating RORγt recruitment to and function at the *Il17a* locus [107].

Amino acid sensing also contributes to T_H17 responses. Neither ASCT2- or LAT1-deficient T cells efficiently differentiate into T_H17 cells in vitro [149, 265, 356]. As noted above, these observations are due in part to glutamine-dependent uptake of leucine that controls mTORC1 activation to induce glycolysis in naïve $CD4^+$ T cells [265]. Interestingly, in contrast to T_H1 cells, ASCT2-deficient T_H17 cells have impaired mTORC1 activity, suggesting that glutamine and/or leucine uptake via ASCT2 helps maintain mTORC1 activity required to sustain the T_H17 program [265]. In support of this, exogenous leucine potentiates T_H17 differentiation in vitro and can restore T_H17 cell responses in the context of ASCT2 deficiency [265]. Because lysosomal-associated mTORC1 can be activated by amino acids [454], these results could indicate a differential requirement for lysosomal function in T_H17 and T_H1 differentiation. In support of this view, Tfam-deficient T cells lose lysosomal function, which is associated with a reduction of T_H17 cell-derived IL-17 production and increase of T_H1 differentiation and function [17]. Whether mTORC1 sensing of lysosomal nutrients, including amino acids, contributes to T_H17 cell functions is currently unknown.

1.5.3.4 T_{FH} Cells

The metabolic programs of T_{FH} cells were recently demonstrated to be divergent from T_H1 cells in the context of a systemic viral infection [324, 427]. LCMV-specific T_{FH} cells are less metabolically active than T_H1 cells [324]. Moreover, T_{FH} cells use more mitochondrial metabolism over glycolysis [324]. How is glycolysis inhibited in T_{FH} cells compared to T_H1 cells? IL-2 signals inhibit the T_{FH} program by modulating AKT and mTOR activities [18, 284, 324]. Indeed, T_{FH} cells display lower levels of AKT and mTOR signaling than T_H1 cells, which is important to suppress glycolysis and support T_{FH} differentiation over T_H1 polarization [324]. The transcription factor Bcl6 is essential for T_{FH} differentiation [180, 273], and it suppresses glycolysis driven by IL-2 signaling in T_H1 cells [285], suggesting that Bcl6 and its regulatory proteins may inhibit the glycolytic program of T_{FH} cells. Consistent with this idea, T cell factor 1 (TCF1, encoded by the *Tcf7* gene) is inhibited by IL-2 signals and promotes Bcl6 expression and subsequent T_{FH} differentiation [71, 427, 429]. *Tcf7* deletion in T_{FH} cells reduces mitochondrial function and the expression of genes associated with multiple branches of metabolism, including branched-chain amino acid degradation, the PPP, fatty acid metabolism, and the TCA cycle [427]. IL-2 signals have also been shown to differentially modulate FOXO1/3a binding to the *Bcl6* locus, in that low IL-2 signals are correlated with higher levels of FOXO1/3a binding [284]; thus, IL-2 signaling might also modulate Bcl6 expression and glycolysis via the FOXOs. However, FOXO1 and its downstream target KLF2 actually antagonize T_{FH} differentiation, in part by suppressing the ICOS-mediated induction of Bcl6 [219, 369, 412, 428]. Recent work has indeed demonstrated that ICOS can control the metabolic reprogramming of T_{FH} cells in an mTOR-dependent manner [443]. Therefore, cross talk between different upstream signals may differentially influence how signaling, transcriptional, and metabolic networks influence T_{FH} differentiation and function in different contexts.

1.5.3.5 T_H9 Cells

T_H9 cells were first characterized as a subset of T_H2 cells capable of producing IL-9 when cultured in the presence of TGF-β, but observations that cells cultured under these conditions rarely co-express GATA3 and IL-9 suggest that T_H9 cells are a unique subset [367, 398]. The differentiation program of these cells is dependent upon the transcription factors, STAT6, interferon-responsive factor 4 (IRF4), and PU.1, and the canonical and noncanonical NF-κB networks [342]. IL-2 signals are indispensable for inducing IL-9 production by $CD4^+$ T cells, with IL-4 and TGF-β potentiating these effects [341]. However, IL-4 is not necessary for T_H9 differentiation if IFN-γ is neutralized [264, 341]. IL-2 induces STAT5 activation, which enhances IL-9 production by $CD4^+$ T cells, via three known mechanisms. First, phosphorylated STAT5 can directly bind to the *Il9* locus [135, 436]. Second, STAT5 induces T_H9 differentiation by increasing IRF4 expression [135]. Third, IL-2 signaling activates Itk, which is important for T_H9 differentiation [135]. However, it should be noted that Itk signaling appears to be more important in the context of limiting TCR-CD28 and/or IL-2R signaling, because excess IL-2 can largely rescue the defective T_H9 differentiation of Itk-deficient T cells [135].

TGF-β induces PU.1 expression, which can subsequently bind the *Il9* locus and also upregulate the acetylation of histone H3 by recruiting histone acetyltransferases to the *Il9* locus [65, 140]. Additionally, the kinase TAK1 inhibits the activity of SIRT1 upon TGF-β stimulation [409]. Thus, TGF-β signaling appears to be an important modulator of epigenetic modifications at the *Il9* locus. How might TGF-β influence the epigenetic landscape of the *Il9* locus to impact T_H9 differentiation? A recent report suggests that SIRT1 antagonizes the mTOR-HIF-1α axis driving glycolysis to suppress T_H9 differentiation [409]. In response to TGF-β and IL-4 stimulation, SIRT1 expression is downregulated. Then, the glycolytic program increases in a mTORC1-HIF-1α-dependent fashion. Consistent with this idea, T_H9 differentiation is enhanced in the absence of SIRT1, but this differentiation is attenuated when glycolysis is inhibited or when mTOR or HIF-1α is deleted. It appears that SIRT1 activity opposes the ability of HIF-1α to bind the *Il9* promoter region. Thus, when SIRT1 activity is antagonized, HIF-1α binds to regions that are also enriched for primed histones (e.g., trimethyl K4 H3). Given that HIF-1α associates with p300/CBP in $CD4^+$ T cells [86], it may cooperate with PU.1 to promote the acetylation of histones at the *Il9* locus. Of note, T_H9 differentiation is not inhibited in the absence of HIF-1α or mTOR [409]. Similarly, T_H9 differentiation is resistant to rapamycin treatment [244]. Thus, future studies will be needed to determine precisely when and how the TAK1-mTOR-HIF-1α controls T_H9 differentiation, but Itk-dependent activation of mTORC1 downstream of IL-2 might help determine the requirements for mTORC1 signaling driving T_H9 differentiation [135].

1.5.4 $CD8^+$ T Cell Differentiation

$CD8^+$ T cell fates are also critically important for the clearance of acute infections and cancers. After encountering antigen and co-stimulatory signals, naïve $CD8^+$ T cells exit quiescence and rapidly proliferate. During this expansion phase, activated T cells integrate environmental antigen, co-stimulatory, and cytokine signals and subsequently become either short-lived effector cells (SLEC) or memory precursor effector cells (MPEC). Over time, antigen-specific $CD8^+$ T cells expand, but most die in a process termed contraction. The remaining memory $CD8^+$ T cells (T_M) are long-lived and have the capacity to rapidly re-expand and acquire effector functions if antigen is reencountered. T_M cells are further subdivided into central memory (T_{CM}), effector memory (T_{EM}), and tissue-resident memory (T_{RM}) cells based upon their expression of different trafficking receptors and functional properties. Several models have been proposed to explain how $CD8^+$ T cells acquire an effector or memory fate [183]. Below, we discuss how metabolic reprogramming contributes to these fate decisions.

Several groups have investigated how metabolic programs differ between T_E and T_M cells (Fig. 1.5a). T_E cells favor glycolysis and glutaminolysis over FAO, whereas T_M cells favor FAO and OXPHOS [63, 370, 393]. Interestingly, although both cell populations have similar cytokine profiles, glucose and glutamine consumption appears higher in, in vitro-derived T_E cells than T_H1 cells [166]. Glycolysis increases in T_E cells because glycolytic enzymes, including aldolase C, α-enolase, and HK2, increase in expression [63, 370], while genes associated with mitochondrial metabolism and FAO are downregulated [370]. Consistent with an important role of glycolysis in T_E function, 2-DG-treated $CD8^+$ T cells produce less IFN-γ and express lower levels of *Prf1* (encodes perforin) and *GzmB* (encodes granzyme B) mRNA than untreated cells [63, 370]. As we discussed above, SCAP-SREBP and/or ACC1-dependent FAS are crucial for $CD8^+$ T cell growth and activation [193, 217]; hence, they support T_E differentiation. Of note, long-chain fatty acid uptake is higher in T_E than naïve T cells or T_M cells due to increased expression of CD36. These lipids are subsequently stored as neutral lipids, such as triacylglycerol or cholesterol esters. Because T_M cells express higher levels of lysosomal acid lipase (LAL) than T_E cells, they can break down these neutral lipids in the lysosome, thus providing T_M cells with an intrinsic substrate for FAO [281]. In addition to increased FAO, T_M cells have enhanced mitochondrial content and have higher spare respiratory capacity (SRC; also called reserve respiratory capacity). Thus, T_M cells have an enhanced capacity to rapidly produce energy via OXPHOS than T_E cells.

The mTOR signaling axis facilitates T_E differentiation and function. T_E cells have high levels of mTORC1 and mTORC2 signaling [166], which is induced by antigenic, co-stimulatory receptor, and IL-12 signals [69, 320, 321]. The activation of mTORC1 positively regulates the transcriptional and metabolic network of T_E cells [10, 321]. How does mTORC1 signaling positively influence T_E differentia-

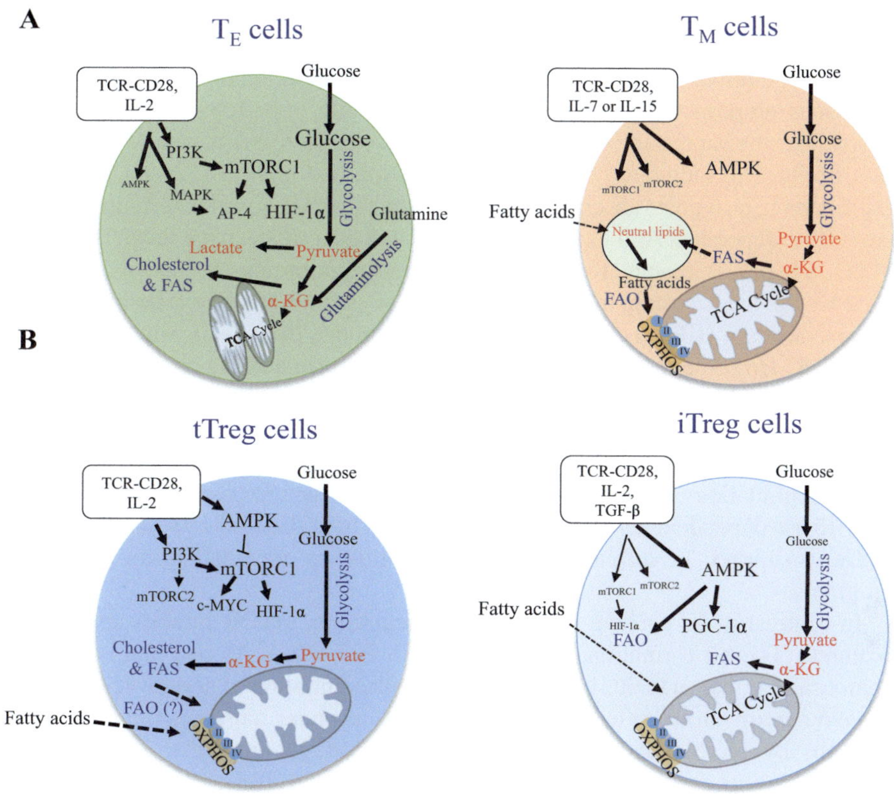

Fig. 1.5 Comparison of the metabolic requirements of CD8⁺ T_E cells and T_M cells versus tTreg cells and iTreg cells

(**a**) $CD8^+$ T_E cells and T_M cells utilize distinct and overlapping metabolic pathways to fuel their specific fates and functions. Glucose and glutaminolysis are the major metabolic programs of T_E cells, wherein these products contribute to α-ketoglutarate (*α-KG*) production, which is used to generate cholesterol and lipid products that support cell growth and proliferation. α-KG can also enter into the TCA to support OXPHOS, but T_E cells have fragmented mitochondria that are not efficient for OXPHOS. The mTORC1 pathway is a crucial regulator of metabolic reprogramming of T_E cells, in part by upregulating the expression of the transcription factors AP4 and HIF-1α. In contrast, T_M cells upregulate AMPK-dependent FAO programs to support their energy metabolism. Mitochondrial biogenesis is also increased downstream of PGC-1α, resulting in more fused mitochondrial networks that support OXPHOS. T_M cells can take up fatty acids to support FAO, but primarily break down de novo-synthesized fatty acids derived from glucose to support their metabolic needs. The reader is referred to Sect. 1.5 for more information related to this topic. (**b**) Several studies have suggested that thymic-derived Treg (*tTreg*) cells and in vitro-derived Treg (*iTreg*) cells use similar metabolic programs as T_E cells and T_M cells, respectively. The mTORC1 pathway is a crucial regulator of tTreg cell functions, and it controls glycolysis and cholesterol and fatty acid synthesis from these cells. The cells also display high levels of AMPK activity that might support fatty acid uptake and FAO under select contexts. Of interest, excessive glycolysis induced by too much PI3K-mTORC2 signaling or mTORC1-HIF-1α or –c-MYC signaling can destabilize tTreg cells, as indicated by the loss of FoxP3 expression (not depicted). Like T_M cells, AMPK-dependent FAO and glycolysis support the differentiation and function of these cells. By contrast, mTORC1 and mTORC2-dependent signals antagonize iTreg cell differentiation and function. The reader should refer to Sect. 1.6 for more information related to these cells

tion? First, mTORC1 signaling is linked to the upregulation of T-bet and EOMES by mechanisms that are not completely understood [313, 321] but may involve epigenetic regulation linked to metabolic programs. Additionally, the mTORC1-c-MYC axis supports the glycolytic program of T_E cells, in part, by upregulating AP4 expression; the expression of this transcription factor increases after c-MYC expression declines to facilitate prolonged proliferation and growth [72]. Of note, AP4 expression is also posttranscriptionally regulated, and MEK and p38 inhibitors suppress IL-2-dependent AP4 expression [72]. These data suggest that the mTORC1-c-MYC network is crucial for early metabolic programming events, whereas the MAPKs-AP4 axis sustain metabolic programs to support T_E cell expansion. Finally, the activation of HIF-1α can also functionally regulate the T_E program downstream of mTOR. *Hif1a*-deficient T_E cells express less perforin and granzyme B than control T_E cells, whereas increased HIF-1 function is linked to the elevated expression of these molecules [76, 98, 117]. However, the terminal differentiation of T_E cells is impaired in the absence of the von-Hippel-Lindau complex (VHL), which drives HIF-1 degradation. This process may be caused by lactate production induced by the HIF-1α-dependent aerobic glycolysis [117]. These data indicate that the mTOR pathway support T_E differentiation and function by modulating the metabolic programs.

In contrast to its positive role in T_E differentiation, mTOR inhibits and AMPK promotes T_M cell formation, respectively [10, 307, 321]. Ahmed and colleagues demonstrated that mTORC1 antagonizes T_M differentiation at several levels. First, in vivo rapamycin treatment during the expansion phase increases the quantity of T_M cells by increasing MPEC generation. Second, rapamycin treatment at the expansion-contraction transition phase improves the survival of the clonal $CD8^+$ T cell pool. Finally, mTORC1 inhibition during the contraction phase improves the quality of the T_M response without affecting the number of cells [10]. Consistent with a role for mTORC1 as an inhibitor of T_M differentiation, RHEB-deficient T cells have defective T_E differentiation but efficiently differentiate into T_M cells in response to a systemic bacterial infection [313]. Given these data, it is not surprising that T cells lacking TSC1 or TSC2 have elevated levels of mTORC1 activity and have impaired T_M cell differentiation [313, 353]. Powell and colleagues recently demonstrated that mTORC2 signaling is also a crucial regulator of T_M differentiation. While RICTOR-deficient T cells display normal T_E differentiation and function, they more efficiently differentiate into T_M cells with enhanced function. Under culture conditions that drive T_E (IL-2) or T_M (IL-7 or IL-15) differentiation, mTORC2 inhibition increases glycolytic metabolism in vitro. However, RICTOR-deficient T cells are resistant to IL-2-induced downregulation of SRC, a metabolic feature of T_M cells [313, 393]. These findings are consistent with observations that FOXO1-deficient T cells fail to downregulate T_E cell-related programs and acquire T_M cell phenotypic features [157, 383]. It has also been reported that TRAF6-deficient T cells cannot efficiently differentiate into T_M cells despite having normal T_E cell responses [307]. This defect is coupled to inefficient upregulation of FAO programs, which are rescued by AMPK activation. Because APMK can increase the uptake of fatty acids [254], it is possible that, in the absence of TRAF6, lipid uptake or accumulation is impaired. The mechanism by which TRAF6 regulates T_M responses still requires further investigation.

How is mTORC1 signaling tuned to influence T_E versus T_M cell fate decisions? Under selective activating conditions, T cells can undergo asymmetric cell division, wherein the APC-proximal daughter cell is more likely to differentiate into a T_E cell, and the APC-distal daughter cell is more likely to become a T_M cell [67]. During the first round of cell division, the metabolic machinery is asymmetrically segregated, wherein the proximal daughter cell acquires higher levels of mTORC1-c-MYC signaling than the distal daughter cell. CD25 and the LAT1-CD98 complex also polarize asymmetrically, and amino acid sensing and glutaminolysis sustain high c-MYC expression in the proximal daughter cell [314, 399]. By contrast, the distal daughter cell has lower levels of mTORC1 activation, higher mitochondrial content, and increased levels of FAO which is likely attributed to increased AMPK activity in the absence of sufficient levels of glucose and/or glutamine [32, 314]. Along with these metabolic features, the distal daughter cell upregulates pro-survival factors that ultimately allow for their long-term survival [314]. While the precise mechanisms contributing to asymmetric mTORC1 partitioning are not completely known, after TCR stimulation, RAGC transiently dissociates from the lysosome and localizes within the cytoplasm before it is redistributed to the lysosome of the proximal daughter cell [314]. This change is accompanied by a rapid, kinase-independent recruitment of mTOR to the lysosome, which is facilitated by LAT1-CD98-dependent amino acid sensing [314]. Then, mTORC1 segregation influences c-MYC expression in the proximal and distal daughter cells to ultimately influence T cell proliferation, metabolism, and differentiation [399].

Pearce and colleagues recently demonstrated that the mitochondria of T_M cells are fused, whereas mitochondrial fission is associated with T_E differentiation [45]. Moreover, $CD8^+$ T cells that have a high content of fused mitochondria display elevated OXPHOS and SRC, signatures of T_M cells [45, 393], whereas $CD8^+$ T cells presenting more discrete or "fissed" mitochondria are more glycolytic [45]. The inner mitochondrial membrane protein Opa1 and the outer mitochondrial membrane proteins, Mfn1 and Mfn2, induce mitochondrial fusion; conversely, the outer mitochondrial protein Drp1 induces fission [392]. The authors found that mitochondrial fusion via Opa1 increases mitochondrial cristae (i.e., inner membrane folding) to support ETC chain activity and increase OXPHOS, whereas mitochondrial fission via Drp1 disrupts the efficiency of the ETC and hence favors glycolysis. Consistent with a role of fusion in T_M cell formation, overexpression of Opa1 in T_E cells increases mitochondrial content, OXPHOS, and T_M cell molecule expression, while Opa1 deletion decreases these parameters. Similarly, inhibiting Drp1 activity diminishes glycolysis. While not investigated in T cells, PI3K-AKT-mTOR signaling has been reported to increase Opa1 expression downstream of the insulin receptor [303], suggesting another potential layer of regulation by mTORC1 in $CD8^+$ T cell differentiation. Interestingly, these fission and fusion events driving differences appears to be independent of cell division, because macrophages and DCs that do not divide display similar differences in mitochondrial morphology [45]. If and how this links to asymmetric mTORC1 signaling and cell division are not known.

It is important to note that the model wherein the loss of mTOR signaling drives T_M differentiation may be overly simplified. In support of this view, it was recently demonstrated that, during chronic viral infections, $CD8^+$ T cells become functionally impaired, or exhausted, due to the loss of PI3K-AKT-mTOR and FOXO1 activity [366]. Further, recent studies have suggested that mTORC1 signaling is also involved in controlling T_M cell tissue distribution. When mice are treated with rapamycin, T_M cell formation is enhanced in the peripheral lymphoid tissues [10, 321, 364], but T_M cells fail to accumulate in mucosal tissues (e.g., the intestines, lung, and vagina) due to decreased cellular trafficking [364]. These phenotypes may be linked to the mTORC1-dependent increase of mevalonate-derived isoprenoids, which can regulate T cell trafficking [385, 434]. The activation of mTORC2 appears to induce T_{RM} differentiation, as these cells express lower levels of the FOXO1-regulated genes, *Klf2* and *S1pr1*. Indeed, these proteins are downregulated in response to cytokines in a PI3K-AKT-dependent manner [359]. Future studies will dissect the precise contributions of mTORC1 and mTORC2 signaling to tissue-infiltrating T_M and T_{RM} differentiation and function and how metabolic programs contribute to the differentiation of these cells.

1.6 Role of Metabolism in Nonconventional T Cells

1.6.1 Overview of Nonconventional T Cells

In addition to conventional αβ T cells that differentiate into effector and memory lineages, several subsets of nonconventional T cells also arise during neonatal development. These include FoxP3-expressing Treg cells, γδ T cells, natural killer T cells (NKT), Tr1 cells, and Th3 cells. These populations serve critical roles in supporting immune and tissue homeostasis, in the resolution of inflammation associated with infections, and can have deleterious roles in autoimmunity or tumor development. Among these populations, the role of cellular and host metabolism is best defined in Treg cells and will hence be discussed below. However, we point our readers to recent studies highlighting roles for metabolic regulators and metabolism in other nonconventional T cells [48, 83, 134, 165, 237, 247, 298, 415, 424, 445].

Treg cells support immune homeostasis and regulate ongoing immune responses in the context of antitumor immunity, autoimmunity, allergic responses, graft-versus-host disease, and pathogen-induced immune responses. Treg cells control immune reactions by suppressing the proliferation, activation, and effector functions of other immune cells. Because it is indispensable for their development and function, mutations in the human and mouse genes encoding FoxP3 drive Treg cell dysfunction and an X-linked lymphoproliferative disease [27, 44, 68, 120, 160, 192, 419]. Most Treg cells arise from the thymus (tTreg cells) but can also be induced in the periphery from naïve T cells (pTreg cells). tTreg cells are essential for immune tolerance and prevent autoimmunity, whereas pTreg cells are locally induced to maintain tissue tolerance, especially at mucosal sites. Activated tTreg cells (also called effector Treg cells) are characterized by low expression of CD62L and high expression of CD44 and are enriched in nonlymphoid tissues, including in the skin, intestines, lung, liver, and adipose tissue [230]. The Treg cells found in tissues express different chemokine receptors and transcription factors and have a different TCR repertoire than Treg cells found in peripheral lymphoid tissues. There is growing emphasis to address what molecular and metabolic requirements tune Treg cell responses in different microenvironments.

The development, homeostasis, and functions of Treg cells are controlled by TCR, CD28, IL-2, and TGF-β signals [70, 146, 181, 196, 222, 227, 340, 344, 447]. TCR recognition of MHC class II-bound peptides and CD28 co-stimulation is essential for tTreg generation [163, 181, 447]. By contrast, low levels of TCR-CD28 signaling combined with IL-2, TGF-β, and all-*trans*-retinoic acid (RA) is required for pTreg induction [181]. Naïve T cells can also be differentiated into in vitro-derived Treg (iTreg) when activated in the presence TGF-β and IL-2. These iTreg cells are more similar to pTreg but do not completely mimic in vivo pTreg cell phenotypes [1]. Mechanistically, these signals drive Treg development or differentiation by inducing or stabilizing FoxP3 expression, in part by promoting epigenetic changes in conversed noncoding DNA sequences (CNS) of the murine *Foxp3* or human *FOXP3* gene. DNA methylation plays an important role in Treg differentiation, where the Treg-specific demethylation region (TSDR) region of the *Foxp3* gene is demethylated in endogenous Treg cells (i.e., already committed to the $CD4^+CD25^+FoxP3^+$ T cell lineage) from humans and mice [286, 449]. By contrast, iTreg cells have predominately methylated regions. Epigenetic modifications like DNA methylation are linked to metabolites [174, 233]. Therefore, the metabolic pathways discussed below likely influence Treg cell development and functions on multiple levels. Figure 1.5b summarizes how metabolic programs regulate Treg cell biology.

1.6.2 FAO and Glycolysis Tune Treg Cell Differentiation and Function

Mitochondrial mass and function are higher in Treg cells than conventional $CD4^+$ T cells ex vivo, and iTreg cells have increased mitochondrial function than other T_H cell subsets in vitro [23, 252, 444]. This increase in respiratory function is important for Treg cell suppressive activity, as it is inhibited upon rotenone treatment [23]. Consistent with this role, deletion of PGC-1α or Tfam impairs selective Treg cell functions, such as their ability to suppress inflammation-triggered immune responses and produce IL-10 [17, 23]. The upregulation of AMPK-dependent FAO driven by TCR-CD28 signals induces FoxP3 expression in the presence of TGF-β. Indeed, the CTP1a inhibitor etomoxir limits murine iTreg cell differentiation, whereas metformin-induced AMPK activation enhances iTreg cell differentiation in vitro and in vivo [252]. Mechanistically, FAO appears to be coupled to IL-2R signaling, because both CD25 expression and STAT5 activation are lower in human iTreg cells generated in the presence of etomoxir [88]. However, despite having reduced CD25 expression on a population level, those Treg cells that upregulate CD25 expression in the presence of etomoxir have higher expression of FoxP3 and increased in vitro suppressive activity than iTreg cells generated in its absence. These data are consistent with observations that the homeostatic functions of Treg cells are independent of AMPKα1 [319]. Thus, the elevated levels of FAO and AMPK activity in ex vivo-isolated Treg cells and iTreg cells only partially accounts for mitochondrial function supporting Treg cell functions [252].

Recent works have also highlighted a role for glycolysis in Treg cell fate decisions and fitness. Human Treg cells express more genes associated with glycolysis than FAO ex vivo. Moreover, upon activation, these ex vivo Treg cells upregulate both FAO and glycolysis, with similar observations made in human iTreg cells. Studies have demonstrated that both glycolysis and FAO are essential to maintain OXPHOS, that iTreg upregulate both glycolysis and FAO programs during their differentiation, and that both FAO and glycolysis contribute to the upregulation of OXPHOS in these cells [88]. Similarly, when human Treg cells are activated ex vivo, they upregulate both FAO- and glycolysis-related enzymes, and human Treg cells analyzed directly ex vivo express more genes associated with glycolysis than FAO [315]. This upregulation of both glycolysis and FAO maintains OXPHOS that supports Treg cell functions [132, 315]. To determine the functional contribution of FAO and glycolysis, the Matarese laboratory compared the differentiation and suppressive activity of human iTreg cells differentiated in the presence or absence of 2-DG or etomoxir. Similar to those generated in the presence of etomoxir, iTreg cells generated in the presence of 2-DG express less CD25 than controls. In contrast to the etomoxir-treated iTreg cells, the $CD25^{hi}$ iTreg cells present in the 2-DG cultures also express lower levels of FoxP3 and activated STAT5 and are hence less suppressive. Similar inhibition of iTreg cell differentiation is observed when ERRα, a transcription factor necessary for glycolytic reprogramming, activity is diminished, although free fatty acids can restore these defects [253]. This observation might be linked to the fact that either glucose or fatty acid-derived α-KG increases the expression Treg cell signature genes via epigenetic regulation. How does glycolysis control iTreg cell differentiation? It was recently demonstrated that, when glycolysis is inhibited, the glycolytic enzyme *enolase*-1 binds to the CNS2 promoter region of the *FOXP3* exon 2 splice variant (*FOXP3*-E2) in human iTreg cells. This binding suppresses *FOXP3-E2* expression, which is important for the suppressive activity of human iTreg cells [88]. Thus, the acquisition of the glycolytic program is critical to support Treg cell functions. It is interesting to note that GLUT1 deletion does impact Treg cell differentiation or functions [240]. Whether this effect is due to the compensatory upregulation of other glucose transporters, such as GLUT3, remains to be elucidated.

Despite the evidence above that glycolysis supports Treg cell differentiation and function, hyperglycolytic responses appear to limit Treg cell fitness. The upregulation of HIF-1-dependent glycolysis was demonstrated to impede iTreg differentiation in favor of T_H17 or T_H1 differentiation [76, 348]; however, HIF-1α also antagonizes iTreg differentiation by driving the degradation of FoxP3 [86]. While still controversial, Treg cells have been reported to lose FoxP3 expression in selected contexts and subsequently increase programs associated with effector T_H cells [94, 257, 331, 450]. The correlation between hyperactivation of glycolysis and Treg cell instability is becoming more evident, with studies demonstrating that inhibition of glycolysis can restore FoxP3 expression in unstable Treg cells [168, 352, 414]. Why might too much glycolysis impact Treg cell stability? As we discussed above, glycolytic metabolites can influence TCR signaling [158]. Treg cells require TCR signals to maintain their homeostasis and functions [222, 340, 391], but recent studies suggest that the TCR-induced signal strength is lower in Treg cells than conventional T cells [343, 431]. Thus, glycolytic metabolites might tune TCR and/or other signaling pathways, so FoxP3 or other Treg suppressive molecule expression is disrupted. Metabolites or metabolic enzymes may also directly or indirectly cooperate with transcription factors to facilitate pro-inflammatory cytokine production by Treg cells and hence impair their immunosuppressive functions. It is unclear whether elevated levels of glycolysis are always detrimental to Treg cell responses. Indeed, recent reports suggest that the glycolytic balance of lymphoid tissue-derived Treg cells might be different than colon Treg cells [74, 182]. Thus, much remains to be uncovered about how glycolysis and FAO cooperatively control Treg cell fate decisions.

It is clear that the PI3K-AKT-mTOR signaling pathway is a crucial regulator of the functional and metabolic signatures of Treg cells. During iTreg differentiation, low levels of PI3K and mTOR signaling are required for maximum differentiation [21, 92, 339]. The requirement for low levels of mTOR signaling is likely multifold, given that both mTORC1 and mTORC2 can affect and influence how efficiently multiple transcription factors, including T-bet, GATA3, and FOXO, are expressed in differentiating T cells [220, 293, 320, 321]. The activation of mTOR is also a crucial determinant of Treg cell proliferation and functions. Both human and murine Treg cells have higher levels of mTORC1 signaling than non-Treg cells [316, 444]. This basal level of mTORC1 activation limits the TCR-CD28-inducible proliferation of Treg cells by suppressing IL-2 production [316]. Indeed, transient mTORC1 inhibition with rapamycin for 1 h increases Treg cell proliferation in vitro and in vivo [316]. Of note, however, activated Treg cells have higher levels of mTOR signaling than quiescent Treg cells [236, 316, 414, 444]. Moreover, *FRAP*-deficient human Treg cells do not robustly proliferate in vitro [316], and *Rptor*-deficient murine Treg cells also fail to proliferate and upregulate effector molecules, including CTLA-4 and ICOS [444]. The suppressive activity of Treg cells is also supported by mTORC1, as mice bearing a conditional deletion of *Rptor* in Treg cells develop a systemic, fatal autoimmune syndrome similar to those reported in mice and humans lacking FoxP3 [27, 44, 68, 120, 160, 192, 316]. Thus, while transient mTORC1 inhibition is beneficial for Treg cell proliferation, reactivation of mTORC1 supports Treg cell proliferation and suppressive functions. However, it should be noted that it is unclear if these effects are characteristic of all Treg cells or if different requirements for mTORC1 signaling exist for tTreg cells and pTreg cells in vivo.

Although mTOR signaling supports Treg cell function, its hyperactivation can also be deleterious to Treg cells. Indeed, studies have shown that deletion of PP2A, TSC1, or Atg7 (upstream inhibitors for mTORC1) increases their ability to produce T_H17- and T_H1-associated cytokines [300, 414]. However, each of these molecules limits the effector-like program in Treg cells by unique mechanisms. PP2A-deficient Treg cells display increases in both glycolysis and respiratory metabolism, yet do not appear to become unstable. In these cells, mTORC1 activity is elevated, which drives Treg cell dysfunction contributing to the development of autoimmunity in mice [8]. By contrast, activated Atg7-deficient Treg cells have dysregulated PI3K-PDK1-mTORC1 signaling that heightens c-MYC-dependent glycolysis, with both contributing to their aberrant production of IFN-γ [414]. Intriguingly, the sustained activation of mTORC2 can also increase T_H1-like responses in PTEN-deficient Treg cells [352]. Changes in the epigenetic program can partially account for the unstable phenotype of PTEN-deficient Treg cells [168]. The precise mechanisms that explain why TSC1-deficient Treg cells acquire T_H17 and T_H1 characteristics are unknown. Because TSC1-deficient T cells have increased mTORC1 signaling but decreased mTORC2 activity [313, 433], the diminished function of mTORC2 might have an influence on the upregulation of the T_H17 and T_H1 programs in TSC1-deficient Treg cells.

It remains unclear how the magnitude and timing of mTOR signaling are controlled in Treg cells. It is possible that the high levels of AMPK activity observed in Treg cells can counterbalance mTORC1 activity such that, in the presence of TCR-CD28 signals, it is downregulated to initiate cell cycling in Treg cells [252, 316]. Then, additional cell-intrinsic signals and cell-extrinsic signals (see below) could continue to tune mTORC1 and/or mTORC2 activation to appropriate levels and support the functions of Treg cells during homeostasis. Ongoing work is required to further elucidate how metabolic and environmental signals regulate mTOR signaling to induce their specific functions in different microenvironments.

1.6.3 Mevalonate-Dependent Metabolism Is a Crucial Regulator of Treg Cell Function

While mTORC1 signaling is a critical regulator of glycolysis and OXPHOS in Treg cells, it was also demonstrated that cholesterol and lipid biosynthesis induced downstream of mTORC1 activity are important for Treg cell function. Mechanistically, mTORC1 induces the mevalonate-dependent lipid and cholesterol synthesis programs in Treg cells to support their growth, proliferation, and expression of effector molecules. In support of this view, RAPTOR-deficient Treg cells have impaired suppressive functions in vivo and in vitro, features of which can be recapitulated by statin treatment in vitro. [444]. The addition of mevalonate completely restores the suppressive activity of these cells. It is presently unclear precisely how cholesterol and lipid biosynthesis regulates Treg cell functions. One possibility is that, like T_M cells, Treg cells catabolize de novo synthesized fatty acids as a means to drive FAO that supports Treg cell functions, although it is noteworthy that LAL appears to be dispensable for Treg cell development and in vitro functions [281, 317]. This possibility might explain why loss of PPAR signaling, which drives fatty acid uptake via CD36- and FAO-related programs, is essential for adipose tissue-resident Treg cell but not splenic Treg cell functions [2, 73]. Further, cholesterol or lipid-derived products might be important for modulating the strength of TCR signaling such that Treg cells appropriately respond to host-derived antigens in different tissues. Indeed, the FoxP3-driven accumulation of ceramides in Treg cells induces the phosphatase activity of PP2A, thus restraining mTORC1 activity to appropriate levels [8]. Whether cholesterol by-products and cholesterol molecules tune Treg cell responses at the level of TCR signaling could be explored [374, 402, 435]. Finally, mevalonate-derived isoprenoids might play an essential role in driving Treg cell recruitment to inflammatory sites or tissues under homeostatic conditions [385].

1.6.4 Cell-Extrinsic Regulation of Treg Cell Metabolism and Function

Several cell-extrinsic regulators that modulate mTOR activity play key roles in modulating Treg cell responses. Leptin is a hormone produced primarily by adipocytes, which controls the release of free fatty acids during conditions of nutrient deprivation [211]. Interestingly, Treg cells constitutively express leptin receptor (LEPR, also known as OBR) and also can produce leptin during homeostasis and after activation. It was found that the leptin-LEPR axis increases mTORC1 signaling in Treg cells to limit their TCR-CD28-induced proliferation. Indeed, like rapamycin treatment, suppressing this axis increased Treg cell proliferation [316]. Interestingly, the loss of LEPR does not appear to inhibit Treg cell suppressive activity in vitro [89]. Vitamins, including RA and the vitamin D metabolite 1,25-dihydroxyvitamin D, are mediators of Treg cell induction and function [78, 138, 176, 184, 309, 371]. Interestingly, these vitamins can modulate mTORC1-related signaling [50, 213, 229, 333]. Treg cells also preferentially express the folate receptor (FR4), and sensing folic acid/folate is crucial for Treg cell survival and maintenance of colon homeostasis [199, 206]. This phenotype is possibly linked to folate-dependent one-carbon metabolism that is likely critical for suppressing the effector T cell program in Treg cells. Future studies will continue to dissect the interplay between vitamin sensing, mTORC1-induced metabolism, and Treg cells, including the role of one-carbon metabolism in these processes.

Commensal bacteria modulate Treg cell responses by triggering intracellular signaling via MyD88-dependent sensors [274, 395, 405]. They also produce the short-chain fatty acids (acetate, butyrate, and propionate) that control the functions and differentiation of Treg cells in mucosal sites [360]. These metabolites signal via the GPR43 receptor expressed on Treg cells to induce their differentiation and function to suppress colitis in mice [360]. These microbial metabolites can also indirectly influence Treg cell programming by binding to intestinal epithelial cells or innate immune cells within the colon. For instance, GPR109a, which is expressed in colon IECs and innate immune cells, only binds butyrate and controls Treg cell programming by upregulating production of IL-10 and retinal dehydrogenases (RALDHs; RA-producing enzymes) in DCs and macrophages [357]. Butyrate can also modulate HDAC activity to induce FoxP3 expression [11, 124]. Moreover, this inhibition of HDAC activity modulates mTORC1-related signaling, since deacetylation of S6K limits its function in T cells [299]. Of note, although CD36 expression appears to be largely restricted to adipose tissue-resident Treg cells, it was recently found that Treg cells in the colon express CD36 and accumulate lipids [73, 182]. How CD36 modulates colon Treg cell responses is unknown.

The availability of amino acids can also influence Treg cell differentiation and functions. Similar to mTOR inhibition, depleting amino acids drives iTreg differentiation over effector T_H cell specialization in vitro [77]. Treg cells can also enforce tolerance at sites of inflammation by driving the localized depletion of selective amino acids, including tryptophan and arginine, which effectively blocks conventional T cell proliferation and allows iTreg/pTreg cells to more readily differentiate. Treg cells drive this depletion through the modulation of DC functions, where specific interactions including CTLA4-CD80/86 increase the expression of enzymes like IDO and arginase in the DCs [77]. This axis also has implications in driving Treg cell accumulation at tumor sites [376]. Of note, LAT1- and ASCT2-deficient naïve T cells are competent to differentiate into iTreg cells [265, 356], so amino acid-dependent modulation of mTORC1 signaling might not entirely account for how amino acid catabolism increases iTreg cell differentiation. In line with this idea, IDO-mediated tryptophan catabolism generates 3-hydroxyanthranilic acid, which activates GCN2 [263, 432]. This protein has been demonstrated to antagonize inflammation and autoimmunity, in part by suppressing T_H1/T_H17 responses and promoting iTreg differentiation [290]. Future studies will continue to explore how tissue microenvironments alter Treg cell responses to promote their specialization in maintaining tissue homeostasis.

1.7 Roles of Metabolism in B Cells

B cells compose the other arm of the adaptive immune system. Activated B cells produce antibodies (also called immunoglobulins) and cytokines and can also serve as APC for T cells. The metabolic requirements underlying B cell functions are only beginning to be understood. Of note, many features of B cell metabolism mirror conventional T cell metabolism. Future work will dissect if specific B cell subsets have different metabolic features related to their functions.

Naïve B cells isolated from the murine spleen utilize both glycolysis and OXPHOS [58]. However, peripheral naïve B cell homeostasis appears to require glucose uptake, as there is a specific loss of mature B cells in the absence of GLUT1 [58]. After BCR engagement, GLUT1 surface expression increases, and glycolysis is upregulated in a PI3K-AKT-dependent manner [58, 101, 421]. BCR engagement also increases OXPHOS, and unlike TCR/CD28-activated T cells that favor glycolysis, BCR-stimulated B cells utilize both glycolysis and OXPHOS [58]. Similarly, LPS treatment also enhances glycolysis and OXPHOS in naïve B cells [58]. This increase is linked to the upregulation of both GLUT1 expression and mitochondrial mass [58, 101], the former of which is PI3K-AKT signaling dependent [101]. Moreover, c-MYC, but not HIF-1α, drives the metabolic reprogramming of activated B cells, but not all metabolic programs are c-MYC dependent. For instance, c-MYC appears to modestly antagonize LPS-induced downregulation of FAO and pyruvate oxidation, while being necessary for LPS-mediated increases in glutamine oxidation [58].

Despite the upregulation of both glycolysis and OXPHOS, glycolysis appears to be the primary pathway supporting B cell proliferation and antibody production. Consistent with this idea, GLUT1-deficient B cells have defects in antibody production [58]. Treatment with 2-DG suppresses LPS-triggered B cell proliferation, as well as IgM and IgG production. Similarly, inhibition of PDHK suppresses antibody production by B cells in vitro and in vivo. This inhibition also suppresses B cell proliferation and antibody production by TLR9-stimulated human B cells. B cell-activating factor (BAFF; also called BLyS, TALL-1, THANK, TNSF13B, and zTNF4) is a crucial pro-growth and pro-survival factor for B cells, and BAFF treatment also upregulates glucose consumption in B cells [421]. Interestingly, in B cells receiving chronic BAFF stimulation, glycolysis is favored over OXPHOS following LPS stimulation, which could partially explain the dependence on glycolysis for antibody production in vivo [58]. Rapamycin treatment inhibits BAFF-induced cell growth [421], suggesting that mTORC1 couples BAFF-dependent signals to increased glucose consumption. In line with this, mTORC1 inhibition blocks B cell proliferation in vitro [228]. However, transient inhibition of mTOR activity can potentiate or inhibit the high-affinity humoral response in vivo [189, 228]. Moreover, the B cell-specific deletion of *Rptor* and *Rictor* antagonizes and promotes class-switch recombination, respectively [228], but hyperactivation of mTORC1 in the context of *Tsc2*-deficient B cells also abrogates germinal center reactions driving antibody production [26]. Thus, mTOR is a complex regulator of B cell responses.

Other receptor systems have also been reported to upregulate glycolysis in B cells. For instance, anti-CD40 antibody stimulation increases glucose consumption by B cells [421]. IL-4-IL-4R signaling triggers glycolysis in a PI3K-AKT-independent but STAT6-dependent manner, which is important for mediating B cell survival [102]. Interestingly, these signals cooperate with BCR, adhesion receptor, and TNF receptor superfamily member signals to drive the T_{FH}-dependent germinal center reaction, suggesting that glycolysis might be important for initiating the germinal B cell program. As with T_{FH} cells, Bcl6 is an important transcription factor for the induction of germinal center B cell reactions [91]. Because Bcl6 antagonizes the glycolytic program in T_H1 cells [285], glycolysis might be temporally regulated during germinal center B cell activation and differentiation. Future work will explore this possibility and whether different metabolic programs influence antibody class switching in the context of various diseases.

The intestinal lamina propria contains IgA^+ plasma cells. Most of these B cells arise from the Peyer's patches, which are specialized gut-associated lymphoid tissues where class switching from IgM^+ naïve B cells to IgA^+-activated B cells occurs. Kiyono and coworkers recently investigated how metabolism differs in naïve B cells and IgA^+ plasma cells from the intestines. They found that naïve B cells and IgA^+ plasma cells produce similar levels of citrate and succinate, but IgA^+ plasma cells express less malate and fumarate than naïve B cells [206]. Moreover, the IgA^+ plasma cells specifically upregulate expression of the glycolytic metabolites, G6P, and fructose bisphosphate; these metabolites are not expressed in either naïve B cells from the intestines or plasma cells from the spleen. Like peripheral naïve B cells, naïve B cells from the intestines generate ATP via the TCA cycle but do not utilize glucose as a source of pyruvate due to low glucose uptake [58, 206]. By contrast, IgA^+ plasma cells in the small intestine shuttle glycolysis-derived pyruvate into the TCA cycle and can generate ATP via both glycolysis and the TCA cycle. However, it is noteworthy that lactate production by IgA^+ B cells is minimal under normal physiological conditions [206], suggesting that the homeostatic energy needs of IgA^+ plasma cells are met via the TCA cycle. The divergent metabolic programs are linked to vitamin B_1 sensing. Naïve B cells in the intestines express the vitamin B_1 transporter, thiamine transporter 1 (THTR1), whereas its expression is absent in IgA^+ plasma cells. Deficiency in vitamin B_1 selectively impairs TCA cycle function without affecting glycolysis, which specifically affects naïve B cell homeostasis [206]. Vitamin B_1 deficiency likely attenuates the conversion of pyruvate to acetyl-CoA and of α-KG to succinyl-CoA, as these reactions require the vitamin B1 derivative, thiamine pyrophosphate (TPP) [96]. Future studies should explore the implications of vitamins as they link to antibody class switching in the context of site-specific germinal center reactions.

1.8 Conclusions and Future Directions

Our understanding of immunometabolism has greatly increased over the last decade. Studies by multiple laboratories increased our mechanistic understanding of how key metabolic pathways are altered during different stages of the immune responses or in different tissues and have provided much insight into how the alterations of these pathways affects immune cell activation and function. Several unanswered questions remain to be fully addressed within the field. For instance, why do effector T cells upregulate similar metabolic pathways but still have profound differences in effector functions? While the transcriptional network induced by polarizing cytokines is strongly linked to these differences, different effector T cell subsets might preferentially accumulate discrete metabolic by-products that serve as second messengers to selectively regulate their differentiation and function. This area of investigation could also be applied to other immune cell subsets that display tremendous phenotypic and functional diversity, such as DCs and B cells. The temporal control of when crucial metabolic second messengers and what upstream receptor-mediated and environmental signals drive the alterations in any such molecules will be important to address in the future.

How do nutrients and metabolites in discrete microenvironments contribute the activation, differentiation, and maintenance of immune cell subsets? It is becoming increasingly clear that immune cells found in different tissues have distinct phenotypic and functional properties than immune cells found in peripheral lymphoid tissues. However, outside of the transcriptional networks that control such cell specialization, it is unclear what molecular signatures control these differences. Given the complex roles that metabolites can play as intracellular messengers, protein modifiers, and epigenetic regulators, it is very likely that metabolism serves important roles in the tissue-specific fate decisions of immune cells. Moreover, nutrient availability in different tissues or in disease states and competition between immune cells are likely to strongly impact immune responses that occur at specific sites. In total, studies investigating how nutrient sensing controls immune reactions will provide crucial insight as to how metabolic pathways can be therapeutically tuned to modulate immune cell responses in the autoimmunity, infectious diseases, and cancers.

Acknowledgments This work was supported by the Hartwell Foundation Biomedical Research Fellowship (N.M.C.) and NIH AI105887, AI101407, CA176624, and NS064599, America Asthma Foundation, and Crohn's & Colitis Foundation of America (to H.C.).

References

1. Abbas AK, Benoist C, Bluestone JA, Campbell DJ, Ghosh S, Hori S, Jiang S, Kuchroo VK, Mathis D, Roncarolo MG, Rudensky A, Sakaguchi S, Shevach EM, Vignali DA, Ziegler SF (2013) Regulatory T cells: recommendations to simplify the nomenclature. Nat Immunol 14(4):307–308. doi:10.1038/ni.2554
2. Ahmadian M, Suh JM, Hah N, Liddle C, Atkins AR, Downes M, Evans RM (2013) PPARgamma signaling and metabolism: the good, the bad and the future. Nat Med 19(5):557–566. doi:10.1038/nm.3159
3. Ai D, Jiang H, Westerterp M, Murphy AJ, Wang M, Ganda A, Abramowicz S, Welch C, Almazan F, Zhu Y, Miller YI, Tall AR (2014) Disruption of mammalian target of rapamycin complex 1 in macrophages decreases chemokine gene expression and atherosclerosis. Circ Res 114(10):1576–1584. doi:10.1161/CIRCRESAHA.114.302313
4. Akula MK, Shi M, Jiang Z, Foster CE, Miao D, Li AS, Zhang X, Gavin RM, Forde SD, Germain G, Carpenter S, Rosadini CV, Gritsman K, Chae JJ, Hampton R, Silverman N, Gravallese EM, Kagan JC, Fitzgerald KA, Kastner DL, Golenbock DT, Bergo MO, Wang D (2016) Control of the innate immune response by the mevalonate pathway. Nat Immunol 17(8):922–929. doi:10.1038/ni.3487

5. Amiel E, Everts B, Freitas TC, King IL, Curtis JD, Pearce EL, Pearce EJ (2012) Inhibition of mechanistic target of rapamycin promotes dendritic cell activation and enhances therapeutic autologous vaccination in mice. J Immunol 189(5):2151–2158. doi:10.4049/jimmunol.1103741
6. Ananieva EA, Patel CH, Drake CH, Powell JD, Hutson SM (2014) Cytosolic branched chain aminotransferase (BCATc) regulates mTORC1 signaling and glycolytic metabolism in CD4+ T cells. J Biol Chem 289(27):18793–18804. doi:10.1074/jbc.M114.554113
7. Angeli V, Hammad H, Staels B, Capron M, Lambrecht BN, Trottein F (2003) Peroxisome proliferator-activated receptor gamma inhibits the migration of dendritic cells: consequences for the immune response. J Immunol 170(10):5295–5301
8. Apostolidis SA, Rodriguez-Rodriguez N, Suarez-Fueyo A, Dioufa N, Ozcan E, Crispin JC, Tsokos MG, Tsokos GC (2016) Phosphatase PP2A is requisite for the function of regulatory T cells. Nat Immunol 17(5):556–564. doi:10.1038/ni.3390
9. Appel S, Mirakaj V, Bringmann A, Weck MM, Grunebach F, Brossart P (2005) PPAR-gamma agonists inhibit toll-like receptor-mediated activation of dendritic cells via the MAP kinase and NF-kappaB pathways. Blood 106(12):3888–3894. doi:10.1182/blood-2004-12-4709
10. Araki K, Turner AP, Shaffer VO, Gangappa S, Keller SA, Bachmann MF, Larsen CP, Ahmed R (2009) mTOR regulates memory CD8 T-cell differentiation. Nature 460(7251):108–112. doi:10.1038/nature08155
11. Arpaia N, Campbell C, Fan X, Dikiy S, van der Veeken J, deRoos P, Liu H, Cross JR, Pfeffer K, Coffer PJ, Rudensky AY (2013) Metabolites produced by commensal bacteria promote peripheral regulatory T-cell generation. Nature 504(7480):451–455. doi:10.1038/nature12726
12. Arranz A, Doxaki C, Vergadi E, Martinez de la Torre Y, Vaporidi K, Lagoudaki ED, Ieronymaki E, Androulidaki A, Venihaki M, Margioris AN, Stathopoulos EN, Tsichlis PN, Tsatsanis C (2012) Akt1 and Akt2 protein kinases differentially contribute to macrophage polarization. Proc Natl Acad Sci U.S.A. 109(24):9517–9522. doi:10.1073/pnas.1119038109
13. Artis D, Spits H (2015) The biology of innate lymphoid cells. Nature 517(7534):293–301. doi:10.1038/nature14189
14. Atkuri KR, Herzenberg LA, Herzenberg LA (2005) Culturing at atmospheric oxygen levels impacts lymphocyte function. Proc Natl Acad Sci U.S.A. 102(10):3756–3759. doi:10.1073/pnas.0409910102
15. Atkuri KR, Herzenberg LA, Niemi AK, Cowan T, Herzenberg LA (2007) Importance of culturing primary lymphocytes at physiological oxygen levels. Proc Natl Acad Sci U.S.A. 104(11):4547–4552. doi:10.1073/pnas.0611732104
16. Baginska J, Viry E, Paggetti J, Medves S, Berchem G, Moussay E, Janji B (2013) The critical role of the tumor microenvironment in shaping natural killer cell-mediated anti-tumor immunity. Front Immunol 4:490. doi:10.3389/fimmu.2013.00490
17. Baixauli F, Acin-Perez R, Villarroya-Beltri C, Mazzeo C, Nunez-Andrade N, Gabande-Rodriguez E, Ledesma MD, Blazquez A, Martin MA, Falcon-Perez JM, Redondo JM, Enriquez JA, Mittelbrunn M (2015) Mitochondrial respiration controls lysosomal function during inflammatory T cell responses. Cell Metab 22(3):485–498. doi:10.1016/j.cmet.2015.07.020
18. Ballesteros-Tato A, Leon B, Graf BA, Moquin A, Adams PS, Lund FE, Randall TD (2012) Interleukin-2 inhibits germinal center formation by limiting T follicular helper cell differentiation. Immunity 36(5):847–856. doi:10.1016/j.immuni.2012.02.012
19. Balsamo M, Manzini C, Pietra G, Raggi F, Blengio F, Mingari MC, Varesio L, Moretta L, Bosco MC, Vitale M (2013) Hypoxia downregulates the expression of activating receptors involved in NK-cell-mediated target cell killing without affecting ADCC. Eur J Immunol 43(10):2756–2764. doi:10.1002/eji.201343448
20. Bar-Peled L, Chantranupong L, Cherniack AD, Chen WW, Ottina KA, Grabiner BC, Spear ED, Carter SL, Meyerson M, Sabatini DM (2013) A tumor suppressor complex with GAP activity for the Rag GTPases that signal amino acid sufficiency to mTORC1. Science 340(6136):1100–1106. doi:10.1126/science.1232044

21. Battaglia M, Stabilini A, Migliavacca B, Horejs-Hoeck J, Kaupper T, Roncarolo MG (2006) Rapamycin promotes expansion of functional CD4+CD25+FOXP3+ regulatory T cells of both healthy subjects and type 1 diabetic patients. J Immunol (Baltimore, Md: 1950) 177(12):8338–8347
22. Bauer DE, Hatzivassiliou G, Zhao F, Andreadis C, Thompson CB (2005) ATP citrate lyase is an important component of cell growth and transformation. Oncogene 24(41):6314–6322. doi:10.1038/sj.onc.1208773
23. Beier UH, Angelin A, Akimova T, Wang L, Liu Y, Xiao H, Koike MA, Hancock SA, Bhatti TR, Han R, Jiao J, Veasey SC, Sims CA, Baur JA, Wallace DC, Hancock WW (2015) Essential role of mitochondrial energy metabolism in Foxp3(+) T-regulatory cell function and allograft survival. FASEB J: Off Publ Fed Am Soc Exp Biol 29(6):2315–2326. doi:10.1096/fj.14-268409
24. Belikov AV, Schraven B, Simeoni L (2015) T cells and reactive oxygen species. J Biomed Sci 22:85. doi:10.1186/s12929-015-0194-3
25. Ben-Shoshan J, Maysel-Auslender S, Mor A, Keren G, George J (2008) Hypoxia controls CD4+CD25+ regulatory T-cell homeostasis via hypoxia-inducible factor-1alpha. Eur J Immunol 38(9):2412–2418. doi:10.1002/eji.200838318
26. Benhamron S, Tirosh B (2011) Direct activation of mTOR in B lymphocytes confers impairment in B-cell maturation andloss of marginal zone B cells. Eur J Immunol 41(8):2390–2396. doi:10.1002/eji.201041336
27. Bennett CL, Christie J, Ramsdell F, Brunkow ME, Ferguson PJ, Whitesell L, Kelly TE, Saulsbury FT, Chance PF, Ochs HD (2001) The immune dysregulation, polyendocrinopathy, enteropathy, X-linked syndrome (IPEX) is caused by mutations of FOXP3. Nat Genet 27(1):20–21. doi:10.1038/83713
28. Bensinger SJ, Bradley MN, Joseph SB, Zelcer N, Janssen EM, Hausner MA, Shih R, Parks JS, Edwards PA, Jamieson BD, Tontonoz P (2008) LXR signaling couples sterol metabolism to proliferation in the acquired immune response. Cell 134(1):97–111. doi:10.1016/j.cell.2008.04.052
29. Bentley J, Itchayanan D, Barnes K, McIntosh E, Tang X, Downes CP, Holman GD, Whetton AD, Owen-Lynch PJ, Baldwin SA (2003) Interleukin-3-mediated cell survival signals include phosphatidylinositol 3-kinase-dependent translocation of the glucose transporter GLUT1 to the cell surface. J Biol Chem 278(41):39337–39348. doi:10.1074/jbc.M305689200
30. Berod L, Friedrich C, Nandan A, Freitag J, Hagemann S, Harmrolfs K, Sandouk A, Hesse C, Castro CN, Bahre H, Tschirner SK, Gorinski N, Gohmert M, Mayer CT, Huehn J, Ponimaskin E, Abraham WR, Muller R, Lochner M, Sparwasser T (2014) De novo fatty acid synthesis controls the fate between regulatory T and T helper 17 cells. Nat Med 20(11):1327–1333. doi:10.1038/nm.3704
31. Bhandari T, Olson J, Johnson RS, Nizet V (2013) HIF-1alpha influences myeloid cell antigen presentation and response to subcutaneous OVA vaccination. J Mol Med 91(10):1199–1205. doi:10.1007/s00109-013-1052-y
32. Blagih J, Coulombe F, Vincent EE, Dupuy F, Galicia-Vazquez G, Yurchenko E, Raissi TC, van der Windt GJ, Viollet B, Pearce EL, Pelletier J, Piccirillo CA, Krawczyk CM, Divangahi M, Jones RG (2015) The energy sensor AMPK regulates T cell metabolic adaptation and effector responses in vivo. Immunity 42(1):41–54. doi:10.1016/j.immuni.2014.12.030
33. Blanc M, Hsieh WY, Robertson KA, Kropp KA, Forster T, Shui G, Lacaze P, Watterson S, Griffiths SJ, Spann NJ, Meljon A, Talbot S, Krishnan K, Covey DF, Wenk MR, Craigon M, Ruzsics Z, Haas J, Angulo A, Griffiths WJ, Glass CK, Wang Y, Ghazal P (2013) The transcription factor STAT-1 couples macrophage synthesis of 25-hydroxycholesterol to the interferon antiviral response. Immunity 38(1):106–118. doi:10.1016/j.immuni.2012.11.004
34. Blanc M, Hsieh WY, Robertson KA, Watterson S, Shui G, Lacaze P, Khondoker M, Dickinson P, Sing G, Rodriguez-Martin S, Phelan P, Forster T, Strobl B, Muller M, Riemersma R, Osborne T, Wenk MR, Angulo A, Ghazal P (2011) Host defense against viral

infection involves interferon mediated down-regulation of sterol biosynthesis. PLoS Biol 9(3):e1000598. doi:10.1371/journal.pbio.1000598
35. Blasius AL, Arnold CN, Georgel P, Rutschmann S, Xia Y, Lin P, Ross C, Li X, Smart NG, Beutler B (2010) Slc15a4, AP-3, and Hermansky-Pudlak syndrome proteins are required for Toll-like receptor signaling in plasmacytoid dendritic cells. Proc Natl Acad Sci U.S.A. 107(46):19973–19978. doi:10.1073/pnas.1014051107
36. Blouin CC, Page EL, Soucy GM, Richard DE (2004) Hypoxic gene activation by lipopolysaccharide in macrophages: implication of hypoxia-inducible factor 1alpha. Blood 103(3):1124–1130. doi:10.1182/blood-2003-07-2427
37. Boor PP, Metselaar HJ, Mancham S, van der Laan LJ, Kwekkeboom J (2013) Rapamycin has suppressive and stimulatory effects on human plasmacytoid dendritic cell functions. Clin Exp Immunol 174(3):389–401. doi:10.1111/cei.12191
38. Bouhlel MA, Derudas B, Rigamonti E, Dievart R, Brozek J, Haulon S, Zawadzki C, Jude B, Torpier G, Marx N, Staels B, Chinetti-Gbaguidi G (2007) PPARgamma activation primes human monocytes into alternative M2 macrophages with anti-inflammatory properties. Cell Metab 6(2):137–143. doi:10.1016/j.cmet.2007.06.010
39. Boxer RB, Stairs DB, Dugan KD, Notarfrancesco KL, Portocarrero CP, Keister BA, Belka GK, Cho H, Rathmell JC, Thompson CB, Birnbaum MJ, Chodosh LA (2006) Isoform-specific requirement for Akt1 in the developmental regulation of cellular metabolism during lactation. Cell Metab 4(6):475–490. doi:10.1016/j.cmet.2006.10.011
40. Brestoff JR, Kim BS, Saenz SA, Stine RR, Monticelli LA, Sonnenberg GF, Thome JJ, Farber DL, Lutfy K, Seale P, Artis D (2015) Group 2 innate lymphoid cells promote beiging of white adipose tissue and limit obesity. Nature 519(7542):242–246. doi:10.1038/nature14115
41. Bronte V, Zanovello P (2005) Regulation of immune responses by L-arginine metabolism. Nat Rev Immunol 5(8):641–654. doi:10.1038/nri1668
42. Brown J, Wang H, Suttles J, Graves DT, Martin M (2011) Mammalian target of rapamycin complex 2 (mTORC2) negatively regulates Toll-like receptor 4-mediated inflammatory response via FoxO1. J Biol Chem 286(52):44295–44305. doi:10.1074/jbc.M111.258053
43. Brunet A, Park J, Tran H, Hu LS, Hemmings BA, Greenberg ME (2001) Protein kinase SGK mediates survival signals by phosphorylating the forkhead transcription factor FKHRL1 (FOXO3a). Mol Cell Biol 21(3):952–965. doi:10.1128/MCB.21.3.952-965.2001
44. Brunkow ME, Jeffery EW, Hjerrild KA, Paeper B, Clark LB, Yasayko SA, Wilkinson JE, Galas D, Ziegler SF, Ramsdell F (2001) Disruption of a new forkhead/winged-helix protein, scurfin, results in the fatal lymphoproliferative disorder of the scurfy mouse. Nat Genet 27(1):68–73. doi:10.1038/83784
45. Buck MD, O'Sullivan D, Klein Geltink RI, Curtis JD, Chang CH, Sanin DE, Qiu J, Kretz O, Braas D, van der Windt GJ, Chen Q, Huang SC, O'Neill CM, Edelson BT, Pearce EJ, Sesaki H, Huber TB, Rambold AS, Pearce EL (2016) Mitochondrial dynamics controls T cell fate through metabolic programming. Cell 166(1):63–76. doi:10.1016/j.cell.2016.05.035
46. Buck MD, O'Sullivan D, Pearce EL (2015) T cell metabolism drives immunity. J Exp Med 212(9):1345–1360. doi:10.1084/jem.20151159
47. Buckley AF, Kuo CT, Leiden JM (2001) Transcription factor LKLF is sufficient to program T cell quiescence via a c-Myc-dependent pathway. Nat Immunol 2(8):698–704. doi:10.1038/90633
48. Bukowski JF, Morita CT, Brenner MB (1999) Human gamma delta T cells recognize alkylamines derived from microbes, edible plants, and tea: implications for innate immunity. Immunity 11(1):57–65
49. Burke B, Tang N, Corke KP, Tazzyman D, Ameri K, Wells M, Lewis CE (2002) Expression of HIF-1alpha by human macrophages: implications for the use of macrophages in hypoxia-regulated cancer gene therapy. J Pathol 196(2):204–212. doi:10.1002/path.1029
50. Busada JT, Chappell VA, Niedenberger BA, Kaye EP, Keiper BD, Hogarth CA, Geyer CB (2015) Retinoic acid regulates Kit translation during spermatogonial differentiation in the mouse. Dev Biol 397(1):140–149

51. Byersdorfer CA, Tkachev V, Opipari AW, Goodell S, Swanson J, Sandquist S, Glick GD, Ferrara JL (2013) Effector T cells require fatty acid metabolism during murine graft-versus-host disease. Blood 122(18):3230–3237. doi:10.1182/blood-2013-04-495515
52. Byles V, Covarrubias AJ, Ben-Sahra I, Lamming DW, Sabatini DM, Manning BD, Horng T (2013) The TSC-mTOR pathway regulates macrophage polarization. Nat Commun 4:2834. doi:10.1038/ncomms3834
53. Caldwell CC, Kojima H, Lukashev D, Armstrong J, Farber M, Apasov SG, Sitkovsky MV (2001) Differential effects of physiologically relevant hypoxic conditions on T lymphocyte development and effector functions. J Immunol 167(11):6140–6149
54. Calnan DR, Brunet A (2008) The FoxO code. Oncogene 27(16):2276–2288. doi:10.1038/onc.2008.21
55. Cao W, Manicassamy S, Tang H, Kasturi SP, Pirani A, Murthy N, Pulendran B (2008) Toll-like receptor-mediated induction of type I interferon in plasmacytoid dendritic cells requires the rapamycin-sensitive PI(3)K-mTOR-p70S6K pathway. Nat Immunol 9(10):1157–1164. doi:10.1038/ni.1645
56. Cao Y, Li H, Liu H, Zheng C, Ji H, Liu X (2010) The serine/threonine kinase LKB1 controls thymocyte survival through regulation of AMPK activation and Bcl-XL expression. Cell Res 20(1):99–108. doi:10.1038/cr.2009.141
57. Carlson CM, Endrizzi BT, Wu J, Ding X, Weinreich MA, Walsh ER, Wani MA, Lingrel JB, Hogquist KA, Jameson SC (2006) Kruppel-like factor 2 regulates thymocyte and T-cell migration. Nature 442(7100):299–302. doi:10.1038/nature04882
58. Caro-Maldonado A, Wang R, Nichols AG, Kuraoka M, Milasta S, Sun LD, Gavin AL, Abel ED, Kelsoe G, Green DR, Rathmell JC (2014) Metabolic reprogramming is required for antibody production that is suppressed in anergic but exaggerated in chronically BAFF-exposed B cells. J Immunol 192(8):3626–3636. doi:10.4049/jimmunol.1302062
59. Carr EL, Kelman A, Wu GS, Gopaul R, Senkevitch E, Aghvanyan A, Turay AM, Frauwirth KA (2010) Glutamine uptake and metabolism are coordinately regulated by ERK/MAPK during T lymphocyte activation. J Immunol 185(2):1037–1044. doi:10.4049/jimmunol.0903586
60. Carroll KC, Viollet B, Suttles J (2013) AMPKalpha1 deficiency amplifies proinflammatory myeloid APC activity and CD40 signaling. J Leukoc Biol 94(6):1113–1121. doi:10.1189/jlb.0313157
61. Castellano E, Downward J (2011) RAS interaction with PI3K: more than just another effector pathway. Genes Cancer 2(3):261–274. doi:10.1177/1947601911408079
62. Chakrabarti R, Engleman EG (1991) Interrelationships between mevalonate metabolism and the mitogenic signaling pathway in T lymphocyte proliferation. J Biol Chem 266(19):12216–12222
63. Cham CM, Gajewski TF (2005) Glucose availability regulates IFN-gamma production and p70S6 kinase activation in CD8+ effector T cells. J Immunol 174(8):4670–4677
64. Chang CH, Curtis JD, Maggi LB Jr, Faubert B, Villarino AV, O'Sullivan D, Huang SC, van der Windt GJ, Blagih J, Qiu J, Weber JD, Pearce EJ, Jones RG, Pearce EL (2013a) Posttranscriptional control of T cell effector function by aerobic glycolysis. Cell 153(6):1239–1251. doi:10.1016/j.cell.2013.05.016
65. Chang HC, Sehra S, Goswami R, Yao W, Yu Q, Stritesky GL, Jabeen R, McKinley C, Ahyi AN, Han L, Nguyen ET, Robertson MJ, Perumal NB, Tepper RS, Nutt SL, Kaplan MH (2010) The transcription factor PU.1 is required for the development of IL-9-producing T cells and allergic inflammation. Nat Immunol 11(6):527–534. doi:10.1038/ni.1867
66. Chang J, Burkett PR, Borges CM, Kuchroo VK, Turka LA, Chang CH (2013b) MyD88 is essential to sustain mTOR activation necessary to promote T helper 17 cell proliferation by linking IL-1 and IL-23 signaling. Proc Natl Acad Sci U.S.A. 110(6):2270–2275. doi:10.1073/pnas.1206048110
67. Chang JT, Palanivel VR, Kinjyo I, Schambach F, Intlekofer AM, Banerjee A, Longworth SA, Vinup KE, Mrass P, Oliaro J, Killeen N, Orange JS, Russell SM, Weninger W, Reiner SL

(2007) Asymmetric T lymphocyte division in the initiation of adaptive immune responses. Science 315(5819):1687–1691. doi:10.1126/science.1139393

68. Chatila TA, Blaeser F, Ho N, Lederman HM, Voulgaropoulos C, Helms C, Bowcock AM (2000) JM2, encoding a fork head-related protein, is mutated in X-linked autoimmunity-allergic disregulation syndrome. J Clin Invest 106(12):R75–R81. doi:10.1172/JCI11679
69. Chi H (2012) Regulation and function of mTOR signalling in T cell fate decisions. Nat Rev Immunol 12(5):325–338. doi:10.1038/nri3198
70. Chinen T, Kannan AK, Levine AG, Fan X, Klein U, Zheng Y, Gasteiger G, Feng Y, Fontenot JD, Rudensky AY (2016) An essential role for the IL-2 receptor in Treg cell function. Nat Immunol. doi:10.1038/ni.3540
71. Choi YS, Gullicksrud JA, Xing S, Zeng Z, Shan Q, Li F, Love PE, Peng W, Xue HH, Crotty S (2015) LEF-1 and TCF-1 orchestrate T(FH) differentiation by regulating differentiation circuits upstream of the transcriptional repressor Bcl6. Nat Immunol 16(9):980–990. doi:10.1038/ni.3226
72. Chou C, Pinto AK, Curtis JD, Persaud SP, Cella M, Lin CC, Edelson BT, Allen PM, Colonna M, Pearce EL, Diamond MS, Egawa T (2014) c-Myc-induced transcription factor AP4 is required for host protection mediated by CD8+ T cells. Nat Immunol 15(9):884–893. doi:10.1038/ni.2943
73. Cipolletta D, Feuerer M, Li A, Kamei N, Lee J, Shoelson SE, Benoist C, Mathis D (2012) PPAR-gamma is a major driver of the accumulation and phenotype of adipose tissue Treg cells. Nature 486(7404):549–553. doi:10.1038/nature11132
74. Clambey ET, McNamee EN, Westrich JA, Glover LE, Campbell EL, Jedlicka P, de Zoeten EF, Cambier JC, Stenmark KR, Colgan SP, Eltzschig HK (2012) Hypoxia-inducible factor-1 alpha-dependent induction of FoxP3 drives regulatory T-cell abundance and function during inflammatory hypoxia of the mucosa. Proc Natl Acad Sci U.S.A. 109(41):E2784–E2793. doi:10.1073/pnas.1202366109
75. Clark RB, Bishop-Bailey D, Estrada-Hernandez T, Hla T, Puddington L, Padula SJ (2000) The nuclear receptor PPAR gamma and immunoregulation: PPAR gamma mediates inhibition of helper T cell responses. J Immunol 164(3):1364–1371
76. Clever D, Roychoudhuri R, Constantinides MG, Askenase MH, Sukumar M, Klebanoff CA, Eil RL, Hickman HD, Yu Z, Pan JH, Palmer DC, Phan AT, Goulding J, Gattinoni L, Goldrath AW, Belkaid Y, Restifo NP (2016) Oxygen sensing by T cells establishes an immunologically tolerant metastatic niche. Cell 166(5):1117–1131. e1114. doi:10.1016/j.cell.2016.07.032
77. Cobbold SP, Adams E, Farquhar CA, Nolan KF, Howie D, Lui KO, Fairchild PJ, Mellor AL, Ron D, Waldmann H (2009) Infectious tolerance via the consumption of essential amino acids and mTOR signaling. Proc Natl Acad Sci U.S.A. 106(29):12055–12060. doi:10.1073/pnas.0903919106
78. Coombes JL, Siddiqui KR, Arancibia-Carcamo CV, Hall J, Sun CM, Belkaid Y, Powrie F (2007) A functionally specialized population of mucosal CD103+ DCs induces Foxp3+ regulatory T cells via a TGF-beta and retinoic acid-dependent mechanism. J Exp Med 204(8):1757–1764. doi:10.1084/jem.20070590
79. Cooper MA, Elliott JM, Keyel PA, Yang L, Carrero JA, Yokoyama WM (2009) Cytokine-induced memory-like natural killer cells. Proc Natl Acad Sci U.S.A. 106(6):1915–1919. doi:10.1073/pnas.0813192106
80. Covarrubias AJ, Aksoylar HI, Horng T (2015) Control of macrophage metabolism and activation by mTOR and Akt signaling. Semin Immunol 27(4):286–296. doi:10.1016/j.smim.2015.08.001
81. Cramer T, Yamanishi Y, Clausen BE, Forster I, Pawlinski R, Mackman N, Haase VH, Jaenisch R, Corr M, Nizet V, Firestein GS, Gerber HP, Ferrara N, Johnson RS (2003) HIF-1alpha is essential for myeloid cell-mediated inflammation. Cell 112(5):645–657
82. Cui G, Qin X, Wu L, Zhang Y, Sheng X, Yu Q, Sheng H, Xi B, Zhang JZ, Zang YQ (2011) Liver X receptor (LXR) mediates negative regulation of mouse and human Th17 differentiation. J Clin Invest 121(2):658–670. doi:10.1172/JCI42974

83. Cui Y, Chen X, Zhang J, Sun X, Liu H, Bai L, Xu C, Liu X (2016) Uhrf1 controls iNKT cell survival and differentiation through the Akt-mTOR Axis. Cell Rep 15(2):256–263. doi:10.1016/j.celrep.2016.03.016
84. Cunningham JT, Rodgers JT, Arlow DH, Vazquez F, Mootha VK, Puigserver P (2007) mTOR controls mitochondrial oxidative function through a YY1-PGC-1alpha transcriptional complex. Nature 450(7170):736–740. doi:10.1038/nature06322
85. Cyster JG, Dang EV, Reboldi A, Yi T (2014) 25-Hydroxycholesterols in innate and adaptive immunity. Nat Rev Immunol 14(11):731–743. doi:10.1038/nri3755
86. Dang EV, Barbi J, Yang HY, Jinasena D, Yu H, Zheng Y, Bordman Z, Fu J, Kim Y, Yen HR, Luo W, Zeller K, Shimoda L, Topalian SL, Semenza GL, Dang CV, Pardoll DM, Pan F (2011) Control of T(H)17/T(reg) balance by hypoxia-inducible factor 1. Cell 146(5):772–784. doi:10.1016/j.cell.2011.07.033
87. de Jong AJ, Kloppenburg M, Toes RE, Ioan-Facsinay A (2014) Fatty acids, lipid mediators, and T-cell function. Front Immunol 5:483. doi:10.3389/fimmu.2014.00483
88. De Rosa V, Galgani M, Porcellini A, Colamatteo A, Santopaolo M, Zuchegna C, Romano A, De Simone S, Procaccini C, La Rocca C, Carrieri PB, Maniscalco GT, Salvetti M, Buscarinu MC, Franzese A, Mozzillo E, La Cava A, Matarese G (2015) Glycolysis controls the induction of human regulatory T cells by modulating the expression of FOXP3 exon 2 splicing variants. Nat Immunol 16(11):1174–1184. doi:10.1038/ni.3269
89. De Rosa V, Procaccini C, Cali G, Pirozzi G, Fontana S, Zappacosta S, La Cava A, Matarese G (2007) A key role of leptin in the control of regulatory T cell proliferation. Immunity 26(2):241–255. doi:10.1016/j.immuni.2007.01.011
90. De Saedeleer CJ, Copetti T, Porporato PE, Verrax J, Feron O, Sonveaux P (2012) Lactate activates HIF-1 in oxidative but not in Warburg-phenotype human tumor cells. PLoS One 7(10):e46571. doi:10.1371/journal.pone.0046571
91. De Silva NS, Klein U (2015) Dynamics of B cells in germinal centres. Nat Rev Immunol 15(3):137–148. doi:10.1038/nri3804
92. Delgoffe GM, Kole TP, Zheng Y, Zarek PE, Matthews KL, Xiao B, Worley PF, Kozma SC, Powell JD (2009) The mTOR kinase differentially regulates effector and regulatory T cell lineage commitment. Immunity 30(6):832–844. doi:10.1016/j.immuni.2009.04.014
93. Delgoffe GM, Pollizzi KN, Waickman AT, Heikamp E, Meyers DJ, Horton MR, Xiao B, Worley PF, Powell JD (2011) The kinase mTOR regulates the differentiation of helper T cells through the selective activation of signaling by mTORC1 and mTORC2. Nat Immunol 12(4):295–303. doi:10.1038/ni.2005
94. Delgoffe GM, Woo SR, Turnis ME, Gravano DM, Guy C, Overacre AE, Bettini ML, Vogel P, Finkelstein D, Bonnevier J, Workman CJ, Vignali DA (2013) Stability and function of regulatory T cells is maintained by a neuropilin-1-semaphorin-4a axis. Nature 501(7466):252–256. doi:10.1038/nature12428
95. Deng Y, Kerdiles Y, Chu J, Yuan S, Wang Y, Chen X, Mao H, Zhang L, Zhang J, Hughes T, Zhang Q, Wang F, Zou X, Liu CG, Freud AG, Li X, Caligiuri MA, Vivier E, Yu J (2015) Transcription factor Foxo1 is a negative regulator of natural killer cell maturation and function. Immunity 42(3):457–470. doi:10.1016/j.immuni.2015.02.006
96. Depeint F, Bruce WR, Shangari N, Mehta R, O'Brien PJ (2006) Mitochondrial function and toxicity: role of B vitamins on the one-carbon transfer pathways. Chem Biol Interact 163(1–2):113–132. doi:10.1016/j.cbi.2006.05.010
97. Desreumaux P, Dubuquoy L, Nutten S, Peuchmaur M, Englaro W, Schoonjans K, Derijard B, Desvergne B, Wahli W, Chambon P, Leibowitz MD, Colombel JF, Auwerx J (2001) Attenuation of colon inflammation through activators of the retinoid X receptor (RXR)/peroxisome proliferator-activated receptor gamma (PPARgamma) heterodimer. A basis for new therapeutic strategies. J Exp Med 193(7):827–838
98. Doedens AL, Phan AT, Stradner MH, Fujimoto JK, Nguyen JV, Yang E, Johnson RS, Goldrath AW (2013) Hypoxia-inducible factors enhance the effector responses of CD8(+) T cells to persistent antigen. Nat Immunol 14(11):1173–1182. doi:10.1038/ni.2714

99. Doege H, Stahl A (2006) Protein-mediated fatty acid uptake: novel insights from in vivo models. Physiology 21:259–268. doi:10.1152/physiol.00014.2006
100. Donnelly RP, Loftus RM, Keating SE, Liou KT, Biron CA, Gardiner CM, Finlay DK (2014) mTORC1-dependent metabolic reprogramming is a prerequisite for NK cell effector function. J Immunol 193(9):4477–4484. doi:10.4049/jimmunol.1401558
101. Doughty CA, Bleiman BF, Wagner DJ, Dufort FJ, Mataraza JM, Roberts MF, Chiles TC (2006) Antigen receptor-mediated changes in glucose metabolism in B lymphocytes: role of phosphatidylinositol 3-kinase signaling in the glycolytic control of growth. Blood 107(11):4458–4465. doi:10.1182/blood-2005-12-4788
102. Dufort FJ, Bleiman BF, Gumina MR, Blair D, Wagner DJ, Roberts MF, Abu-Amer Y, Chiles TC (2007) Cutting edge: IL-4-mediated protection of primary B lymphocytes from apoptosis via Stat6-dependent regulation of glycolytic metabolism. J Immunol 179(8):4953–4957
103. Dunn SE, Youssef S, Goldstein MJ, Prod'homme T, Weber MS, Zamvil SS, Steinman L (2006) Isoprenoids determine Th1/Th2 fate in pathogenic T cells, providing a mechanism of modulation of autoimmunity by atorvastatin. J Exp Med 203(2):401–412. doi:10.1084/jem.20051129
104. Ecker J, Liebisch G, Englmaier M, Grandl M, Robenek H, Schmitz G (2010) Induction of fatty acid synthesis is a key requirement for phagocytic differentiation of human monocytes. Proc Natl Acad Sci U.S.A. 107(17):7817–7822. doi:10.1073/pnas.0912059107
105. Egan DF, Shackelford DB, Mihaylova MM, Gelino S, Kohnz RA, Mair W, Vasquez DS, Joshi A, Gwinn DM, Taylor R, Asara JM, Fitzpatrick J, Dillin A, Viollet B, Kundu M, Hansen M, Shaw RJ (2011) Phosphorylation of ULK1 (hATG1) by AMP-activated protein kinase connects energy sensing to mitophagy. Science 331(6016):456–461. doi:10.1126/science.1196371
106. Eijkelenboom A, Burgering BM (2013) FOXOs: signalling integrators for homeostasis maintenance. Nat Rev Mol Cell Biol 14(2):83–97. doi:10.1038/nrm3507
107. Endo Y, Asou HK, Matsugae N, Hirahara K, Shinoda K, Tumes DJ, Tokuyama H, Yokote K, Nakayama T (2015) Obesity drives Th17 cell differentiation by inducing the lipid metabolic kinase, ACC1. Cell Rep 12(6):1042–1055. doi:10.1016/j.celrep.2015.07.014
108. Everts B, Amiel E, Huang SC, Smith AM, Chang CH, Lam WY, Redmann V, Freitas TC, Blagih J, van der Windt GJ, Artyomov MN, Jones RG, Pearce EL, Pearce EJ (2014) TLR-driven early glycolytic reprogramming via the kinases TBK1-IKKvarepsilon supports the anabolic demands of dendritic cell activation. Nat Immunol 15(4):323–332. doi:10.1038/ni.2833
109. Everts B, Amiel E, van der Windt GJ, Freitas TC, Chott R, Yarasheski KE, Pearce EL, Pearce EJ (2012) Commitment to glycolysis sustains survival of NO-producing inflammatory dendritic cells. Blood 120(7):1422–1431. doi:10.1182/blood-2012-03-419747
110. Fan W, Morinaga H, Kim JJ, Bae E, Spann NJ, Heinz S, Glass CK, Olefsky JM (2010) FoxO1 regulates Tlr4 inflammatory pathway signalling in macrophages. EMBO J 29(24):4223–4236. doi:10.1038/emboj.2010.268
111. Fang C, Yu J, Luo Y, Chen S, Wang W, Zhao C, Sun Z, Wu W, Guo W, Han Z, Hu X, Liao F, Feng X (2015) Tsc1 is a critical regulator of macrophage survival and function. Cell Physiol Biochem 36(4):1406–1418. doi:10.1159/000430306
112. Faveeuw C, Fougeray S, Angeli V, Fontaine J, Chinetti G, Gosset P, Delerive P, Maliszewski C, Capron M, Staels B, Moser M, Trottein F (2000) Peroxisome proliferator-activated receptor gamma activators inhibit interleukin-12 production in murine dendritic cells. FEBS Lett 486(3):261–266
113. Feng X, Wang H, Takata H, Day TJ, Willen J, Hu H (2011) Transcription factor Foxp1 exerts essential cell-intrinsic regulation of the quiescence of naive T cells. Nat Immunol 12(6):544–550. doi:10.1038/ni.2034
114. Fernandez-Marcos PJ, Auwerx J (2011) Regulation of PGC-1alpha, a nodal regulator of mitochondrial biogenesis. Am J Clin Nutr 93(4):884S–8890. doi:10.3945/ajcn.110.001917

115. Festuccia WT, Pouliot P, Bakan I, Sabatini DM, Laplante M (2014) Myeloid-specific Rictor deletion induces M1 macrophage polarization and potentiates in vivo pro-inflammatory response to lipopolysaccharide. PLoS One 9(4):e95432. doi:10.1371/journal.pone.0095432
116. Finlay DK (2012) Regulation of glucose metabolism in T cells: new insight into the role of Phosphoinositide 3-kinases. Front Immunol 3:247. doi:10.3389/fimmu.2012.00247
117. Finlay DK, Rosenzweig E, Sinclair LV, Feijoo-Carnero C, Hukelmann JL, Rolf J, Panteleyev AA, Okkenhaug K, Cantrell DA (2012) PDK1 regulation of mTOR and hypoxia-inducible factor 1 integrate metabolism and migration of CD8+ T cells. J Exp Med 209(13):2441–2453. doi:10.1084/jem.20112607
118. Finlay DK, Sinclair LV, Feijoo C, Waugh CM, Hagenbeek TJ, Spits H, Cantrell DA (2009) Phosphoinositide-dependent kinase 1 controls migration and malignant transformation but not cell growth and proliferation in PTEN-null lymphocytes. J Exp Med 206(11):2441–2454. doi:10.1084/jem.20090219
119. Fletcher M, Ramirez ME, Sierra RA, Raber P, Thevenot P, Al-Khami AA, Sanchez-Pino D, Hernandez C, Wyczechowska DD, Ochoa AC, Rodriguez PC (2015) L-arginine depletion blunts antitumor T-cell responses by inducing myeloid-derived suppressor cells. Cancer Res 75(2):275–283. doi:10.1158/0008-5472.can-14-1491
120. Fontenot JD, Rasmussen JP, Williams LM, Dooley JL, Farr AG, Rudensky AY (2005) Regulatory T cell lineage specification by the forkhead transcription factor foxp3. Immunity 22(3):329–341. doi:10.1016/j.immuni.2005.01.016
121. Frauwirth KA, Riley JL, Harris MH, Parry RV, Rathmell JC, Plas DR, Elstrom RL, June CH, Thompson CB (2002) The CD28 signaling pathway regulates glucose metabolism. Immunity 16(6):769–777
122. Freemerman AJ, Johnson AR, Sacks GN, Milner JJ, Kirk EL, Troester MA, Macintyre AN, Goraksha-Hicks P, Rathmell JC, Makowski L (2014) Metabolic reprogramming of macrophages: glucose transporter 1 (GLUT1)-mediated glucose metabolism drives a proinflammatory phenotype. J Biol Chem 289(11):7884–7896. doi:10.1074/jbc.M113.522037
123. Fullerton MD, Galic S, Marcinko K, Sikkema S, Pulinilkunnil T, Chen ZP, O'Neill HM, Ford RJ, Palanivel R, O'Brien M, Hardie DG, Macaulay SL, Schertzer JD, Dyck JR, van Denderen BJ, Kemp BE, Steinberg GR (2013) Single phosphorylation sites in Acc1 and Acc2 regulate lipid homeostasis and the insulin-sensitizing effects of metformin. Nat Med 19(12):1649–1654. doi:10.1038/nm.3372
124. Furusawa Y, Obata Y, Fukuda S, Endo TA, Nakato G, Takahashi D, Nakanishi Y, Uetake C, Kato K, Kato T, Takahashi M, Fukuda NN, Murakami S, Miyauchi E, Hino S, Atarashi K, Onawa S, Fujimura Y, Lockett T, Clarke JM, Topping DL, Tomita M, Hori S, Ohara O, Morita T, Koseki H, Kikuchi J, Honda K, Hase K, Ohno H (2013) Commensal microbe-derived butyrate induces the differentiation of colonic regulatory T cells. Nature 504(7480):446–450. doi:10.1038/nature12721
125. Galic S, Fullerton MD, Schertzer JD, Sikkema S, Marcinko K, Walkley CR, Izon D, Honeyman J, Chen ZP, van Denderen BJ, Kemp BE, Steinberg GR (2011) Hematopoietic AMPK beta1 reduces mouse adipose tissue macrophage inflammation and insulin resistance in obesity. J Clin Invest 121(12):4903–4915. doi:10.1172/JCI58577
126. Galvan-Pena S, O'Neill LA (2014) Metabolic reprograming in macrophage polarization. Front Immunol 5:420. doi:10.3389/fimmu.2014.00420
127. Gan X, Wang J, Su B, Wu D (2011) Evidence for direct activation of mTORC2 kinase activity by phosphatidylinositol 3,4,5-trisphosphate. J Biol Chem 286(13):10998–11002. doi:10.1074/jbc.M110.195016
128. Ganeshan K, Chawla A (2014) Metabolic regulation of immune responses. Annu Rev Immunol 32:609–634. doi:10.1146/annurev-immunol-032713-120236
129. Ganguly D, Haak S, Sisirak V, Reizis B (2013) The role of dendritic cells in autoimmunity. Nat Rev Immunol 13(8):566–577. doi:10.1038/nri3477

130. Ganley IG, Lam du H, Wang J, Ding X, Chen S, Jiang X (2009) ULK1.ATG13.FIP200 complex mediates mTOR signaling and is essential for autophagy. J Biol Chem 284(18):12297–12305. doi:10.1074/jbc.M900573200
131. Gehart H, Kumpf S, Ittner A, Ricci R (2010) MAPK signalling in cellular metabolism: stress or wellness? EMBO Rep 11(11):834–840. doi:10.1038/embor.2010.160
132. Gerriets VA, Kishton RJ, Nichols AG, Macintyre AN, Inoue M, Ilkayeva O, Winter PS, Liu X, Priyadharshini B, Slawinska ME, Haeberli L, Huck C, Turka LA, Wood KC, Hale LP, Smith PA, Schneider MA, MacIver NJ, Locasale JW, Newgard CB, Shinohara ML, Rathmell JC (2015) Metabolic programming and PDHK1 control CD4+ T cell subsets and inflammation. J Clin Invest 125(1):194–207. doi:10.1172/JCI76012
133. Giandomenico V, Simonsson M, Gronroos E, Ericsson J (2003) Coactivator-dependent acetylation stabilizes members of the SREBP family of transcription factors. Mol Cell Biol 23(7):2587–2599
134. Gober HJ, Kistowska M, Angman L, Jeno P, Mori L, De Libero G (2003) Human T cell receptor gammadelta cells recognize endogenous mevalonate metabolites in tumor cells. J Exp Med 197(2):163–168
135. Gomez-Rodriguez J, Meylan F, Handon R, Hayes ET, Anderson SM, Kirby MR, Siegel RM, Schwartzberg PL (2016) Itk is required for Th9 differentiation via TCR-mediated induction of IL-2 and IRF4. Nat Commun 7:10857. doi:10.1038/ncomms10857
136. Gomez-Rodriguez J, Wohlfert EA, Handon R, Meylan F, Wu JZ, Anderson SM, Kirby MR, Belkaid Y, Schwartzberg PL (2014) Itk-mediated integration of T cell receptor and cytokine signaling regulates the balance between Th17 and regulatory T cells. J Exp Med 211(3):529–543. doi:10.1084/jem.20131459
137. Gorentla BK, Wan CK, Zhong XP (2011) Negative regulation of mTOR activation by diacylglycerol kinases. Blood 117(15):4022–4031. doi:10.1182/blood-2010-08-300731
138. Gorman S, Kuritzky LA, Judge MA, Dixon KM, McGlade JP, Mason RS, Finlay-Jones JJ, Hart PH (2007) Topically applied 1,25-dihydroxyvitamin D3 enhances the suppressive activity of CD4+CD25+ cells in the draining lymph nodes. J Immunol 179(9):6273–6283
139. Gosset P, Charbonnier AS, Delerive P, Fontaine J, Staels B, Pestel J, Tonnel AB, Trottein F (2001) Peroxisome proliferator-activated receptor gamma activators affect the maturation of human monocyte-derived dendritic cells. Eur J Immunol 31(10):2857–2865
140. Goswami R, Kaplan MH (2012) Gcn5 is required for PU.1-dependent IL-9 induction in Th9 cells. J Immunol 189(6):3026–3033. doi:10.4049/jimmunol.1201496
141. Greenwood J, Steinman L, Zamvil SS (2006) Statin therapy and autoimmune disease: from protein prenylation to immunomodulation. Nat Rev Immunol 6(5):358–370. doi:10.1038/nri1839
142. Gross DN, van den Heuvel AP, Birnbaum MJ (2008) The role of FoxO in the regulation of metabolism. Oncogene 27(16):2320–2336. doi:10.1038/onc.2008.25
143. Gualdoni GA, Mayer KA, Goschl L, Boucheron N, Ellmeier W, Zlabinger GJ (2016) The AMP analog AICAR modulates the Treg/Th17 axis through enhancement of fatty acid oxidation. FASEB J: Off Publ Fed Am Soc Exp Biol. doi:10.1096/fj.201600522R
144. Gubbels Bupp MR, Edwards B, Guo C, Wei D, Chen G, Wong B, Masteller E, Peng SL (2009) T cells require Foxo1 to populate the peripheral lymphoid organs. Eur J Immunol 39(11):2991–2999. doi:10.1002/eji.200939427
145. Gulen MF, Kang Z, Bulek K, Youzhong W, Kim TW, Chen Y, Altuntas CZ, Sass Bak-Jensen K, McGeachy MJ, Do JS, Xiao H, Delgoffe GM, Min B, Powell JD, Tuohy VK, Cua DJ, Li X (2010) The receptor SIGIRR suppresses Th17 cell proliferation via inhibition of the interleukin-1 receptor pathway and mTOR kinase activation. Immunity 32(1):54–66. doi:10.1016/j.immuni.2009.12.003
146. Guo F, Iclozan C, Suh WK, Anasetti C, Yu XZ (2008) CD28 controls differentiation of regulatory T cells from naive CD4 T cells. J Immunol 181(4):2285–2291
147. Haas R, Smith J, Rocher-Ros V, Nadkarni S, Montero-Melendez T, D'Acquisto F, Bland EJ, Bombardieri M, Pitzalis C, Perretti M, Marelli-Berg FM, Mauro C (2015) Lactate regulates

metabolic and pro-inflammatory circuits in control of T cell migration and effector functions. PLoS Biol 13(7):e1002202. doi:10.1371/journal.pbio.1002202
148. Hagenbeek TJ, Naspetti M, Malergue F, Garcon F, Nunes JA, Cleutjens KB, Trapman J, Krimpenfort P, Spits H (2004) The loss of PTEN allows TCR alphabeta lineage thymocytes to bypass IL-7 and Pre-TCR-mediated signaling. J Exp Med 200(7):883–894. doi:10.1084/jem.20040495
149. Haghikia A, Jorg S, Duscha A, Berg J, Manzel A, Waschbisch A, Hammer A, Lee DH, May C, Wilck N, Balogh A, Ostermann AI, Schebb NH, Akkad DA, Grohme DA, Kleinewietfeld M, Kempa S, Thone J, Demir S, Muller DN, Gold R, Linker RA (2015) Dietary fatty acids directly impact central nervous system autoimmunity via the small intestine. Immunity 43(4):817–829. doi:10.1016/j.immuni.2015.09.007
150. Haidinger M, Poglitsch M, Geyeregger R, Kasturi S, Zeyda M, Zlabinger GJ, Pulendran B, Horl WH, Saemann MD, Weichhart T (2010) A versatile role of mammalian target of rapamycin in human dendritic cell function and differentiation. J Immunol 185(7):3919–3931. doi:10.4049/jimmunol.1000296
151. Hamilton KS, Phong B, Corey C, Cheng J, Gorentla B, Zhong X, Shiva S, Kane LP (2014) T cell receptor-dependent activation of mTOR signaling in T cells is mediated by Carma1 and MALT1, but not Bcl10. Sci Signal 7(329):ra55. doi:10.1126/scisignal.2005169
152. Hard GC (1970) Some biochemical aspects of the immune macrophage. Br J Exp Pathol 51(1):97–105
153. Harms M, Seale P (2013) Brown and beige fat: development, function and therapeutic potential. Nat Med 19(10):1252–1263. doi:10.1038/nm.3361
154. Haschemi A, Kosma P, Gille L, Evans CR, Burant CF, Starkl P, Knapp B, Haas R, Schmid JA, Jandl C, Amir S, Lubec G, Park J, Esterbauer H, Bilban M, Brizuela L, Pospisilik JA, Otterbein LE, Wagner O (2012) The sedoheptulose kinase CARKL directs macrophage polarization through control of glucose metabolism. Cell Metab 15(6):813–826. doi:10.1016/j.cmet.2012.04.023
155. Hawley SA, Davison M, Woods A, Davies SP, Beri RK, Carling D, Hardie DG (1996) Characterization of the AMP-activated protein kinase kinase from rat liver and identification of threonine 172 as the major site at which it phosphorylates AMP-activated protein kinase. J Biol Chem 271(44):27879–27887
156. Hedrick SM, Hess Michelini R, Doedens AL, Goldrath AW, Stone EL (2012) FOXO transcription factors throughout T cell biology. Nat Rev Immunol 12(9):649–661. doi:10.1038/nri3278
157. Hess Michelini R, Doedens AL, Goldrath AW, Hedrick SM (2013) Differentiation of CD8 memory T cells depends on Foxo1. J Exp Med 210(6):1189–1200. doi:10.1084/jem.20130392
158. Ho PC, Bihuniak JD, Macintyre AN, Staron M, Liu X, Amezquita R, Tsui YC, Cui G, Micevic G, Perales JC, Kleinstein SH, Abel ED, Insogna KL, Feske S, Locasale JW, Bosenberg MW, Rathmell JC, Kaech SM (2015) Phosphoenolpyruvate is a metabolic checkpoint of anti-tumor T cell responses. Cell 162(6):1217–1228. doi:10.1016/j.cell.2015.08.012
159. Hontecillas R, Bassaganya-Riera J (2007) Peroxisome proliferator-activated receptor gamma is required for regulatory CD4+ T cell-mediated protection against colitis. J Immunol 178(5):2940–2949
160. Hori S, Nomura T, Sakaguchi S (2003) Control of regulatory T cell development by the transcription factor Foxp3. Science 299(5609):1057–1061. doi:10.1126/science.1079490
161. Hosios AM, Hecht VC, Danai LV, Johnson MO, Rathmell JC, Steinhauser ML, Manalis SR, Vander Heiden MG (2016) Amino acids rather than glucose account for the majority of cell mass in proliferating mammalian cells. Dev Cell 36(5):540–549. doi:10.1016/j.devcel.2016.02.012
162. Housley WJ, Adams CO, Vang AG, Brocke S, Nichols FC, LaCombe M, Rajan TV, Clark RB (2011) Peroxisome proliferator-activated receptor gamma is required for CD4+ T cell-mediated lymphopenia-associated autoimmunity. J Immunol 187(8):4161–4169. doi:10.4049/jimmunol.1101731

163. Hsieh CS, Lee HM, Lio CW (2012) Selection of regulatory T cells in the thymus. Nat Rev Immunol 12(3):157–167. doi:10.1038/nri3155
164. Huang D, Li T, Li X, Zhang L, Sun L, He X, Zhong X, Jia D, Song L, Semenza GL, Gao P, Zhang H (2014) HIF-1-mediated suppression of acyl-CoA dehydrogenases and fatty acid oxidation is critical for cancer progression. Cell Rep 8(6):1930–1942. doi:10.1016/j.celrep.2014.08.028
165. Huh JY, Kim JI, Park YJ, Hwang IJ, Lee YS, Sohn JH, Lee SK, Alfadda AA, Kim SS, Choi SH, Lee DS, Park SH, Seong RH, Choi CS, Kim JB (2013) A novel function of adipocytes in lipid antigen presentation to iNKT cells. Mol Cell Biol 33(2):328–339. doi:10.1128/MCB.00552-12
166. Hukelmann JL, Anderson KE, Sinclair LV, Grzes KM, Murillo AB, Hawkins PT, Stephens LR, Lamond AI, Cantrell DA (2016) The cytotoxic T cell proteome and its shaping by the kinase mTOR. Nat Immunol 17(1):104–112. doi:10.1038/ni.3314
167. Hussaarts L, Smits HH, Schramm G, van der Ham AJ, van der Zon GC, Haas H, Guigas B, Yazdanbakhsh M (2013) Rapamycin and omega-1: mTOR-dependent and -independent Th2 skewing by human dendritic cells. Immunol Cell Biol 91(7):486–489. doi:10.1038/icb.2013.31
168. Huynh A, DuPage M, Priyadharshini B, Sage PT, Quiros J, Borges CM, Townamchai N, Gerriets VA, Rathmell JC, Sharpe AH, Bluestone JA, Turka LA (2015) Control of PI(3) kinase in Treg cells maintains homeostasis and lineage stability. Nat Immunol 16(2):188–196. doi:10.1038/ni.3077
169. Ikeda S, Saijo S, Murayama MA, Shimizu K, Akitsu A, Iwakura Y (2014) Excess IL-1 signaling enhances the development of Th17 cells by downregulating TGF-beta-induced Foxp3 expression. J Immunol 192(4):1449–1458. doi:10.4049/jimmunol.1300387
170. Im SS, Yousef L, Blaschitz C, Liu JZ, Edwards RA, Young SG, Raffatellu M, Osborne TF (2011) Linking lipid metabolism to the innate immune response in macrophages through sterol regulatory element binding protein-1a. Cell Metab 13(5):540–549. doi:10.1016/j.cmet.2011.04.001
171. Imtiyaz HZ, Williams EP, Hickey MM, Patel SA, Durham AC, Yuan LJ, Hammond R, Gimotty PA, Keith B, Simon MC (2010) Hypoxia-inducible factor 2alpha regulates macrophage function in mouse models of acute and tumor inflammation. J Clin Invest 120(8):2699–2714. doi:10.1172/JCI39506
172. Infantino V, Convertini P, Cucci L, Panaro MA, Di Noia MA, Calvello R, Palmieri F, Iacobazzi V (2011) The mitochondrial citrate carrier: a new player in inflammation. Biochem J 438(3):433–436. doi:10.1042/BJ20111275
173. Jacobs SR, Herman CE, Maciver NJ, Wofford JA, Wieman HL, Hammen JJ, Rathmell JC (2008) Glucose uptake is limiting in T cell activation and requires CD28-mediated Akt-dependent and independent pathways. J Immunol 180(7):4476–4486
174. Janke R, Dodson AE, Rine J (2015) Metabolism and epigenetics. Annu Rev Cell Dev Biol 31:473–496. doi:10.1146/annurev-cellbio-100814-125544
175. Jantsch J, Chakravortty D, Turza N, Prechtel AT, Buchholz B, Gerlach RG, Volke M, Glasner J, Warnecke C, Wiesener MS, Eckardt KU, Steinkasserer A, Hensel M, Willam C (2008) Hypoxia and hypoxia-inducible factor-1 alpha modulate lipopolysaccharide-induced dendritic cell activation and function. J Immunol 180(7):4697–4705
176. Jeffery LE, Burke F, Mura M, Zheng Y, Qureshi OS, Hewison M, Walker LS, Lammas DA, Raza K, Sansom DM (2009) 1,25-Dihydroxyvitamin D3 and IL-2 combine to inhibit T cell production of inflammatory cytokines and promote development of regulatory T cells expressing CTLA-4 and FoxP3. J Immunol 183(9):5458–5467. doi:10.4049/jimmunol.0803217
177. Jewell JL, Russell RC, Guan KL (2013) Amino acid signalling upstream of mTOR. Nat Rev Mol Cell Biol 14(3):133–139. doi:10.1038/nrm3522
178. Ji G, Zhang Y, Yang Q, Cheng S, Hao J, Zhao X, Jiang Z (2012) Genistein suppresses LPS-induced inflammatory response through inhibiting NF-kappaB following AMP kinase activation in RAW 264.7 macrophages. PloS One 7(12):e53101. doi:10.1371/journal.pone.0053101

179. Jiang H, Westerterp M, Wang C, Zhu Y, Ai D (2014) Macrophage mTORC1 disruption reduces inflammation and insulin resistance in obese mice. Diabetologia 57(11):2393–2404. doi:10.1007/s00125-014-3350-5
180. Johnston RJ, Poholek AC, DiToro D, Yusuf I, Eto D, Barnett B, Dent AL, Craft J, Crotty S (2009) Bcl6 and blimp-1 are reciprocal and antagonistic regulators of T follicular helper cell differentiation. Science 325(5943):1006–1010. doi:10.1126/science.1175870
181. Josefowicz SZ, Lu LF, Rudensky AY (2012) Regulatory T cells: mechanisms of differentiation and function. Annu Rev Immunol 30:531–564. doi:10.1146/annurev.immunol.25.022106.141623
182. Kabat AM, Harrison OJ, Riffelmacher T, Moghaddam AE, Pearson CF, Laing A, Abeler-Dorner L, Forman SP, Grencis RK, Sattentau Q, Simon AK, Pott J, Maloy KJ (2016) The autophagy gene Atg16l1 differentially regulates Treg and TH2 cells to control intestinal inflammation. eLife 5:e12444. doi:10.7554/eLife.12444
183. Kaech SM, Cui W (2012) Transcriptional control of effector and memory CD8+ T cell differentiation. Nat Rev Immunol 12(11):749–761. doi:10.1038/nri3307
184. Kang SG, Lim HW, Andrisani OM, Broxmeyer HE, Kim CH (2007) Vitamin A metabolites induce gut-homing FoxP3+ regulatory T cells. J Immunol 179(6):3724–3733
185. Kanno Y, Vahedi G, Hirahara K, Singleton K, O'Shea JJ (2012) Transcriptional and epigenetic control of T helper cell specification: molecular mechanisms underlying commitment and plasticity. Annu Rev Immunol 30:707–731. doi:10.1146/annurev-immunol-020711-075058
186. Kaplan MH, Hufford MM, Olson MR (2015) The development and in vivo function of T helper 9 cells. Nat Rev Immunol 15(5):295–307. doi:10.1038/nri3824
187. Kawanaka T, Kubo A, Ikushima H, Sano T, Takegawa Y, Nishitani H (2008) Prognostic significance of HIF-2alpha expression on tumor infiltrating macrophages in patients with uterine cervical cancer undergoing radiotherapy. J Med Invest: JMI 55(1–2):78–86
188. Kc W, Satpathy AT, Rapaport AS, Briseno CG, Wu X, Albring JC, Russler-Germain EV, Kretzer NM, Durai V, Persaud SP, Edelson BT, Loschko J, Cella M, Allen PM, Nussenzweig MC, Colonna M, Sleckman BP, Murphy TL, Murphy KM (2014) L-Myc expression by dendritic cells is required for optimal T-cell priming. Nature 507(7491):243–247. doi:10.1038/nature12967
189. Keating R, Hertz T, Wehenkel M, Harris TL, Edwards BA, McClaren JL, Brown SA, Surman S, Wilson ZS, Bradley P, Hurwitz J, Chi H, Doherty PC, Thomas PG, McGargill MA (2013) The kinase mTOR modulates the antibody response to provide cross-protective immunity to lethal infection with influenza virus. Nat Immunol 14(12):1266–1276. doi:10.1038/ni.2741
190. Keppel MP, Saucier N, Mah AY, Vogel TP, Cooper MA (2015) Activation-specific metabolic requirements for NK cell IFN-gamma production. J Immunol 194(4):1954–1962. doi:10.4049/jimmunol.1402099
191. Kerdiles YM, Beisner DR, Tinoco R, Dejean AS, Castrillon DH, DePinho RA, Hedrick SM (2009) Foxo1 links homing and survival of naive T cells by regulating L-selectin, CCR7 and interleukin 7 receptor. Nat Immunol 10(2):176–184. doi:10.1038/ni.1689
192. Khattri R, Cox T, Yasayko SA, Ramsdell F (2003) An essential role for Scurfin in CD4+CD25+ T regulatory cells. Nat Immunol 4(4):337–342. doi:10.1038/ni909
193. Kidani Y, Elsaesser H, Hock MB, Vergnes L, Williams KJ, Argus JP, Marbois BN, Komisopoulou E, Wilson EB, Osborne TF, Graeber TG, Reue K, Brooks DG, Bensinger SJ (2013) Sterol regulatory element-binding proteins are essential for the metabolic programming of effector T cells and adaptive immunity. Nat Immunol 14(5):489–499. doi:10.1038/ni.2570
194. Kim I, He YY (2013) Targeting the AMP-activated protein kinase for cancer prevention and therapy. Front Oncol 3:175. doi:10.3389/fonc.2013.00175
195. Kim J, Kundu M, Viollet B, Guan KL (2011) AMPK and mTOR regulate autophagy through direct phosphorylation of Ulk1. Nat Cell Biol 13(2):132–141. doi:10.1038/ncb2152
196. Kim JK, Klinger M, Benjamin J, Xiao Y, Erle DJ, Littman DR, Killeen N (2009) Impact of the TCR signal on regulatory T cell homeostasis, function, and trafficking. PLoS One 4(8):e6580. doi:10.1371/journal.pone.0006580

197. Kim JW, Tchernyshyov I, Semenza GL, Dang CV (2006) HIF-1-mediated expression of pyruvate dehydrogenase kinase: a metabolic switch required for cellular adaptation to hypoxia. Cell Metab 3(3):177–185. doi:10.1016/j.cmet.2006.02.002
198. Kimura A, Naka T, Nohara K, Fujii-Kuriyama Y, Kishimoto T (2008) Aryl hydrocarbon receptor regulates Stat1 activation and participates in the development of Th17 cells. Proc Natl Acad Sci U.S.A. 105(28):9721–9726. doi:10.1073/pnas.0804231105
199. Kinoshita M, Kayama H, Kusu T, Yamaguchi T, Kunisawa J, Kiyono H, Sakaguchi S, Takeda K (2012) Dietary folic acid promotes survival of Foxp3+ regulatory T cells in the colon. J Immunol 189(6):2869–2878. doi:10.4049/jimmunol.1200420
200. Kleinewietfeld M, Manzel A, Titze J, Kvakan H, Yosef N, Linker RA, Muller DN, Hafler DA (2013) Sodium chloride drives autoimmune disease by the induction of pathogenic TH17 cells. Nature 496(7446):518–522. doi:10.1038/nature11868
201. Klotz L, Dani I, Edenhofer F, Nolden L, Evert B, Paul B, Kolanus W, Klockgether T, Knolle P, Diehl L (2007) Peroxisome proliferator-activated receptor gamma control of dendritic cell function contributes to development of CD4+ T cell anergy. J Immunol 178(4):2122–2131
202. Kobayashi T, Shimabukuro-Demoto S, Yoshida-Sugitani R, Furuyama-Tanaka K, Karyu H, Sugiura Y, Shimizu Y, Hosaka T, Goto M, Kato N, Okamura T, Suematsu M, Yokoyama S, Toyama-Sorimachi N (2014) The histidine transporter SLC15A4 coordinates mTOR-dependent inflammatory responses and pathogenic antibody production. Immunity 41(3):375–388. doi:10.1016/j.immuni.2014.08.011
203. Krawczyk CM, Holowka T, Sun J, Blagih J, Amiel E, DeBerardinis RJ, Cross JR, Jung E, Thompson CB, Jones RG, Pearce EJ (2010) Toll-like receptor-induced changes in glycolytic metabolism regulate dendritic cell activation. Blood 115(23):4742–4749. doi:10.1182/blood-2009-10-249540
204. Kress TR, Sabo A, Amati B (2015) MYC: connecting selective transcriptional control to global RNA production. Nat Rev Cancer 15(10):593–607. doi:10.1038/nrc3984
205. Kryczek I, Zhao E, Liu Y, Wang Y, Vatan L, Szeliga W, Moyer J, Klimczak A, Lange A, Zou W (2011) Human TH17 cells are long-lived effector memory cells. Sci Transl Med 3(104):104ra100. doi:10.1126/scitranslmed.3002949
206. Kunisawa J, Hashimoto E, Ishikawa I, Kiyono H (2012) A pivotal role of vitamin B9 in the maintenance of regulatory T cells in vitro and in vivo. PLoS One 7(2):e32094. doi:10.1371/journal.pone.0032094
207. Kuo CT, Veselits ML, Leiden JM (1997) LKLF: a transcriptional regulator of single-positive T cell quiescence and survival. Science 277(5334):1986–1990
208. Kurebayashi Y, Nagai S, Ikejiri A, Ohtani M, Ichiyama K, Baba Y, Yamada T, Egami S, Hoshii T, Hirao A, Matsuda S, Koyasu S (2012) PI3K-Akt-mTORC1-S6K1/2 axis controls Th17 differentiation by regulating Gfi1 expression and nuclear translocation of RORgamma. Cell Rep 1(4):360–373. doi:10.1016/j.celrep.2012.02.007
209. Kurihara T, Arimochi H, Bhuyan ZA, Ishifune C, Tsumura H, Ito M, Ito Y, Kitamura A, Maekawa Y, Yasutomo K (2015) CD98 heavy chain is a potent positive regulator of CD4+ T cell proliferation and interferon-gamma production in vivo. PLoS One 10(10):e0139692. doi:10.1371/journal.pone.0139692
210. Kurobe H, Urata M, Ueno M, Ueki M, Ono S, Izawa-Ishizawa Y, Fukuhara Y, Lei Y, Ripen AM, Kanbara T, Aihara K, Ishizawa K, Akaike M, Gonzalez FJ, Tamaki T, Takahama Y, Yoshizumi M, Kitagawa T, Tomita S (2010) Role of hypoxia-inducible factor 1alpha in T cells as a negative regulator in development of vascular remodeling. Arterioscler Thromb Vasc Biol 30(2):210–217. doi:10.1161/ATVBAHA.109.192666
211. La Cava A, Matarese G (2004) The weight of leptin in immunity. Nat Rev Immunol 4(5):371–379. doi:10.1038/nri1350
212. Laderoute KR, Amin K, Calaoagan JM, Knapp M, Le T, Orduna J, Foretz M, Viollet B (2006) 5′-AMP-activated protein kinase (AMPK) is induced by low-oxygen and glucose deprivation conditions found in solid-tumor microenvironments. Mol Cell Biol 26(14):5336–5347. doi:10.1128/MCB.00166-06

213. Lal L, Li Y, Smith J, Sassano A, Uddin S, Parmar S, Tallman MS, Minucci S, Hay N, Platanias LC (2005) Activation of the p70 S6 kinase by all-trans-retinoic acid in acute promyelocytic leukemia cells. Blood 105(4):1669–1677. doi:10.1182/blood-2004-06-2078
214. Laplante M, Sabatini DM (2009) An emerging role of mTOR in lipid biosynthesis. Curr Biol: CB 19(22):R1046–R1052. doi:10.1016/j.cub.2009.09.058
215. Laplante M, Sabatini DM (2012) mTOR signaling in growth control and disease. Cell 149(2):274–293. doi:10.1016/j.cell.2012.03.017
216. Larbi A, Zelba H, Goldeck D, Pawelec G (2010) Induction of HIF-1alpha and the glycolytic pathway alters apoptotic and differentiation profiles of activated human T cells. J Leukoc Biol 87(2):265–273. doi:10.1189/jlb.0509304
217. Lee J, Walsh MC, Hoehn KL, James DE, Wherry EJ, Choi Y (2014) Regulator of fatty acid metabolism, acetyl coenzyme a carboxylase 1, controls T cell immunity. J Immunol 192(7):3190–3199. doi:10.4049/jimmunol.1302985
218. Lee JE, Walsh MC, Hoehn KL, James DE, Wherry EJ, Choi Y (2015a) Acetyl CoA carboxylase 2 is dispensable for CD8+ T cell responses. PLoS One 10(9):e0137776. doi:10.1371/journal.pone.0137776
219. Lee JY, Skon CN, Lee YJ, Oh S, Taylor JJ, Malhotra D, Jenkins MK, Rosenfeld MG, Hogquist KA, Jameson SC (2015b) The transcription factor KLF2 restrains CD4(+) T follicular helper cell differentiation. Immunity 42(2):252–264. doi:10.1016/j.immuni.2015.01.013
220. Lee K, Gudapati P, Dragovic S, Spencer C, Joyce S, Killeen N, Magnuson MA, Boothby M (2010) Mammalian target of rapamycin protein complex 2 regulates differentiation of Th1 and Th2 cell subsets via distinct signaling pathways. Immunity 32(6):743–753. doi:10.1016/j.immuni.2010.06.002
221. Leek RD, Talks KL, Pezzella F, Turley H, Campo L, Brown NS, Bicknell R, Taylor M, Gatter KC, Harris AL (2002) Relation of hypoxia-inducible factor-2 alpha (HIF-2 alpha) expression in tumor-infiltrative macrophages to tumor angiogenesis and the oxidative thymidine phosphorylase pathway in human breast cancer. Cancer Res 62(5):1326–1329
222. Levine AG, Arvey A, Jin W, Rudensky AY (2014) Continuous requirement for the TCR in regulatory T cell function. Nat Immunol 15(11):1070–1078. doi:10.1038/ni.3004
223. Li B, Reynolds JM, Stout RD, Bernlohr DA, Suttles J (2009) Regulation of Th17 differentiation by epidermal fatty acid-binding protein. J Immunol 182(12):7625–7633. doi:10.4049/jimmunol.0804192
224. Li H, Shi B (2015) Tolerogenic dendritic cells and their applications in transplantation. Cell Mol Immunol 12(1):24–30. doi:10.1038/cmi.2014.52
225. Li Y, Xu S, Mihaylova MM, Zheng B, Hou X, Jiang B, Park O, Luo Z, Lefai E, Shyy JY, Gao B, Wierzbicki M, Verbeuren TJ, Shaw RJ, Cohen RA, Zang M (2011) AMPK phosphorylates and inhibits SREBP activity to attenuate hepatic steatosis and atherosclerosis in diet-induced insulin-resistant mice. Cell Metab 13(4):376–388. doi:10.1016/j.cmet.2011.03.009
226. Liang SL, Liu H, Zhou A (2006) Lovastatin-induced apoptosis in macrophages through the Rac1/Cdc42/JNK pathway. J Immunol 177(1):651–656
227. Liang Y, Cucchetti M, Roncagalli R, Yokosuka T, Malzac A, Bertosio E, Imbert J, Nijman IJ, Suchanek M, Saito T, Wulfing C, Malissen B, Malissen M (2013) The lymphoid lineage-specific actin-uncapping protein Rltpr is essential for costimulation via CD28 and the development of regulatory T cells. Nat Immunol 14(8):858–866. doi:10.1038/ni.2634
228. Limon JJ, So L, Jellbauer S, Chiu H, Corado J, Sykes SM, Raffatellu M, Fruman DA (2014) mTOR kinase inhibitors promote antibody class switching via mTORC2 inhibition. Proc Natl Acad Sci U.S.A. 111(47):E5076–E5085. doi:10.1073/pnas.1407104111
229. Lisse TS, Hewison M (2011) Vitamin D: a new player in the world of mTOR signaling. Cell Cycle 10(12):1888–1889. doi:10.4161/cc.10.12.15620
230. Liston A, Gray DH (2014) Homeostatic control of regulatory T cell diversity. Nat Rev Immunol 14(3):154–165. doi:10.1038/nri3605
231. Liu SY, Aliyari R, Chikere K, Li G, Marsden MD, Smith JK, Pernet O, Guo H, Nusbaum R, Zack JA, Freiberg AN, Su L, Lee B, Cheng G (2013) Interferon-inducible cholesterol-25-

hydroxylase broadly inhibits viral entry by production of 25-hydroxycholesterol. Immunity 38(1):92–105. doi:10.1016/j.immuni.2012.11.005
232. Locasale JW (2013) Serine, glycine and one-carbon units: cancer metabolism in full circle. Nat Rev Cancer 13(8):572–583. doi:10.1038/nrc3557
233. Lu C, Thompson CB (2012) Metabolic regulation of epigenetics. Cell Metab 16(1):9–17. doi:10.1016/j.cmet.2012.06.001
234. Lu H, Forbes RA, Verma A (2002) Hypoxia-inducible factor 1 activation by aerobic glycolysis implicates the Warburg effect in carcinogenesis. J Biol Chem 277(26):23111–23115. doi:10.1074/jbc.M202487200
235. Lukashev D, Klebanov B, Kojima H, Grinberg A, Ohta A, Berenfeld L, Wenger RH, Ohta A, Sitkovsky M (2006) Cutting edge: hypoxia-inducible factor 1alpha and its activation-inducible short isoform I.1 negatively regulate functions of CD4+ and CD8+ T lymphocytes. J Immunol 177(8):4962–4965
236. Luo CT, Liao W, Dadi S, Toure A, Li MO (2016) Graded Foxo1 activity in Treg cells differentiates tumour immunity from spontaneous autoimmunity. Nature 529(7587):532–536. doi:10.1038/nature16486
237. Lynch L, Nowak M, Varghese B, Clark J, Hogan AE, Toxavidis V, Balk SP, O'Shea D, O'Farrelly C, Exley MA (2012) Adipose tissue invariant NKT cells protect against diet-induced obesity and metabolic disorder through regulatory cytokine production. Immunity 37(3):574–587. doi:10.1016/j.immuni.2012.06.016
238. Ma Y, Xu L, Rodriguez-Agudo D, Li X, Heuman DM, Hylemon PB, Pandak WM, Ren S (2008) 25-hydroxycholesterol-3-sulfate regulates macrophage lipid metabolism via the LXR/SREBP-1 signaling pathway. Am J Phys Endocrinol Metab 295(6):E1369–E1379. doi:10.1152/ajpendo.90555.2008
239. Macintyre AN, Finlay D, Preston G, Sinclair LV, Waugh CM, Tamas P, Feijoo C, Okkenhaug K, Cantrell DA (2011) Protein kinase B controls transcriptional programs that direct cytotoxic T cell fate but is dispensable for T cell metabolism. Immunity 34(2):224–236. doi:10.1016/j.immuni.2011.01.012
240. Macintyre AN, Gerriets VA, Nichols AG, Michalek RD, Rudolph MC, Deoliveira D, Anderson SM, Abel ED, Chen BJ, Hale LP, Rathmell JC (2014) The glucose transporter Glut1 is selectively essential for CD4 T cell activation and effector function. Cell Metab 20(1):61–72. doi:10.1016/j.cmet.2014.05.004
241. MacIver NJ, Blagih J, Saucillo DC, Tonelli L, Griss T, Rathmell JC, Jones RG (2011) The liver kinase B1 is a central regulator of T cell development, activation, and metabolism. J Immunol 187(8):4187–4198. doi:10.4049/jimmunol.1100367
242. MacIver NJ, Michalek RD, Rathmell JC (2013) Metabolic regulation of T lymphocytes. Annu Rev Immunol 31:259–283. doi:10.1146/annurev-immunol-032712-095956
243. Malinarich F, Duan K, Hamid RA, Bijin A, Lin WX, Poidinger M, Fairhurst AM, Connolly JE (2015) High mitochondrial respiration and glycolytic capacity represent a metabolic phenotype of human tolerogenic dendritic cells. J Immunol 194(11):5174–5186. doi:10.4049/jimmunol.1303316
244. Mangus CW, Massey PR, Fowler DH, Amarnath S (2013) Rapamycin resistant murine th9 cells have a stable in vivo phenotype and inhibit graft-versus-host reactivity. PLoS One 8(8):e72305. doi:10.1371/journal.pone.0072305
245. Marcais A, Cherfils-Vicini J, Viant C, Degouve S, Viel S, Fenis A, Rabilloud J, Mayol K, Tavares A, Bienvenu J, Gangloff YG, Gilson E, Vivier E, Walzer T (2014) The metabolic checkpoint kinase mTOR is essential for IL-15 signaling during the development and activation of NK cells. Nat Immunol 15(8):749–757. doi:10.1038/ni.2936
246. Marko AJ, Miller RA, Kelman A, Frauwirth KA (2010) Induction of glucose metabolism in stimulated T lymphocytes is regulated by mitogen-activated protein kinase signaling. PLoS One 5(11):e15425. doi:10.1371/journal.pone.0015425
247. Mascanfroni ID, Takenaka MC, Yeste A, Patel B, Wu Y, Kenison JE, Siddiqui S, Basso AS, Otterbein LE, Pardoll DM, Pan F, Priel A, Clish CB, Robson SC, Quintana FJ (2015)

Metabolic control of type 1 regulatory T cell differentiation by AHR and HIF1-alpha. Nat Med 21(6):638–646. doi:10.1038/nm.3868
248. Masson N, Ratcliffe PJ (2014) Hypoxia signaling pathways in cancer metabolism: the importance of co-selecting interconnected physiological pathways. Cancer Metab 2(1):3. doi:10.1186/2049-3002-2-3
249. Masui K, Tanaka K, Akhavan D, Babic I, Gini B, Matsutani T, Iwanami A, Liu F, Villa GR, Gu Y, Campos C, Zhu S, Yang H, Yong WH, Cloughesy TF, Mellinghoff IK, Cavenee WK, Shaw RJ, Mischel PS (2013) mTOR complex 2 controls glycolytic metabolism in glioblastoma through FoxO acetylation and upregulation of c-Myc. Cell Metab 18(5):726–739. doi:10.1016/j.cmet.2013.09.013
250. Mazumdar J, Hickey MM, Pant DK, Durham AC, Sweet-Cordero A, Vachani A, Jacks T, Chodosh LA, Kissil JL, Simon MC, Keith B (2010) HIF-2alpha deletion promotes Kras-driven lung tumor development. Proc Natl Acad Sci U.S.A. 107(32):14182–14187. doi:10.1073/pnas.1001296107
251. McNamee EN, Korns Johnson D, Homann D, Clambey ET (2013) Hypoxia and hypoxia-inducible factors as regulators of T cell development, differentiation, and function. Immunol Res 55(1–3):58–70. doi:10.1007/s12026-012-8349-8
252. Michalek RD, Gerriets VA, Jacobs SR, Macintyre AN, MacIver NJ, Mason EF, Sullivan SA, Nichols AG, Rathmell JC (2011a) Cutting edge: distinct glycolytic and lipid oxidative metabolic programs are essential for effector and regulatory CD4+ T cell subsets. J Immunol 186(6):3299–3303. doi:10.4049/jimmunol.1003613
253. Michalek RD, Gerriets VA, Nichols AG, Inoue M, Kazmin D, Chang CY, Dwyer MA, Nelson ER, Pollizzi KN, Ilkayeva O, Giguere V, Zuercher WJ, Powell JD, Shinohara ML, McDonnell DP, Rathmell JC (2011b) Estrogen-related receptor-alpha is a metabolic regulator of effector T-cell activation and differentiation. Proc Natl Acad Sci U.S.A. 108(45):18348–18353. doi:10.1073/pnas.1108856108
254. Mihaylova MM, Shaw RJ (2011) The AMPK signalling pathway coordinates cell growth, autophagy and metabolism. Nat Cell Biol 13(9):1016–1023. doi:10.1038/ncb2329
255. Milasta S, Dillon CP, Sturm OE, Verbist KC, Brewer TL, Quarato G, Brown SA, Frase S, Janke LJ, Perry SS, Thomas PG, Green DR (2016) Apoptosis-inducing-factor-dependent mitochondrial function is required for T cell but not B cell function. Immunity 44(1):88–102. doi:10.1016/j.immuni.2015.12.002
256. Mirebeau-Prunier D, Le Pennec S, Jacques C, Gueguen N, Poirier J, Malthiery Y, Savagner F (2010) Estrogen-related receptor alpha and PGC-1-related coactivator constitute a novel complex mediating the biogenesis of functional mitochondria. FEBS J 277(3):713–725. doi:10.1111/j.1742-4658.2009.07516.x
257. Miyao T, Floess S, Setoguchi R, Luche H, Fehling HJ, Waldmann H, Huehn J, Hori S (2012) Plasticity of Foxp3(+) T cells reflects promiscuous Foxp3 expression in conventional T cells but not reprogramming of regulatory T cells. Immunity 36(2):262–275. doi:10.1016/j.immuni.2011.12.012
258. Monticelli LA, Buck MD, Flamar AL, Saenz SA, Tait Wojno ED, Yudanin NA, Osborne LC, Hepworth MR, Tran SV, Rodewald HR, Shah H, Cross JR, Diamond JM, Cantu E, Christie JD, Pearce EL, Artis D (2016) Arginase 1 is an innate lymphoid-cell-intrinsic metabolic checkpoint controlling type 2 inflammation. Nat Immunol 17(6):656–665. doi:10.1038/ni.3421
259. Moore KJ, Rosen ED, Fitzgerald ML, Randow F, Andersson LP, Altshuler D, Milstone DS, Mortensen RM, Spiegelman BM, Freeman MW (2001) The role of PPAR-gamma in macrophage differentiation and cholesterol uptake. Nat Med 7(1):41–47. doi:10.1038/83328
260. Morrison DK (2012) MAP kinase pathways. Cold Spring Harb Perspect Biol 4(11). doi:10.1101/cshperspect.a011254
261. Mounier R, Theret M, Arnold L, Cuvellier S, Bultot L, Goransson O, Sanz N, Ferry A, Sakamoto K, Foretz M, Viollet B, Chazaud B (2013) AMPKalpha1 regulates macrophage skewing at the time of resolution of inflammation during skeletal muscle regeneration. Cell Metab 18(2):251–264. doi:10.1016/j.cmet.2013.06.017

262. Mungai PT, Waypa GB, Jairaman A, Prakriya M, Dokic D, Ball MK, Schumacker PT (2011) Hypoxia triggers AMPK activation through reactive oxygen species-mediated activation of calcium release-activated calcium channels. Mol Cell Biol 31(17):3531–3545. doi:10.1128/MCB.05124-11
263. Munn DH, Sharma MD, Baban B, Harding HP, Zhang Y, Ron D, Mellor AL (2005) GCN2 kinase in T cells mediates proliferative arrest and anergy induction in response to indoleamine 2,3-dioxygenase. Immunity 22(5):633–642. doi:10.1016/j.immuni.2005.03.013
264. Murugaiyan G, Beynon V, Pires Da Cunha A, Joller N, Weiner HL (2012) IFN-gamma limits Th9-mediated autoimmune inflammation through dendritic cell modulation of IL-27. J Immunol 189(11):5277–5283. doi:10.4049/jimmunol.1200808
265. Nakaya M, Xiao Y, Zhou X, Chang JH, Chang M, Cheng X, Blonska M, Lin X, Sun SC (2014) Inflammatory T cell responses rely on amino acid transporter ASCT2 facilitation of glutamine uptake and mTORC1 kinase activation. Immunity 40(5):692–705. doi:10.1016/j.immuni.2014.04.007
266. Nath N, Khan M, Paintlia MK, Singh I, Hoda MN, Giri S (2009a) Metformin attenuated the autoimmune disease of the central nervous system in animal models of multiple sclerosis. J Immunol 182(12):8005–8014. doi:10.4049/jimmunol.0803563
267. Nath N, Khan M, Rattan R, Mangalam A, Makkar RS, de Meester C, Bertrand L, Singh I, Chen Y, Viollet B, Giri S (2009b) Loss of AMPK exacerbates experimental autoimmune encephalomyelitis disease severity. Biochem Biophys Res Commun 386(1):16–20. doi:10.1016/j.bbrc.2009.05.106
268. Nencioni A, Grunebach F, Zobywlaski A, Denzlinger C, Brugger W, Brossart P (2002) Dendritic cell immunogenicity is regulated by peroxisome proliferator-activated receptor gamma. J Immunol 169(3):1228–1235
269. Newsholme P, Curi R, Gordon S, Newsholme EA (1986) Metabolism of glucose, glutamine, long-chain fatty acids and ketone bodies by murine macrophages. Biochem J 239(1):121–125
270. Newsholme P, Gordon S, Newsholme EA (1987) Rates of utilization and fates of glucose, glutamine, pyruvate, fatty acids and ketone bodies by mouse macrophages. Biochem J 242(3):631–636
271. Newsholme P, Newsholme EA (1989) Rates of utilization of glucose, glutamine and oleate and formation of end-products by mouse peritoneal macrophages in culture. Biochem J 261(1):211–218
272. Nishi K, Oda T, Takabuchi S, Oda S, Fukuda K, Adachi T, Semenza GL, Shingu K, Hirota K (2008) LPS induces hypoxia-inducible factor 1 activation in macrophage-differentiated cells in a reactive oxygen species-dependent manner. Antioxid Redox Signal 10(5):983–995. doi:10.1089/ars.2007.1825
273. Nurieva RI, Chung Y, Martinez GJ, Yang XO, Tanaka S, Matskevitch TD, Wang YH, Dong C (2009) Bcl6 mediates the development of T follicular helper cells. Science 325(5943):1001–1005. doi:10.1126/science.1176676
274. Nyirenda MH, Sanvito L, Darlington PJ, O'Brien K, Zhang GX, Constantinescu CS, Bar-Or A, Gran B (2011) TLR2 stimulation drives human naive and effector regulatory T cells into a Th17-like phenotype with reduced suppressive function. J Immunol 187(5):2278–2290. doi:10.4049/jimmunol.1003715
275. O'Brien TF, Gorentla BK, Xie D, Srivatsan S, McLeod IX, He YW, Zhong XP (2011) Regulation of T-cell survival and mitochondrial homeostasis by TSC1. Eur J Immunol 41(11):3361–3370. doi:10.1002/eji.201141411
276. O'Leary JG, Goodarzi M, Drayton DL, von Andrian UH (2006) T cell- and B cell-independent adaptive immunity mediated by natural killer cells. Nat Immunol 7(5):507–516. doi:10.1038/ni1332
277. O'Neill HM, Maarbjerg SJ, Crane JD, Jeppesen J, Jorgensen SB, Schertzer JD, Shyroka O, Kiens B, van Denderen BJ, Tarnopolsky MA, Kemp BE, Richter EA, Steinberg GR (2011) AMP-activated protein kinase (AMPK) beta1beta2 muscle null mice reveal an essential role for AMPK in maintaining mitochondrial content and glucose uptake during exercise. Proc Natl Acad Sci U.S.A. 108(38):16092–16097. doi:10.1073/pnas.1105062108

278. O'Neill LA, Golenbock D, Bowie AG (2013) The history of toll-like receptors – redefining innate immunity. Nat Rev Immunol 13(6):453–460. doi:10.1038/nri3446
279. O'Neill LA, Hardie DG (2013) Metabolism of inflammation limited by AMPK and pseudo-starvation. Nature 493(7432):346–355. doi:10.1038/nature11862
280. O'Rourke AM, Rider CC (1989) Glucose, glutamine and ketone body utilisation by resting and concanavalin A activated rat splenic lymphocytes. Biochim Biophys Acta 1010(3):342–345
281. O'Sullivan D, van der Windt GJ, Huang SC, Curtis JD, Chang CH, Buck MD, Qiu J, Smith AM, Lam WY, DiPlato LM, Hsu FF, Birnbaum MJ, Pearce EJ, Pearce EL (2014) Memory CD8(+) T cells use cell-intrinsic lipolysis to support the metabolic programming necessary for development. Immunity 41(1):75–88. doi:10.1016/j.immuni.2014.06.005
282. O'Sullivan TE, Johnson LR, Kang HH, Sun JC (2015) BNIP3- and BNIP3L-mediated mitophagy promotes the generation of natural killer cell memory. Immunity 43(2):331–342. doi:10.1016/j.immuni.2015.07.012
283. Odegaard JI, Ricardo-Gonzalez RR, Goforth MH, Morel CR, Subramanian V, Mukundan L, Red Eagle A, Vats D, Brombacher F, Ferrante AW, Chawla A (2007) Macrophage-specific PPARgamma controls alternative activation and improves insulin resistance. Nature 447(7148):1116–1120. doi:10.1038/nature05894
284. Oestreich KJ, Mohn SE, Weinmann AS (2012) Molecular mechanisms that control the expression and activity of Bcl-6 in TH1 cells to regulate flexibility with a TFH-like gene profile. Nat Immunol 13(4):405–411. doi:10.1038/ni.2242
285. Oestreich KJ, Read KA, Gilbertson SE, Hough KP, McDonald PW, Krishnamoorthy V, Weinmann AS (2014) Bcl-6 directly represses the gene program of the glycolysis pathway. Nat Immunol 15(10):957–964. doi:10.1038/ni.2985
286. Ohkura N, Hamaguchi M, Morikawa H, Sugimura K, Tanaka A, Ito Y, Osaki M, Tanaka Y, Yamashita R, Nakano N, Huehn J, Fehling HJ, Sparwasser T, Nakai K, Sakaguchi S (2012) T cell receptor stimulation-induced epigenetic changes and Foxp3 expression are independent and complementary events required for Treg cell development. Immunity 37(5):785–799. doi:10.1016/j.immuni.2012.09.010
287. Ohtani M, Nagai S, Kondo S, Mizuno S, Nakamura K, Tanabe M, Takeuchi T, Matsuda S, Koyasu S (2008) Mammalian target of rapamycin and glycogen synthase kinase 3 differentially regulate lipopolysaccharide-induced interleukin-12 production in dendritic cells. Blood 112(3):635–643. doi:10.1182/blood-2008-02-137430
288. Okkenhaug K, Bilancio A, Farjot G, Priddle H, Sancho S, Peskett E, Pearce W, Meek SE, Salpekar A, Waterfield MD, Smith AJ, Vanhaesebroeck B (2002) Impaired B and T cell antigen receptor signaling in p110delta PI 3-kinase mutant mice. Science 297(5583):1031–1034. doi:10.1126/science.1073560
289. Okkenhaug K, Patton DT, Bilancio A, Garcon F, Rowan WC, Vanhaesebroeck B (2006) The p110delta isoform of phosphoinositide 3-kinase controls clonal expansion and differentiation of Th cells. J Immunol 177(8):5122–5128
290. Orsini H, Araujo LP, Maricato JT, Guereschi MG, Mariano M, Castilho BA, Basso AS (2014) GCN2 kinase plays an important role triggering the remission phase of experimental autoimmune encephalomyelitis (EAE) in mice. Brain Behav Immun 37:177–186. doi:10.1016/j.bbi.2013.12.012
291. Oshiro N, Rapley J, Avruch J (2014) Amino acids activate mammalian target of rapamycin (mTOR) complex 1 without changing Rag GTPase guanyl nucleotide charging. J Biol Chem 289(5):2658–2674. doi:10.1074/jbc.M113.528505
292. Ouyang W, Beckett O, Flavell RA, Li MO (2009) An essential role of the Forkhead-box transcription factor Foxo1 in control of T cell homeostasis and tolerance. Immunity 30(3):358–371. doi:10.1016/j.immuni.2009.02.003
293. Ouyang W, Liao W, Luo CT, Yin N, Huse M, Kim MV, Peng M, Chan P, Ma Q, Mo Y, Meijer D, Zhao K, Rudensky AY, Atwal G, Zhang MQ, Li MO (2012) Novel Foxo1-dependent transcriptional programs control T(reg) cell function. Nature 491(7425):554–559. doi:10.1038/nature11581

294. Owen OE, Kalhan SC, Hanson RW (2002) The key role of anaplerosis and cataplerosis for citric acid cycle function. J Biol Chem 277(34):30409–30412. doi:10.1074/jbc.R200006200
295. Panday A, Sahoo MK, Osorio D, Batra S (2015) NADPH oxidases: an overview from structure to innate immunity-associated pathologies. Cell Mol Immunol 12(1):5–23. doi:10.1038/cmi.2014.89
296. Pantel A, Teixeira A, Haddad E, Wood EG, Steinman RM, Longhi MP (2014) Direct type I IFN but not MDA5/TLR3 activation of dendritic cells is required for maturation and metabolic shift to glycolysis after poly IC stimulation. PLoS Biol 12(1):e1001759. doi:10.1371/journal.pbio.1001759
297. Papandreou I, Cairns RA, Fontana L, Lim AL, Denko NC (2006) HIF-1 mediates adaptation to hypoxia by actively downregulating mitochondrial oxygen consumption. Cell Metab 3(3):187–197. doi:10.1016/j.cmet.2006.01.012
298. Park H, Tsang M, Iritani BM, Bevan MJ (2014) Metabolic regulator Fnip1 is crucial for iNKT lymphocyte development. Proc Natl Acad Sci U.S.A. 111(19):7066–7071. doi:10.1073/pnas.1406473111
299. Park J, Kim M, Kang SG, Jannasch AH, Cooper B, Patterson J, Kim CH (2015) Short-chain fatty acids induce both effector and regulatory T cells by suppression of histone deacetylases and regulation of the mTOR-S6K pathway. Mucosal Immunol 8(1):80–93. doi:10.1038/mi.2014.44
300. Park Y, Jin HS, Lopez J, Elly C, Kim G, Murai M, Kronenberg M, Liu YC (2013) TSC1 regulates the balance between effector and regulatory T cells. J Clin Invest 123(12):5165–5178. doi:10.1172/JCI69751
301. Park YH, Wood G, Kastner DL, Chae JJ (2016) Pyrin inflammasome activation and RhoA signaling in the autoinflammatory diseases FMF and HIDS. Nat Immunol 17(8):914–921. doi:10.1038/ni.3457
302. Parmigiani A, Nourbakhsh A, Ding B, Wang W, Kim YC, Akopiants K, Guan KL, Karin M, Budanov AV (2014) Sestrins inhibit mTORC1 kinase activation through the GATOR complex. Cell Rep 9(4):1281–1291. doi:10.1016/j.celrep.2014.10.019
303. Parra V, Verdejo HE, Iglewski M, Del Campo A, Troncoso R, Jones D, Zhu Y, Kuzmicic J, Pennanen C, Lopez-Crisosto C, Jana F, Ferreira J, Noguera E, Chiong M, Bernlohr DA, Klip A, Hill JA, Rothermel BA, Abel ED, Zorzano A, Lavandero S (2014) Insulin stimulates mitochondrial fusion and function in cardiomyocytes via the Akt-mTOR-NFkappaB-Opa-1 signaling pathway. Diabetes 63(1):75–88. doi:10.2337/db13-0340
304. Pearce EJ, Everts B (2015) Dendritic cell metabolism. Nat Rev Immunol 15(1):18–29. doi:10.1038/nri3771
305. Pearce EL, Pearce EJ (2013) Metabolic pathways in immune cell activation and quiescence. Immunity 38(4):633–643. doi:10.1016/j.immuni.2013.04.005
306. Pearce EL, Poffenberger MC, Chang CH, Jones RG (2013) Fueling immunity: insights into metabolism and lymphocyte function. Science 342(6155):1242454. doi:10.1126/science.1242454
307. Pearce EL, Walsh MC, Cejas PJ, Harms GM, Shen H, Wang LS, Jones RG, Choi Y (2009) Enhancing CD8 T-cell memory by modulating fatty acid metabolism. Nature 460(7251):103–107. doi:10.1038/nature08097
308. Peng M, Yin N, Li MO (2014) Sestrins function as guanine nucleotide dissociation inhibitors for Rag GTPases to control mTORC1 signaling. Cell 159(1):122–133. doi:10.1016/j.cell.2014.08.038
309. Penna G, Roncari A, Amuchastegui S, Daniel KC, Berti E, Colonna M, Adorini L (2005) Expression of the inhibitory receptor ILT3 on dendritic cells is dispensable for induction of CD4+Foxp3+ regulatory T cells by 1,25-dihydroxyvitamin D3. Blood 106(10):3490–3497. doi:10.1182/blood-2005-05-2044
310. Peyssonnaux C, Cejudo-Martin P, Doedens A, Zinkernagel AS, Johnson RS, Nizet V (2007a) Cutting edge: essential role of hypoxia inducible factor-1alpha in development of lipopolysaccharide-induced sepsis. J Immunol 178(12):7516–7519

311. Peyssonnaux C, Zinkernagel AS, Schuepbach RA, Rankin E, Vaulont S, Haase VH, Nizet V, Johnson RS (2007b) Regulation of iron homeostasis by the hypoxia-inducible transcription factors (HIFs). J Clin Invest 117(7):1926–1932. doi:10.1172/JCI31370
312. Pochini L, Scalise M, Galluccio M, Indiveri C (2014) Membrane transporters for the special amino acid glutamine: structure/function relationships and relevance to human health. Front Chem 2:61. doi:10.3389/fchem.2014.00061
313. Pollizzi KN, Patel CH, Sun IH, Oh MH, Waickman AT, Wen J, Delgoffe GM, Powell JD (2015) mTORC1 and mTORC2 selectively regulate CD8(+) T cell differentiation. J Clin Invest 125(5):2090–2108. doi:10.1172/JCI77746
314. Pollizzi KN, Sun IH, Patel CH, Lo YC, Oh MH, Waickman AT, Tam AJ, Blosser RL, Wen J, Delgoffe GM, Powell JD (2016) Asymmetric inheritance of mTORC1 kinase activity during division dictates CD8(+) T cell differentiation. Nat Immunol 17(6):704–711. doi:10.1038/ni.3438
315. Procaccini C, Carbone F, Di Silvestre D, Brambilla F, De Rosa V, Galgani M, Faicchia D, Marone G, Tramontano D, Corona M, Alviggi C, Porcellini A, La Cava A, Mauri P, Matarese G (2016) The proteomic landscape of human ex vivo regulatory and conventional T cells reveals specific metabolic requirements. Immunity 44(2):406–421. doi:10.1016/j.immuni.2016.01.028
316. Procaccini C, De Rosa V, Galgani M, Abanni L, Calì G, Porcellini A, Carbone F, Fontana S, Horvath TL, La Cava A, Matarese G (2010) An oscillatory switch in mTOR kinase activity sets regulatory T cell responsiveness. Immunity 33(6):929–941. doi:10.1016/j.immuni.2010.11.024
317. Qu P, Du H, Wilkes DS, Yan C (2009) Critical roles of lysosomal acid lipase in T cell development and function. Am J Pathol 174(3):944–956. doi:10.2353/ajpath.2009.080562
318. Raikwar HP, Muthian G, Rajasingh J, Johnson C, Bright JJ (2005) PPARgamma antagonists exacerbate neural antigen-specific Th1 response and experimental allergic encephalomyelitis. J Neuroimmunol 167(1–2):99–107. doi:10.1016/j.jneuroim.2005.06.026
319. Rao E, Zhang Y, Zhu G, Hao J, Persson XM, Egilmez NK, Suttles J, Li B (2015) Deficiency of AMPK in CD8+ T cells suppresses their anti-tumor function by inducing protein phosphatase-mediated cell death. Oncotarget 6(10):7944–7958. doi:10.18632/oncotarget.3501
320. Rao RR, Li Q, Gubbels Bupp MR, Shrikant PA (2012) Transcription factor Foxo1 represses T-bet-mediated effector functions and promotes memory CD8(+) T cell differentiation. Immunity 36(3):374–387. doi:10.1016/j.immuni.2012.01.015
321. Rao RR, Li Q, Odunsi K, Shrikant PA (2010) The mTOR kinase determines effector versus memory CD8+ T cell fate by regulating the expression of transcription factors T-bet and Eomesodermin. Immunity 32(1):67–78. doi:10.1016/j.immuni.2009.10.010
322. Rathmell JC, Vander Heiden MG, Harris MH, Frauwirth KA, Thompson CB (2000) In the absence of extrinsic signals, nutrient utilization by lymphocytes is insufficient to maintain either cell size or viability. Mol Cell 6(3):683–692
323. Ravitz MJ, Chen L, Lynch M, Schmidt EV (2007) C-myc repression of TSC2 contributes to control of translation initiation and Myc-induced transformation. Cancer Res 67(23):11209–11217. doi:10.1158/0008-5472.CAN-06-4351
324. Ray JP, Staron MM, Shyer JA, Ho PC, Marshall HD, Gray SM, Laidlaw BJ, Araki K, Ahmed R, Kaech SM, Craft J (2015) The interleukin-2-mTORc1 kinase axis defines the signaling, differentiation, and metabolism of T helper 1 and follicular B helper T cells. Immunity 43(4):690–702. doi:10.1016/j.immuni.2015.08.017
325. Reboldi A, Dang EV, McDonald JG, Liang G, Russell DW, Cyster JG (2014) Inflammation. 25-Hydroxycholesterol suppresses interleukin-1-driven inflammation downstream of type I interferon. Science 345(6197):679–684. doi:10.1126/science.1254790
326. Rocher C, Singla DK (2013) SMAD-PI3K-Akt-mTOR pathway mediates BMP-7 polarization of monocytes into M2 macrophages. PLoS One 8(12):e84009. doi:10.1371/journal.pone.0084009

327. Rodriguez-Prados JC, Traves PG, Cuenca J, Rico D, Aragones J, Martin-Sanz P, Cascante M, Bosca L (2010) Substrate fate in activated macrophages: a comparison between innate, classic, and alternative activation. J Immunol 185(1):605–614. doi:10.4049/jimmunol.0901698
328. Rolf J, Zarrouk M, Finlay DK, Foretz M, Viollet B, Cantrell DA (2013) AMPKalpha1: a glucose sensor that controls CD8 T-cell memory. Eur J Immunol 43(4):889–896. doi:10.1002/eji.201243008
329. Roos D, Loos JA (1973) Changes in the carbohydrate metabolism of mitogenically stimulated human peripheral lymphocytes. II Relative importance of glycolysis and oxidative phosphorylation on phytohaemagglutinin stimulation. Exp Cell Res 77(1):127–135
330. Rubic T, Lametschwandtner G, Jost S, Hinteregger S, Kund J, Carballido-Perrig N, Schwarzler C, Junt T, Voshol H, Meingassner JG, Mao X, Werner G, Rot A, Carballido JM (2008) Triggering the succinate receptor GPR91 on dendritic cells enhances immunity. Nat Immunol 9(11):1261–1269. doi:10.1038/ni.1657
331. Rubtsov YP, Niec RE, Josefowicz S, Li L, Darce J, Mathis D, Benoist C, Rudensky AY (2010) Stability of the regulatory T cell lineage in vivo. Science 329(5999):1667–1671. doi:10.1126/science.1191996
332. Sag D, Carling D, Stout RD, Suttles J (2008) Adenosine 5′-monophosphate-activated protein kinase promotes macrophage polarization to an anti-inflammatory functional phenotype. J Immunol 181(12):8633–8641
333. Salles J, Chanet A, Giraudet C, Patrac V, Pierre P, Jourdan M, Luiking YC, Verlaan S, Migne C, Boirie Y, Walrand S (2013) 1,25(OH)2-vitamin D3 enhances the stimulating effect of leucine and insulin on protein synthesis rate through Akt/PKB and mTOR mediated pathways in murine C2C12 skeletal myotubes. Mol Nutr Food Res 57(12):2137–2146. doi:10.1002/mnfr.201300074
334. Sancak Y, Bar-Peled L, Zoncu R, Markhard AL, Nada S, Sabatini DM (2010) Ragulator-Rag complex targets mTORC1 to the lysosomal surface and is necessary for its activation by amino acids. Cell 141(2):290–303. doi:10.1016/j.cell.2010.02.024
335. Sancak Y, Peterson TR, Shaul YD, Lindquist RA, Thoreen CC, Bar-Peled L, Sabatini DM (2008) The Rag GTPases bind raptor and mediate amino acid signaling to mTORC1. Science 320(5882):1496–1501. doi:10.1126/science.1157535
336. Sang N, Stiehl DP, Bohensky J, Leshchinsky I, Srinivas V, Caro J (2003) MAPK signaling up-regulates the activity of hypoxia-inducible factors by its effects on p300. J Biol Chem 278(16):14013–14019. doi:10.1074/jbc.M209702200
337. Sarkar S, Germeraad WT, Rouschop KM, Steeghs EM, van Gelder M, Bos GM, Wieten L (2013) Hypoxia induced impairment of NK cell cytotoxicity against multiple myeloma can be overcome by IL-2 activation of the NK cells. PLoS One 8(5):e64835. doi:10.1371/journal.pone.0064835
338. Sauer K, Cooke MP (2010) Regulation of immune cell development through soluble inositol-1,3,4,5-tetrakisphosphate. Nat Rev Immunol 10(4):257–271. doi:10.1038/nri2745
339. Sauer S, Bruno L, Hertweck A, Finlay D, Leleu M, Spivakov M, Knight ZA, Cobb BS, Cantrell D, O'Connor E, Shokat KM, Fisher AG, Merkenschlager M (2008) T cell receptor signaling controls Foxp3 expression via PI3K, Akt, and mTOR. Proc Natl Acad Sci U.S.A 105(22):7797–7802. doi:10.1073/pnas.0800928105
340. Schmidt AM, Lu W, Sindhava VJ, Huang Y, Burkhardt JK, Yang E, Riese MJ, Maltzman JS, Jordan MS, Kambayashi T (2015) Regulatory T cells require TCR signaling for their suppressive function. J Immunol 194(9):4362–4370. doi:10.4049/jimmunol.1402384
341. Schmitt E, Germann T, Goedert S, Hoehn P, Huels C, Koelsch S, Kuhn R, Muller W, Palm N, Rude E (1994) IL-9 production of naive CD4+ T cells depends on IL-2, is synergistically enhanced by a combination of TGF-beta and IL-4, and is inhibited by IFN-gamma. J Immunol 153(9):3989–3996
342. Schmitt E, Klein M, Bopp T (2014) Th9 cells, new players in adaptive immunity. Trends Immunol 35(2):61–68. doi:10.1016/j.it.2013.10.004

343. Schonle A, Hartl FA, Mentzel J, Noltner T, Rauch KS, Prestipino A, Wohlfeil SA, Apostolova P, Hechinger AK, Melchinger W, Fehrenbach K, Guadamillas MC, Follo M, Prinz G, Ruess AK, Pfeifer D, del Pozo MA, Schmitt-Graeff A, Duyster J, Hippen KI, Blazar BR, Schachtrup K, Minguet S, Zeiser R (2016) Caveolin-1 regulates TCR signal strength and regulatory T-cell differentiation into alloreactive T cells. Blood 127(15):1930–1939. doi:10.1182/blood-2015-09-672428
344. Semple K, Nguyen A, Yu Y, Wang H, Anasetti C, Yu XZ (2011) Strong CD28 costimulation suppresses induction of regulatory T cells from naive precursors through Lck signaling. Blood 117(11):3096–3103. doi:10.1182/blood-2010-08-301275
345. Sena LA, Li S, Jairaman A, Prakriya M, Ezponda T, Hildeman DA, Wang CR, Schumacker PT, Licht JD, Perlman H, Bryce PJ, Chandel NS (2013) Mitochondria are required for antigen-specific T cell activation through reactive oxygen species signaling. Immunity 38(2):225–236. doi:10.1016/j.immuni.2012.10.020
346. Shao W, Espenshade PJ (2012) Expanding roles for SREBP in metabolism. Cell Metab 16(4):414–419. doi:10.1016/j.cmet.2012.09.002
347. Shen CJ, Yuan ZH, Liu YX, Hu GY (2012) Increased numbers of T helper 17 cells and the correlation with clinicopathological characteristics in multiple myeloma. J Int Med Res 40(2):556–564
348. Shi LZ, Wang R, Huang G, Vogel P, Neale G, Green DR, Chi H (2011) HIF1alpha-dependent glycolytic pathway orchestrates a metabolic checkpoint for the differentiation of TH17 and Treg cells. J Exp Med 208(7):1367–1376. doi:10.1084/jem.20110278
349. Shimobayashi M, Hall MN (2014) Making new contacts: the mTOR network in metabolism and signalling crosstalk. Nat Rev Mol Cell Biol 15(3):155–162. doi:10.1038/nrm3757
350. Shimobayashi M, Hall MN (2016) Multiple amino acid sensing inputs to mTORC1. Cell Res 26(1):7–20. doi:10.1038/cr.2015.146
351. Shin I, Yakes FM, Rojo F, Shin NY, Bakin AV, Baselga J, Arteaga CL (2002) PKB/Akt mediates cell-cycle progression by phosphorylation of p27(Kip1) at threonine 157 and modulation of its cellular localization. Nat Med 8(10):1145–1152. doi:10.1038/nm759
352. Shrestha S, Yang K, Guy C, Vogel P, Neale G, Chi H (2015) Treg cells require the phosphatase PTEN to restrain TH1 and TFH cell responses. Nat Immunol 16(2):178–187. doi:10.1038/ni.3076
353. Shrestha S, Yang K, Wei J, Karmaus PWF, Neale G, Chi H (2014) Tsc1 promotes the differentiation of memory CD8+ T cells via orchestrating the transcriptional and metabolic programs. Proc Natl Acad Sci 111(41):14858–14863. doi:10.1073/pnas.1404264111
354. Silverstein RL, Febbraio M (2009) CD36, a scavenger receptor involved in immunity, metabolism, angiogenesis, and behavior. Sci Signal 2(72):re3. doi:10.1126/scisignal.272re3
355. Sim WJ, Ahl PJ, Connolly JE (2016) Metabolism is central to tolerogenic dendritic cell function. Mediat Inflamm 2016:2636701. doi:10.1155/2016/2636701
356. Sinclair LV, Rolf J, Emslie E, Shi Y-B, Taylor PM, Cantrell DA (2013) Control of amino-acid transport by antigen receptors coordinates the metabolic reprogramming essential for T cell differentiation. Nat Immunol 14(5):500–508. doi:10.1038/ni.2556
357. Singh N, Gurav A, Sivaprakasam S, Brady E, Padia R, Shi H, Thangaraju M, Prasad PD, Manicassamy S, Munn DH, Lee JR, Offermanns S, Ganapathy V (2014) Activation of Gpr109a, receptor for niacin and the commensal metabolite butyrate, suppresses colonic inflammation and carcinogenesis. Immunity 40(1):128–139. doi:10.1016/j.immuni.2013.12.007
358. Singh R, Xiang Y, Wang Y, Baikati K, Cuervo AM, Luu YK, Tang Y, Pessin JE, Schwartz GJ, Czaja MJ (2009) Autophagy regulates adipose mass and differentiation in mice. J Clin Invest 119(11):3329–3339. doi:10.1172/JCI39228
359. Skon CN, Lee JY, Anderson KG, Masopust D, Hogquist KA, Jameson SC (2013) Transcriptional downregulation of S1pr1 is required for the establishment of resident memory CD8+ T cells. Nat Immunol 14(12):1285–1293. doi:10.1038/ni.2745
360. Smith PM, Howitt MR, Panikov N, Michaud M, Gallini CA, Bohlooly YM, Glickman JN, Garrett WS (2013) The microbial metabolites, short-chain fatty acids, regulate colonic Treg cell homeostasis. Science 341(6145):569–573. doi:10.1126/science.1241165

361. Sonnenberg GF, Artis D (2015) Innate lymphoid cells in the initiation, regulation and resolution of inflammation. Nat Med 21(7):698–708. doi:10.1038/nm.3892
362. Sonoda J, Laganiere J, Mehl IR, Barish GD, Chong LW, Li X, Scheffler IE, Mock DC, Bataille AR, Robert F, Lee CH, Giguere V, Evans RM (2007) Nuclear receptor ERR alpha and coactivator PGC-1 beta are effectors of IFN-gamma-induced host defense. Genes Dev 21(15):1909–1920. doi:10.1101/gad.1553007
363. Sonveaux P, Copetti T, De Saedeleer CJ, Vegran F, Verrax J, Kennedy KM, Moon EJ, Dhup S, Danhier P, Frerart F, Gallez B, Ribeiro A, Michiels C, Dewhirst MW, Feron O (2012) Targeting the lactate transporter MCT1 in endothelial cells inhibits lactate-induced HIF-1 activation and tumor angiogenesis. PLoS One 7(3):e33418. doi:10.1371/journal.pone.0033418
364. Sowell RT, Rogozinska M, Nelson CE, Vezys V, Marzo AL (2014) Cutting edge: generation of effector cells that localize to mucosal tissues and form resident memory CD8 T cells is controlled by mTOR. J Immunol 193(5):2067–2071. doi:10.4049/jimmunol.1400074
365. Spirig R, Djafarzadeh S, Regueira T, Shaw SG, von Garnier C, Takala J, Jakob SM, Rieben R, Lepper PM (2010) Effects of TLR agonists on the hypoxia-regulated transcription factor HIF-1alpha and dendritic cell maturation under normoxic conditions. PLoS One 5(6):e0010983. doi:10.1371/journal.pone.0010983
366. Staron MM, Gray SM, Marshall HD, Parish IA, Chen JH, Perry CJ, Cui G, Li MO, Kaech SM (2014) The transcription factor FoxO1 sustains expression of the inhibitory receptor PD-1 and survival of antiviral CD8(+) T cells during chronic infection. Immunity 41(5):802–814. doi:10.1016/j.immuni.2014.10.013
367. Staudt V, Bothur E, Klein M, Lingnau K, Reuter S, Grebe N, Gerlitzki B, Hoffmann M, Ulges A, Taube C, Dehzad N, Becker M, Stassen M, Steinborn A, Lohoff M, Schild H, Schmitt E, Bopp T (2010) Interferon-regulatory factor 4 is essential for the developmental program of T helper 9 cells. Immunity 33(2):192–202. doi:10.1016/j.immuni.2010.07.014
368. Steinberg GR, Michell BJ, van Denderen BJ, Watt MJ, Carey AL, Fam BC, Andrikopoulos S, Proietto J, Gorgun CZ, Carling D, Hotamisligil GS, Febbraio MA, Kay TW, Kemp BE (2006) Tumor necrosis factor alpha-induced skeletal muscle insulin resistance involves suppression of AMP-kinase signaling. Cell Metab 4(6):465–474. doi:10.1016/j.cmet.2006.11.005
369. Stone EL, Pepper M, Katayama CD, Kerdiles YM, Lai CY, Emslie E, Lin YC, Yang E, Goldrath AW, Li MO, Cantrell DA, Hedrick SM (2015) ICOS coreceptor signaling inactivates the transcription factor FOXO1 to promote Tfh cell differentiation. Immunity 42(2):239–251. doi:10.1016/j.immuni.2015.01.017
370. Sukumar M, Liu J, Ji Y, Subramanian M, Crompton JG, Yu Z, Roychoudhuri R, Palmer DC, Muranski P, Karoly ED, Mohney RP, Klebanoff CA, Lal A, Finkel T, Restifo NP, Gattinoni L (2013) Inhibiting glycolytic metabolism enhances CD8+ T cell memory and antitumor function. J Clin Invest 123(10):4479–4488. doi:10.1172/JCI69589
371. Sun CM, Hall JA, Blank RB, Bouladoux N, Oukka M, Mora JR, Belkaid Y (2007) Small intestine lamina propria dendritic cells promote de novo generation of Foxp3 T reg cells via retinoic acid. J Exp Med 204(8):1775–1785. doi:10.1084/jem.20070602
372. Sun JC, Beilke JN, Lanier LL (2009) Adaptive immune features of natural killer cells. Nature 457(7229):557–561. doi:10.1038/nature07665
373. Sun X, Kanwar JR, Leung E, Lehnert K, Wang D, Krissansen GW (2001) Gene transfer of antisense hypoxia inducible factor-1 alpha enhances the therapeutic efficacy of cancer immunotherapy. Gene Ther 8(8):638–645. doi:10.1038/sj.gt.3301388
374. Swamy M, Beck-Garcia K, Beck-Garcia E, Hartl FA, Morath A, Yousefi OS, Dopfer EP, Molnar E, Schulze AK, Blanco R, Borroto A, Martin-Blanco N, Alarcon B, Hofer T, Minguet S, Schamel WW (2016) A cholesterol-based allostery model of T cell receptor phosphorylation. Immunity 44(5):1091–1101. doi:10.1016/j.immuni.2016.04.011
375. Szatmari I, Gogolak P, Im JS, Dezso B, Rajnavolgyi E, Nagy L (2004) Activation of PPARgamma specifies a dendritic cell subtype capable of enhanced induction of iNKT cell expansion. Immunity 21(1):95–106. doi:10.1016/j.immuni.2004.06.003

376. Takamatsu M, Hirata A, Ohtaki H, Hoshi M, Ando T, Ito H, Hatano Y, Tomita H, Kuno T, Saito K, Seishima M, Hara A (2015) Inhibition of indoleamine 2,3-dioxygenase 1 expression alters immune response in colon tumor microenvironment in mice. Cancer Sci 106(8):1008–1015. doi:10.1111/cas.12705
377. Takeda N, O'Dea EL, Doedens A, Kim JW, Weidemann A, Stockmann C, Asagiri M, Simon MC, Hoffmann A, Johnson RS (2010) Differential activation and antagonistic function of HIF-{alpha} isoforms in macrophages are essential for NO homeostasis. Genes Dev 24(5):491–501. doi:10.1101/gad.1881410
378. Talks KL, Turley H, Gatter KC, Maxwell PH, Pugh CW, Ratcliffe PJ, Harris AL (2000) The expression and distribution of the hypoxia-inducible factors HIF-1alpha and HIF-2alpha in normal human tissues, cancers, and tumor-associated macrophages. Am J Pathol 157(2):411–421
379. Tamas P, Hawley SA, Clarke RG, Mustard KJ, Green K, Hardie DG, Cantrell DA (2006) Regulation of the energy sensor AMP-activated protein kinase by antigen receptor and Ca2+ in T lymphocytes. J Exp Med 203(7):1665–1670. doi:10.1084/jem.20052469
380. Tamas P, Macintyre A, Finlay D, Clarke R, Feijoo-Carnero C, Ashworth A, Cantrell D (2010) LKB1 is essential for the proliferation of T-cell progenitors and mature peripheral T cells. Eur J Immunol 40(1):242–253. doi:10.1002/eji.200939677
381. Tannahill GM, Curtis AM, Adamik J, Palsson-McDermott EM, McGettrick AF, Goel G, Frezza C, Bernard NJ, Kelly B, Foley NH, Zheng L, Gardet A, Tong Z, Jany SS, Corr SC, Haneklaus M, Caffrey BE, Pierce K, Walmsley S, Beasley FC, Cummins E, Nizet V, Whyte M, Taylor CT, Lin H, Masters SL, Gottlieb E, Kelly VP, Clish C, Auron PE, Xavier RJ, O'Neill LA (2013) Succinate is an inflammatory signal that induces IL-1beta through HIF-1alpha. Nature 496(7444):238–242. doi:10.1038/nature11986
382. Tato I, Bartrons R, Ventura F, Rosa JL (2011) Amino acids activate mammalian target of rapamycin complex 2 (mTORC2) via PI3K/Akt signaling. J Biol Chem 286(8):6128–6142. doi:10.1074/jbc.M110.166991
383. Tejera MM, Kim EH, Sullivan JA, Plisch EH, Suresh M (2013) FoxO1 controls effector-to-memory transition and maintenance of functional CD8 T cell memory. J Immunol 191(1):187–199. doi:10.4049/jimmunol.1300331
384. Thiel M, Caldwell CC, Kreth S, Kuboki S, Chen P, Smith P, Ohta A, Lentsch AB, Lukashev D, Sitkovsky MV (2007) Targeted deletion of HIF-1alpha gene in T cells prevents their inhibition in hypoxic inflamed tissues and improves septic mice survival. PLoS One 2(9):e853. doi:10.1371/journal.pone.0000853
385. Thurnher M, Gruenbacher G (2015) T lymphocyte regulation by mevalonate metabolism. Sci Signal 8(370):re4. doi:10.1126/scisignal.2005970
386. Toschi A, Lee E, Gadir N, Ohh M, Foster DA (2008) Differential dependence of hypoxia-inducible factors 1 alpha and 2 alpha on mTORC1 and mTORC2. J Biol Chem 283(50):34495–34499. doi:10.1074/jbc.C800170200
387. Tripmacher R, Gaber T, Dziurla R, Haupl T, Erekul K, Grutzkau A, Tschirschmann M, Scheffold A, Radbruch A, Burmester GR, Buttgereit F (2008) Human CD4(+) T cells maintain specific functions even under conditions of extremely restricted ATP production. Eur J Immunol 38(6):1631–1642. doi:10.1002/eji.200738047
388. Troutman TD, Hu W, Fulenchek S, Yamazaki T, Kurosaki T, Bazan JF, Pasare C (2012) Role for B-cell adapter for PI3K (BCAP) as a signaling adapter linking toll-like receptors (TLRs) to serine/threonine kinases PI3K/Akt. Proc Natl Acad Sci U.S.A. 109(1):273–278. doi:10.1073/pnas.1118579109
389. Tsun ZY, Bar-Peled L, Chantranupong L, Zoncu R, Wang T, Kim C, Spooner E, Sabatini DM (2013) The folliculin tumor suppressor is a GAP for the RagC/D GTPases that signal amino acid levels to mTORC1. Mol Cell 52(4):495–505. doi:10.1016/j.molcel.2013.09.016
390. Turcotte LP, Raney MA, Todd MK (2005) ERK1/2 inhibition prevents contraction-induced increase in plasma membrane FAT/CD36 content and FA uptake in rodent muscle. Acta Physiol Scand 184(2):131–139. doi:10.1111/j.1365-201X.2005.01445.x

391. Vahl JC, Drees C, Heger K, Heink S, Fischer JC, Nedjic J, Ohkura N, Morikawa H, Poeck H, Schallenberg S, Riess D, Hein MY, Buch T, Polic B, Schonle A, Zeiser R, Schmitt-Graff A, Kretschmer K, Klein L, Korn T, Sakaguchi S, Schmidt-Supprian M (2014) Continuous T cell receptor signals maintain a functional regulatory T cell pool. Immunity 41(5):722–736. doi:10.1016/j.immuni.2014.10.012
392. van der Bliek AM, Shen Q, Kawajiri S (2013) Mechanisms of mitochondrial fission and fusion. Cold Spring Harb Perspect Biol 5(6). doi:10.1101/cshperspect.a011072
393. van der Windt GJ, Everts B, Chang CH, Curtis JD, Freitas TC, Amiel E, Pearce EJ, Pearce EL (2012) Mitochondrial respiratory capacity is a critical regulator of CD8+ T cell memory development. Immunity 36(1):68–78. doi:10.1016/j.immuni.2011.12.007
394. Vander Heiden MG, Locasale JW, Swanson KD, Sharfi H, Heffron GJ, Amador-Noguez D, Christofk HR, Wagner G, Rabinowitz JD, Asara JM, Cantley LC (2010) Evidence for an alternative glycolytic pathway in rapidly proliferating cells. Science 329(5998):1492–1499. doi:10.1126/science.1188015
395. Vasanthakumar A, Moro K, Xin A, Liao Y, Gloury R, Kawamoto S, Fagarasan S, Mielke LA, Afshar-Sterle S, Masters SL, Nakae S, Saito H, Wentworth JM, Li P, Liao W, Leonard WJ, Smyth GK, Shi W, Nutt SL, Koyasu S, Kallies A (2015) The transcriptional regulators IRF4, BATF and IL-33 orchestrate development and maintenance of adipose tissue-resident regulatory T cells. Nat Immunol 16(3):276–285. doi:10.1038/ni.3085
396. Vats D, Mukundan L, Odegaard JI, Zhang L, Smith KL, Morel CR, Wagner RA, Greaves DR, Murray PJ, Chawla A (2006) Oxidative metabolism and PGC-1beta attenuate macrophage-mediated inflammation. Cell Metab 4(1):13–24. doi:10.1016/j.cmet.2006.05.011
397. Veldhoen M, Hirota K, Westendorf AM, Buer J, Dumoutier L, Renauld JC, Stockinger B (2008a) The aryl hydrocarbon receptor links TH17-cell-mediated autoimmunity to environmental toxins. Nature 453(7191):106–109. doi:10.1038/nature06881
398. Veldhoen M, Uyttenhove C, van Snick J, Helmby H, Westendorf A, Buer J, Martin B, Wilhelm C, Stockinger B (2008b) Transforming growth factor-beta 'reprograms' the differentiation of T helper 2 cells and promotes an interleukin 9-producing subset. Nat Immunol 9(12):1341–1346. doi:10.1038/ni.1659
399. Verbist KC, Guy CS, Milasta S, Liedmann S, Kaminski MM, Wang R, Green DR (2016) Metabolic maintenance of cell asymmetry following division in activated T lymphocytes. Nature 532(7599):389–393. doi:10.1038/nature17442
400. Vousden KH, Lane DP (2007) p53 in health and disease. Nat Rev Mol Cell Biol 8(4):275–283. doi:10.1038/nrm2147
401. Wahli W, Michalik L (2012) PPARs at the crossroads of lipid signaling and inflammation. Trends Endocrinol Metab 23(7):351–363. doi:10.1016/j.tem.2012.05.001
402. Wang F, Alain T, Szretter KJ, Stephenson K, Pol JG, Atherton MJ, Hoang HD, Fonseca BD, Zakaria C, Chen L, Rangwala Z, Hesch A, Chan ES, Tuinman C, Suthar MS, Jiang Z, Ashkar AA, Thomas G, Kozma SC, Gale M Jr, Fitzgerald KA, Diamond MS, Mossman K, Sonenberg N, Wan Y, Lichty BD (2016a) S6K-STING interaction regulates cytosolic DNA-mediated activation of the transcription factor IRF3. Nat Immunol 17(5):514–522. doi:10.1038/ni.3433
403. Wang R, Dillon CP, Shi LZ, Milasta S, Carter R, Finkelstein D, McCormick LL, Fitzgerald P, Chi H, Munger J, Green DR (2011) The transcription factor Myc controls metabolic reprogramming upon T lymphocyte activation. Immunity 35(6):871–882. doi:10.1016/j.immuni.2011.09.021
404. Wang R, Green DR (2012) Metabolic checkpoints in activated T cells. Nat Immunol 13(10):907–915. doi:10.1038/ni.2386
405. Wang S, Charbonnier LM, Noval Rivas M, Georgiev P, Li N, Gerber G, Bry L, Chatila TA (2015a) MyD88 adaptor-dependent microbial sensing by regulatory T cells promotes mucosal tolerance and enforces commensalism. Immunity 43(2):289–303. doi:10.1016/j.immuni.2015.06.014
406. Wang S, Tsun ZY, Wolfson RL, Shen K, Wyant GA, Plovanich ME, Yuan ED, Jones TD, Chantranupong L, Comb W, Wang T, Bar-Peled L, Zoncu R, Straub C, Kim C, Park J, Sabatini

BL, Sabatini DM (2015b) Metabolism. Lysosomal amino acid transporter SLC38A9 signals arginine sufficiency to mTORC1. Science 347(6218):188–194. doi:10.1126/science.1257132
407. Wang S, Xia P, Huang G, Zhu P, Liu J, Ye B, Du Y, Fan Z (2016b) FoxO1-mediated autophagy is required for NK cell development and innate immunity. Nat Commun 7:11023. doi:10.1038/ncomms11023
408. Wang X, Wang W, Xu J, Wu S, Le Q (2015c) All-trans retinoid acid promotes allogeneic corneal graft survival in mice by regulating Treg-Th17 balance in the presence of TGF-beta. BMC Immunol 16:17. doi:10.1186/s12865-015-0082-3
409. Wang Y, Bi Y, Chen X, Li C, Li Y, Zhang Z, Wang J, Lu Y, Yu Q, Su H, Yang H, Liu G (2016c) Histone deacetylase SIRT1 negatively regulates the differentiation of interleukin-9-producing CD4(+) T cells. Immunity 44(6):1337–1349. doi:10.1016/j.immuni.2016.05.009
410. Wang Y, Huang G, Zeng H, Yang K, Lamb RF, Chi H (2013) Tuberous sclerosis 1 (Tsc1)-dependent metabolic checkpoint controls development of dendritic cells. Proc Natl Acad Sci USA 110(50):E4894–E4903. doi:10.1073/pnas.1308905110
411. Waugh C, Sinclair L, Finlay D, Bayascas JR, Cantrell D (2009) Phosphoinositide (3,4,5)-triphosphate binding to phosphoinositide-dependent kinase 1 regulates a protein kinase B/Akt signaling threshold that dictates T-cell migration, not proliferation. Mol Cell Biol 29(21):5952–5962. doi:10.1128/MCB.00585-09
412. Weber JP, Fuhrmann F, Feist RK, Lahmann A, Al Baz MS, Gentz LJ, Vu Van D, Mages HW, Haftmann C, Riedel R, Grun JR, Schuh W, Kroczek RA, Radbruch A, Mashreghi MF, Hutloff A (2015) ICOS maintains the T follicular helper cell phenotype by down-regulating Kruppel-like factor 2. J Exp Med 212(2):217–233. doi:10.1084/jem.20141432
413. Wei H, Geng J, Shi B, Liu Z, Wang YH, Stevens AC, Sprout SL, Yao M, Wang H, Hu H (2016a) Cutting edge: Foxp1 controls naive CD8+ T cell quiescence by simultaneously repressing key pathways in cellular metabolism and cell cycle progression. J Immunol 196(9):3537–3541. doi:10.4049/jimmunol.1501896
414. Wei J, Long L, Yang K, Guy C, Shrestha S, Chen Z, Wu C, Vogel P, Neale G, Green DR, Chi H (2016b) Autophagy enforces functional integrity of regulatory T cells by coupling environmental cues and metabolic homeostasis. Nat Immunol 17(3):277–285. doi:10.1038/ni.3365
415. Wei J, Yang K, Chi H (2014) Cutting edge: discrete functions of mTOR signaling in invariant NKT cell development and NKT17 fate decision. J Immunol 193(9):4297–4301. doi:10.4049/jimmunol.1402042
416. Wellen KE, Hatzivassiliou G, Sachdeva UM, Bui TV, Cross JR, Thompson CB (2009) ATP-citrate lyase links cellular metabolism to histone acetylation. Science 324(5930):1076–1080. doi:10.1126/science.1164097
417. West AP, Shadel GS, Ghosh S (2011) Mitochondria in innate immune responses. Nat Rev Immunol 11(6):389–402. doi:10.1038/nri2975
418. Wieman HL, Wofford JA, Rathmell JC (2007) Cytokine stimulation promotes glucose uptake via phosphatidylinositol-3 kinase/Akt regulation of Glut1 activity and trafficking. Mol Biol Cell 18(4):1437–1446. doi:10.1091/mbc.E06-07-0593
419. Wildin RS, Ramsdell F, Peake J, Faravelli F, Casanova JL, Buist N, Levy-Lahad E, Mazzella M, Goulet O, Perroni L, Bricarelli FD, Byrne G, McEuen M, Proll S, Appleby M, Brunkow ME (2001) X-linked neonatal diabetes mellitus, enteropathy and endocrinopathy syndrome is the human equivalent of mouse scurfy. Nat Genet 27(1):18–20. doi:10.1038/83707
420. Wofford JA, Wieman HL, Jacobs SR, Zhao Y, Rathmell JC (2008) IL-7 promotes Glut1 trafficking and glucose uptake via STAT5-mediated activation of Akt to support T-cell survival. Blood 111(4):2101–2111. doi:10.1182/blood-2007-06-096297
421. Woodland RT, Fox CJ, Schmidt MR, Hammerman PS, Opferman JT, Korsmeyer SJ, Hilbert DM, Thompson CB (2008) Multiple signaling pathways promote B lymphocyte stimulator dependent B-cell growth and survival. Blood 111(2):750–760. doi:10.1182/blood-2007-03-077222
422. Wu C, Yosef N, Thalhamer T, Zhu C, Xiao S, Kishi Y, Regev A, Kuchroo VK (2013a) Induction of pathogenic TH17 cells by inducible salt-sensing kinase SGK1. Nature 496(7446):513–517. doi:10.1038/nature11984

423. Wu D, Sanin DE, Everts B, Chen Q, Qiu J, Buck MD, Patterson A, Smith AM, Chang CH, Liu Z, Artyomov MN, Pearce EL, Cella M, Pearce EJ (2016) Type 1 interferons induce changes in core metabolism that are critical for immune function. Immunity 44(6):1325–1336. doi:10.1016/j.immuni.2016.06.006
424. Wu J, Yang J, Yang K, Wang H, Gorentla B, Shin J, Qiu Y, Que LG, Foster WM, Xia Z, Chi H, Zhong XP (2014) iNKT cells require TSC1 for terminal maturation and effector lineage fate decisions. J Clin Invest 124(4):1685–1698. doi:10.1172/JCI69780
425. Wu N, Zheng B, Shaywitz A, Dagon Y, Tower C, Bellinger G, Shen CH, Wen J, Asara J, McGraw TE, Kahn BB, Cantley LC (2013b) AMPK-dependent degradation of TXNIP upon energy stress leads to enhanced glucose uptake via GLUT1. Mol Cell 49(6):1167–1175. doi:10.1016/j.molcel.2013.01.035
426. Wu Q, Liu Y, Chen C, Ikenoue T, Qiao Y, Li CS, Li W, Guan KL, Liu Y, Zheng P (2011) The tuberous sclerosis complex-mammalian target of rapamycin pathway maintains the quiescence and survival of naive T cells. J Immunol 187(3):1106–1112. doi:10.4049/jimmunol.1003968
427. Wu T, Shin HM, Moseman EA, Ji Y, Huang B, Harly C, Sen JM, Berg LJ, Gattinoni L, McGavern DB, Schwartzberg PL (2015) TCF1 is required for the T follicular helper cell response to viral infection. Cell Rep 12(12):2099–2110. doi:10.1016/j.celrep.2015.08.049
428. Xiao N, Eto D, Elly C, Peng G, Crotty S, Liu YC (2014) The E3 ubiquitin ligase Itch is required for the differentiation of follicular helper T cells. Nat Immunol 15(7):657–666. doi:10.1038/ni.2912
429. Xu L, Cao Y, Xie Z, Huang Q, Bai Q, Yang X, He R, Hao Y, Wang H, Zhao T, Fan Z, Qin A, Ye J, Zhou X, Ye L, Wu Y (2015) The transcription factor TCF-1 initiates the differentiation of T(FH) cells during acute viral infection. Nat Immunol 16(9):991–999. doi:10.1038/ni.3229
430. Yamagata K, Daitoku H, Takahashi Y, Namiki K, Hisatake K, Kako K, Mukai H, Kasuya Y, Fukamizu A (2008) Arginine methylation of FOXO transcription factors inhibits their phosphorylation by Akt. Mol Cell 32(2):221–231. doi:10.1016/j.molcel.2008.09.013
431. Yan D, Farache J, Mingueneau M, Mathis D, Benoist C (2015) Imbalanced signal transduction in regulatory T cells expressing the transcription factor FoxP3. Proc Natl Acad Sci U.S.A. 112(48):14942–14947. doi:10.1073/pnas.1520393112
432. Yan Y, Zhang GX, Gran B, Fallarino F, Yu S, Li H, Cullimore ML, Rostami A, Xu H (2010) IDO upregulates regulatory T cells via tryptophan catabolite and suppresses encephalitogenic T cell responses in experimental autoimmune encephalomyelitis. J Immunol 185(10):5953–5961. doi:10.4049/jimmunol.1001628
433. Yang K, Neale G, Green DR, He W, Chi H (2011) The tumor suppressor Tsc1 enforces quiescence of naive T cells to promote immune homeostasis and function. Nat Immunol 12(9):888–897. doi:10.1038/ni.2068
434. Yang K, Shrestha S, Zeng H, Karmaus PW, Neale G, Vogel P, Guertin DA, Lamb RF, Chi H (2013a) T cell exit from quiescence and differentiation into Th2 cells depend on Raptor-mTORC1-mediated metabolic reprogramming. Immunity 39(6):1043–1056. doi:10.1016/j.immuni.2013.09.015
435. Yang W, Bai Y, Xiong Y, Zhang J, Chen S, Zheng X, Meng X, Li L, Wang J, Xu C, Yan C, Wang L, Chang CC, Chang TY, Zhang T, Zhou P, Song BL, Liu W, Sun SC, Liu X, Li BL, Xu C (2016) Potentiating the antitumour response of CD8(+) T cells by modulating cholesterol metabolism. Nature 531(7596):651–655. doi:10.1038/nature17412
436. Yang XO, Zhang H, Kim BS, Niu X, Peng J, Chen Y, Kerketta R, Lee YH, Chang SH, Corry DB, Wang D, Watowich SS, Dong C (2013b) The signaling suppressor CIS controls proallergic T cell development and allergic airway inflammation. Nat Immunol 14(7):732–740. doi:10.1038/ni.2633
437. Yang XY, Wang LH, Chen T, Hodge DR, Resau JH, DaSilva L, Farrar WL (2000) Activation of human T lymphocytes is inhibited by peroxisome proliferator-activated receptor gamma (PPARgamma) agonists. PPARgamma co-association with transcription factor NFAT. J Biol Chem 275(7):4541–4544

438. Yang Y, Lovett-Racke AE, Racke MK (2010a) Regulation of immune responses and autoimmune encephalomyelitis by PPARs. PPAR Res 2010:104705. doi:10.1155/2010/104705
439. Yang Z, Kahn BB, Shi H, Xue BZ (2010b) Macrophage alpha1 AMP-activated protein kinase (alpha1AMPK) antagonizes fatty acid-induced inflammation through SIRT1. J Biol Chem 285(25):19051–19059. doi:10.1074/jbc.M110.123620
440. Ye J, DeBose-Boyd RA (2011) Regulation of cholesterol and fatty acid synthesis. Cold Spring Harb Perspect Biol 3(7). doi:10.1101/cshperspect.a004754
441. Yin Y, Choi SC, Xu Z, Perry DJ, Seay H, Croker BP, Sobel ES, Brusko TM, Morel L (2015) Normalization of CD4+ T cell metabolism reverses lupus. Sci Transl Med 7(274):274ra218. doi:10.1126/scitranslmed.aaa0835
442. York AG, Williams KJ, Argus JP, Zhou QD, Brar G, Vergnes L, Gray EE, Zhen A, Wu NC, Yamada DH, Cunningham CR, Tarling EJ, Wilks MQ, Casero D, Gray DH, Yu AK, Wang ES, Brooks DG, Sun R, Kitchen SG, Wu TT, Reue K, Stetson DB, Bensinger SJ (2015) Limiting cholesterol biosynthetic flux spontaneously engages type I IFN signaling. Cell 163(7):1716–1729. doi:10.1016/j.cell.2015.11.045
443. Zeng H, Cohen S, Guy C, Shrestha S, Neale G, Brown SA, Cloer C, Kishton RJ, Youngblood B, Do M, Li MO, Locasale JW, Rathmell JC, Chi H (2016) mTORC1 and mTORC2 kinase signaling and glucose metabolism drive follicular helper T cell differentitiation. Immunity 45:540
444. Zeng H, Yang K, Cloer C, Neale G, Vogel P, Chi H (2013) mTORC1 couples immune signals and metabolic programming to establish T(reg)-cell function. Nature 499(7459):485–490. doi:10.1038/nature12297
445. Zhang L, Tschumi BO, Corgnac S, Ruegg MA, Hall MN, Mach JP, Romero P, Donda A (2014) Mammalian target of rapamycin complex 1 orchestrates invariant NKT cell differentiation and effector function. J Immunol 193(4):1759–1765. doi:10.4049/jimmunol.1400769
446. Zhang L, Zhang H, Li L, Xiao Y, Rao E, Miao Z, Chen H, Sun L, Li H, Liu G, Zhao Y (2012) TSC1/2 signaling complex is essential for peripheral naive CD8+ T cell survival and homeostasis in mice. PLoS One 7(2):e30592. doi:10.1371/journal.pone.0030592
447. Zhang R, Huynh A, Whitcher G, Chang J, Maltzman JS, Turka LA (2013) An obligate cell-intrinsic function for CD28 in Tregs. J Clin Invest 123(2):580–593. doi:10.1172/JCI65013
448. Zheng Y, Delgoffe GM, Meyer CF, Chan W, Powell JD (2009) Anergic T cells are metabolically anergic. J Immunol 183(10):6095–6101. doi:10.4049/jimmunol.0803510
449. Zheng Y, Josefowicz S, Chaudhry A, Peng XP, Forbush K, Rudensky AY (2010) Role of conserved non-coding DNA elements in the Foxp3 gene in regulatory T-cell fate. Nature 463(7282):808–812. doi:10.1038/nature08750
450. Zhou X, Bailey-Bucktrout SL, Jeker LT, Penaranda C, Martinez-Llordella M, Ashby M, Nakayama M, Rosenthal W, Bluestone JA (2009) Instability of the transcription factor Foxp3 leads to the generation of pathogenic memory T cells in vivo. Nat Immunol 10(9):1000–1007. doi:10.1038/ni.1774
451. Zhu L, Yang T, Li L, Sun L, Hou Y, Hu X, Zhang L, Tian H, Zhao Q, Peng J, Zhang H, Wang R, Yang Z, Zhao Y (2014) TSC1 controls macrophage polarization to prevent inflammatory disease. Nat Commun 5:4696. doi:10.1038/ncomms5696
452. Zinzalla V, Stracka D, Oppliger W, Hall MN (2011) Activation of mTORC2 by association with the ribosome. Cell 144(5):757–768. doi:10.1016/j.cell.2011.02.014
453. Zizzo G, Cohen PL (2015) The PPAR-gamma antagonist GW9662 elicits differentiation of M2c-like cells and upregulation of the MerTK/Gas6 axis: a key role for PPAR-gamma in human macrophage polarization. J Inflamm 12:36. doi:10.1186/s12950-015-0081-4
454. Zoncu R, Bar-Peled L, Efeyan A, Wang S, Sancak Y, Sabatini DM (2011) mTORC1 senses lysosomal amino acids through an inside-out mechanism that requires the vacuolar H(+)-ATPase. Science 334(6056):678–683. doi:10.1126/science.1207056
455. Zong Y, Sun L, Liu B, Deng YS, Zhan D, Chen YL, He Y, Liu J, Zhang ZJ, Sun J, Lu D (2012) Resveratrol inhibits LPS-induced MAPKs activation via activation of the phosphatidylinositol 3-kinase pathway in murine RAW 264.7 macrophage cells. PloS One 7(8):e44107. doi:10.1371/journal.pone.0044107

Chapter 2
Metabolic Regulation of T Cell Immunity

Paolo D.A. Vignali, Joseph Barbi, and Fan Pan

Abstract It is becoming increasingly clear that cellular metabolism plays a critical role in the propagation of appropriate, effective, and pathologic immune responses. In this chapter, we detail the metabolic pathways involved in T cell activation and differentiation, highlighting specific factors responsible for directing the processes that lead to metabolic programming at important stages in the dynamic life cycle of this immune cell lineage. Additionally, this chapter will discuss how key metabolites are acquired, touching on the factors and conditions regulating the expression of crucial transporter molecules in response to activation and pathological states.

Keywords Metabolism • T cell • Regulation • mTOR • Myc • AMPK • HIF-1 • PPAR • Pathways • Activation • Differentiation

2.1 Introduction to Cellular Metabolism

The diverse cell types of the immune system perform a range of specialized functions that are critical for host defense. They also actively participate in the processes of wound healing, tissue remodeling, regulation of human metabolism, and restraint of potentially harmful (overzealous or misdirected) immune responses. These many and distinct roles are largely orchestrated through signaling cascades that are triggered by the ligation of extracellular and intracellular receptors. Such stimuli may dictate the migration of immune cells to drastically different microenvironments or the initiation of cellular proliferation and differentiation. As such, proper leukocyte function requires adaptation to varying nutrient levels and a flexible metabolism capable of accommodating changes in cellular demands for energy and biosynthesis. Therefore, the metabolic profile of leukocytes is central to multiple aspects of

P.D.A. Vignali • F. Pan (✉)
Immunology and Hematopoiesis Division, Department of Oncology, Sidney Kimmel Comprehensive Cancer Center, Johns Hopkins University School of Medicine, Baltimore, MD 21287, USA
e-mail: fpan1@jhmi.edu

J. Barbi
Department of Immunology, Roswell Park Cancer Institute, Buffalo, NY 14263, USA

B. Li, F. Pan (eds.), *Immune Metabolism in Health and Tumor*, Advances in Experimental Medicine and Biology 1011, DOI 10.1007/978-94-024-1170-6_2

their function. Recently, it has become clear that elements of cellular metabolism are not only integral to immune cell development. Indeed, they may act as distinct switches capable of reshaping the nature of an immune response. As will be discussed in this chapter, pathways that program cellular metabolism also regulate significant signaling cascades that control the effector functions of immune cells.

The T lymphocyte cell type is a heterogeneous population encompassing an expansive number of subsets capable of highly specialized functions in response to diverse stimuli. Yet some characteristics are widely shared among these and other leukocyte populations. Among these commonalities is the ability to detect and respond to extracellular and extraorganismal threats. When a mature, yet antigen-inexperienced, T cell emerges from the thymus to enter the circulation, they are relatively inert or quiescent. Prior to receiving activating signals, the modest energetic and biosynthetic needs of naïve T cells are met chiefly through the acquisition and metabolism of pyruvate derived from glucose (via mitochondrial oxidative phosphorylation) or fatty acids (via fatty acid oxidation) to generate adenosine triphosphate (ATP) through the tricarboxylic acid (TCA) cycle. When a naïve T cell encounters activating signals, a dramatic metabolic reprogramming occurs, allowing for the increased production of energy and biological raw materials – e.g., amino acids, fatty acids, and nucleotides [1] (Fig. 2.1). The significance of this metabolic reprogramming cannot be understated, as the robust production of effector molecules (e.g., cytokines, tissue adhesion molecules, cytotoxic factors) and cell division is as costly in terms of cellular resources as it is necessary for the establishment of an effective immune response. Moreover, specific alterations to a naïve T cell's metabolic profile – or inadequacies therein – during activation may affect the ability of a cell to differentiate into the appropriate effector T cell lineage capable of responding to a specific biological or chemical insults.

2.1.1 T Cell Differentiation

Engagement of the T cell receptor (TCR) by antigen/MHC complexes and the interaction of costimulatory receptor and ligands (e.g., CD28 and B7 family members) bring about T cell activation. Stimulation of naïve $CD4^+$ T cells not only leads to expansion of these cells but also to their acquisition of highly specialized effector functions accompanying their commitment to defined T helper (Th) lineages. Th differentiation is driven by lineage-specific cytokines present during naïve $CD4^+$ T cell activation. Signaling events downstream of cytokine-cytokine receptor interactions promote the expression of "master regulator" transcription factors responsible for establishing and enforcing Th-specific programs of gene expression. These, in turn, underlie the unique effector molecules and functions of Th cells [2].

For example, Th1 cells are well known for their production of interferon-γ (IFNγ) which, along with other characteristic functions, is driven by the transcription factor T-bet. These cells are important for the cell-mediated immunity necessary to resist intracellular viral, bacterial, and parasitic infections. A robust Th1

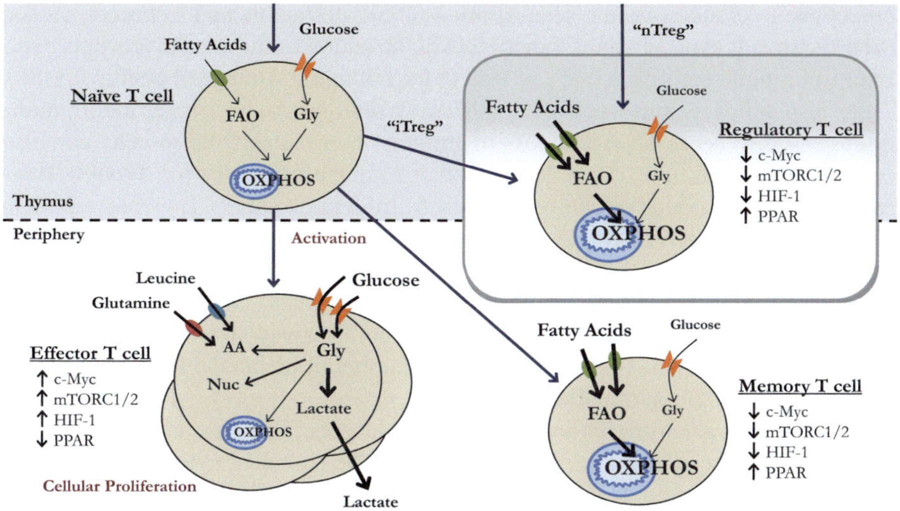

Fig. 2.1 Metabolic reprogramming in T cell activation/differentiation. The characteristically quiescent naïve T cell consumes meager amounts of glucose and fatty acids, processing their catabolic byproducts through OXPHOS. Upon TCR ligation and costimulation, T cells engage a rapidly proliferative profile, increasing the surface expression of glucose and amino acid transporters to accommodate this increased demand for products for cellular growth and effector function. Aerobic glycolysis is prioritized at the expense of OXPHOS. During the contraction phase of an immune response, memory lineages persist, relying on a metabolic profile dominated by the metabolism of fatty acids through OXPHOS. Tregs possess a metabolic profile similar to Tmem. Acronyms: *FAO* fatty acid oxidation, *Gly* glycolysis, *OXPHOS* oxidative phosphorylation, *AA* amino acids, *Nuc* nucleotides, *iTreg* peripherally induced Treg, *nTreg* thymically derived natural Treg

response is also needed for effective antitumor immune responses. Cells of the Th2 lineage, on the other hand, produce interleukin-4 (IL-4) and direct immunity to extracellular parasites (e.g., helminthes). GATA3 is a critical regulator of Th2-accociated gene expression [2]. Meanwhile Th17 cell differentiation is driven by STAT3-activating cytokines (such as IL-6) and the transcriptional regulator RORγt. Th17 cells are responsible for fighting extracellular bacterial and fungal infections through secretion of the characteristic cytokines IL-17A, IL-17F, and IL-22. These cells have considerable inflammatory potential and their involvement in autoimmune and inflammatory diseases is well established [3].

Another important subset of $CD4^+$ T cells is known for their ability to suppress the activity of other leukocytes. While several types of T cells can exert regulatory functions, the most recognized and arguably the most important of these is $CD4^+$ T cell subset marked by constitutively high expression of the transcription factor Foxp3 [4]. $Foxp3^+$ regulatory T cells (Tregs) utilize a variety of mechanisms to carry out this suppressive function. These include the production of anti-inflammatory cytokines (e.g., IL-10, TGFβ, and IL-35), the expression of coinhibitory molecules (e.g., CTLA-4 and LAG3), subversion of antigen-presenting cell activity to perpetuate tolerance, and the sequestration of growth factors and resources needed for effector cell expansion and function [5]. The action of these cells prevents over-zealous immune activation that can lead to the collateral damage of healthy tissues. Tregs also suppress autoreactive T cells that would otherwise trigger autoimmune disease. While necessary mediators of immune homeostasis, these cells can also oppose the mounting of effective anti-tumor immune responses [6]. Foxp3^{+} Tregs can arise in the thymus or can be induced from Foxp3^{-} naïve CD4+ precursors in peripheral tissues or ex vivo upon activation in the presence of the cytokines IL-2 and TGF-β [7] (Fig. 2.1). Just as with the aforementioned Th subsets, cellular metabolism plays a role both in the differentiation of these suppressor cells. While cytokines and transcription factors hold incredible sway over the Th decision-making process, as will be discussed later, a number of metabolic factors also influence the differentiation of naïve CD4^{+} T cells.

CD8^{+} T cells also differentiate into distinct subsets. These include effector cells capable of killing infected or transformed cells (the so-called cytotoxic T lymphocytes or CTLs) as well as long-lived memory CD8^{+} T cells that respond with the kinetics of innate immune cells to repeat encounters with antigens. Metabolic factors play a role in the distinct biology of these subsets as well.

2.2 Key Metabolic Pathways of T Cells

2.2.1 Glycolysis and the Tricarboxylic Acid Cycle

Within T cells, production of ATP largely results from the catabolism (breakdown) of glucose or fatty acids. There are a multitude of potential fates of glucose once it is transported into the cell, although many of the initial processing steps of the pathway termed *glycolysis* are shared by these paths. Upon entering the cell, glucose is rapidly phosphorylated at the sixth carbon position by the enzyme, *hexokinase*, which expends a molecule ATP and produces glucose-6-phosphate. In glycolysis, subsequent steps of rearrangement, isomerization, and bond-breaking follow resulting in the generation of the high-energy molecules, nicotinamide adenine dinucleotide (NADH), and ATP, as well as two identical three-carbon pyruvate molecules. Additionally, intermediate products of this process may be shuttled off to the pentose phosphate or serine biosynthesis pathways, the β-oxidation pathway, or the glycogenesis pathway, allowing for the production of nucleotides, fatty acids, or the energy storage molecule, glycogen, respectively (Fig. 2.2).

Once pyruvate is generated, it can be shuttled into the mitochondria where further breakdown occurs via the tricarboxylic acid (TCA) cycle – also known as the citric acid or Krebs cycle [8]. Inside the mitochondrial matrix, the three-carbon molecule is either carboxylated or decarboxylated and bound to coenzyme a (CoA), yielding the four-carbon molecule, *oxaloacetate*, or the two-carbon molecule, *acetyl-CoA*, respectively. Molecules of oxaloacetate and acetyl-CoA, generated

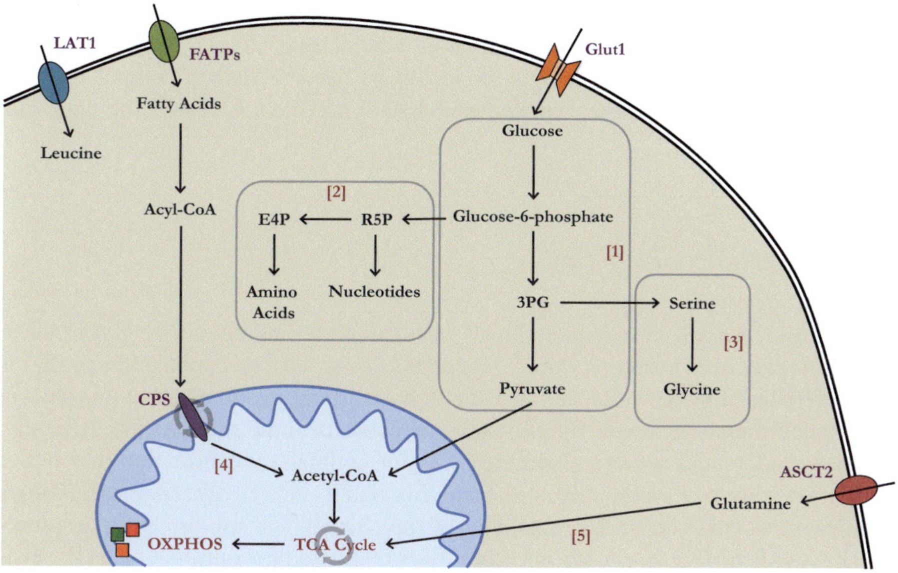

Fig. 2.2 Pathways in T cell metabolism. This flowchart details the acquisition of extracellular nutrients and the de novo synthesis of anabolic products. Many of these processes are readily reversible, despite their being represented with an unidirectional *arrow*. Key intermediates of these pathways and enzymes targeted for regulation are highlighted; some intermediates are excluded for simplicity's sake. *1* Glycolysis; *2* Pentose Phosphate Pathway; *3* Serine Biosynthesis Pathway; *4* β-Oxidation; *5* Glutaminolysis. Acronyms: *3PG* 3-phosphoglyceric acid, *R5P* ribose-5-phosphate, *E4P* erythrose-4-phosphate, *FATPs:fCPS* carnitine palmitoyltransferase system

separately from molecules of pyruvate, are bound through the action of *citrate synthase*, producing the six-carbon molecule, *citrate*. Seven successive steps of the TCA cycle follow, expelling two-carbon molecules such as carbon dioxide (CO_2) and regenerating oxaloacetate to be rebound to a new molecule of acetyl-CoA. Two reducing agents are generated through this cycle, NADH and flavin adenine dinucleotide ($FADH_2$), which donate electrons to cytochromes of the electron transport chain, resulting in the robust production of ATP through oxidative phosphorylation (OXPHOS). Three molecules of NADH (roughly equal to ~2.5 molecules of ATP) and one of $FADH_2$ (~1.5 molecules of ATP) are produced from the consumption of one molecule of acetyl-CoA yielding ~9 molecules of ATP (~30 to 32 molecules of ATP are produced per molecule of glucose through this entire pathway). This process relies on the availability of both NAD^+ and $FADH^+$ to oxidize the byproducts of glucose and, importantly, oxygen to serve as the final acceptor of electrons. Additionally, a molecule of GTP is proceeded during one step of the TCA cycle, which, like ATP, contains a high-energy phosphate bond that is broken to supply energy or to activate/deactivate enzymatic processes.

Alternatively, in a pathway that avoids mitochondrial involvement, pyruvate can be processed into lactate through the action of lactate dehydrogenase. The reaction consumes a molecule of NADH and bypasses the requirement for oxygen as the terminal electron acceptor, thereby allowing for a rapid, albeit less efficient, production of ATP from glucose in environments where little oxygen may be present. When oxygen is scarce, glycolysis alone can be used to sustain cellular requirements for ATP, with lactate expelled and NAD^+ recycled to restart the catabolic process.

2.2.2 Aerobic Glycolysis

Certain immune cell subsets including T cells can preferentially utilize glycolysis to generate ATP, even when oxygen is abundant. This particular metabolic profile is characteristic of tumor cells, and in 1956, Otto Warburg postulated that this shift in cancer cell metabolism was the fundamental cause of tumorigenesis [9]. Although we now understand a more elaborate basis for cellular transformation and tumor formation, this metabolic shift still bears his name, termed *Warburg metabolism* (also, aerobic glycolysis; the remainder of the chapter will use the latter terminology). As will be discussed, aerobic glycolysis is utilized by subsets of T cells post-activation to support their effector functions and characteristically rapid proliferation, a phenotype similar to the rampant cellular replication of tumor cells [10, 11].

Although aerobic glycolysis is less efficient for generating ATP from a single molecule of glucose, the use of these pathways confers a significant survival advantage to the cells that employ in it – namely, through the generation of molecular "raw materials" for biosynthetic processes. Two intermediates of glycolysis, namely, glucose-6-phosphate and 3-phosphoglyceric acid (3PG), can be fed into the pentose phosphate or serine biosynthesis pathways, respectively [1, 12, 13]. The resulting products from these pathways provide the cell with essential precursors of anabolic (synthetic) processes. The pentose phosphate pathway can be divided into two phases, an oxidative phase and a subsequent non-oxidative phase, each producing distinct precursors for biosynthetic and effector processes. The first generates the reducing agent, nicotinamide adenine dinucleotide phosphate (NADPH; distinct from NADH), which catalyzes numerous anabolic processes in lipid and cholesterol biosynthesis pathways and facilitates the production of superoxide anions (later, free radicals) within macrophages and neutrophils (see the section on reactive oxygen signaling below) [14]. The non-oxidative phase results in the production of the sugars, ribose-5-phosphate and erythrose-4-phosphate. This allows for the synthesis of nucleotides and nonessential aromatic amino acids, respectively. The serine biosynthesis pathway is essential for the de novo production of serine and glycine for protein and nucleotide biosynthesis. 3PG is converted in three steps to serine, which in turn allosterically inhibits the first enzyme of the series, aptly named, *3PG dehydrogenase*. 3PG dehydrogenase requires the cofactor NAD^+, which accepts a hydride from 3PG to become NADH. One final step in the process synthesizes glycine from serine, in a readily reversible fashion.

Many activated immune cells – e.g., effector T cells (Teff; namely, Th1, Th2, and Th17 $CD4^+$ subsets and cytotoxic $CD8^+$ cells), neutrophils, M1 macrophages, and some DC subsets – will predominantly utilize aerobic glycolysis to provide energy to the cell. Here the majority of pyruvate molecules produced are rapidly converted into lactate and expelled (see below for distinct regulatory and memory T cell metabolism) [12]. During this process, glycolytic intermediates are diverted through the pentose phosphate and serine biosynthesis pathways, promoting the generation of amino acids, nucleotides, and NADPH. Within $CD4^+$ Teff cells, there is evidence suggesting that a portion of the pyruvate generated during activation-induced glycolysis is still metabolized through the TCA cycle and OXPHOS. $CD8^+$ cells, on the other hand, do not appear to enhance OXPHOS following activation [15–17].

Reflecting the importance of glycolytic metabolism in activated Teff cells, many studies have shown that small-molecule inhibitors interfering with this process can be immunomodulatory. The glucose analogue, 2-deoxyglucose (2-DG) actively competes for the active site of hexokinase II (one of the three known isoforms of hexokinase), the enzyme that catalyzes the initial phosphorylation of glucose to G6P. 2-DG treatment selectively inhibits the glycolysis-dependent development of Th17 cells while reciprocally promoting the generation of $Foxp3^+$ regulatory T cells (Tregs) in vitro. Furthermore, adoptive transfer of 2-DG-conditioned Th17 cells results in a less efficient induction of neuropathology in the murine autoimmune model of experimental autoimmune encephalomyelitis (EAE) as compared to untreated Th17 cells [18, 19]. EAE mimics multiple sclerosis, with mice developing ascending paralysis as self-reactive Th1 and Th17 cells assault the myelin sheaths of neurons within the central nervous system (CNS). These results demonstrate the importance of glycolysis for (in this case pathologic) Teff responses. Memory T cells, on the other hand, do not rely on glycolytic metabolism. Reflecting this, 2-DG treatment augments the generation of memory T cells in mice and promotes the execution of memory T cell-mediated antitumor immunity [20].

2.2.3 Fatty Acid Oxidation (FAO)

By mass, the oxidation of fatty acids is the most energetically efficient metabolic process described in this chapter. Due to the negative charge on the polar head group, fatty acids must first be transported across the cellular membrane before they can be metabolized [21]. Once in the cytosol, ligases facilitate the association of a fatty acid to a molecule of CoA at the proximal carbon (C-1; α carbon) of the fatty acid chain. In the case of long chain fatty acids, the newly formed *acyl-CoA* undergoes a temporary subunit switch, in which CoA is replaced with a carnitine molecule to facilitate its transport across the mitochondrial matrix. Within the matrix, the reverse reaction occurs and further catabolism of the fatty acid proceeds. This reaction is catalyzed by the *carnitine palmitoyltransferase system* and is essential for the efficient metabolism of fatty acids. Once within the mitochondrial matrix, *acyl-CoA dehydrogenase* catalyzes the initial catabolic step of the process termed *β-oxidation*, where the bond between C-2 (β carbon) and C-3 is manipulated by successive oxidations and is eventually cleaved. The process results in the isolation of the two-carbon molecule, acetyl-CoA, from the fatty acid chain and the production of two high-energy reducing molecules, NADH and flavin adenine dinucleotide ($FADH_2$). If a single acetyl-CoA molecule were to continue through the TCA cycle and OXPHOS, it would produce ~10 molecules of ATP. Should the consumption of the NADH and $FADH_2$ molecules produced by β-oxidation be included in this calculation, ~14 molecules of ATP would be produced from a single β-oxidation step. The most common naturally occurring fatty acids are saturated and even numbered, between 4 and 28 carbons in length. Thus, a single fatty acid molecule from this category can result in the production of 28–196 molecules of ATP. Two by two, a fatty acid chain can be shrunk down, each cycle turning over molecules that can be shuttled through OXPHOS to produce ATP, until its last acetyl-CoA molecule remains. Mechanisms exist to account for odd-numbered and unsaturated fatty acids in this process, but these topics are beyond the scope of this chapter. Besides serving as high-energy metabolites, fatty acids also provide the key components of the cellular membranes, namely, phospholipids and steroids. In T cells and a great many other cell types, fatty acid synthesis is used to promote both cellular growth and organelle biogenesis [22].

Tregs and memory T cells (Tmem), as well as M2 macrophages, adopt a vastly different immune profile compared to their naïve and effector counterparts, primarily metabolizing lipids through fatty acid oxidation (FAO; β-oxidation) to support both their function and survival [17, 23] (Fig. 2.1). Tregs are critical for the maintenance of immune homeostasis as they counteract the milieu of proinflammatory immune cells through suppression of leukocyte activation and function. As previously mentioned with the glycolytic small-molecule inhibitor, 2-DG, impairment of glucose metabolism can promote a Treg fate and hinder proinflammatory Teff differentiation. Excess exogenous fatty acids in vitro have a decidedly negative effect on acquisition of the Th17 phenotype, illustrating a metabolic mechanism for the reciprocal regulation of Treg and Th17 cell fates. Indeed, forced reliance on FAO during in vitro T cell differentiation favors the generation of Tregs (as evidenced by the heightened induction of Foxp3 expression and the suppression of responder cell proliferation) at the marked expense of Teff differentiation [23]. The genetic or chemical inhibition of the metabolic regulator, mechanistic target of rapamycin (mTOR; also, FRAP1, which will be discussed in detail below), during $CD4^+$ T cell activation decreases aerobic glycolysis and augments FAO favoring the generation of $Foxp3^+$ Tregs [24] (see the mTOR section below for more details). However, when FAO is simultaneously interrupted through the etomoxir-mediated inhibition of the carnitine palmitoyltransferase system (and thus inhibition of fatty acid transport), T cells are no longer preferentially shunted toward a Treg fate by suppressed mTOR activity.

While Teff cells expand clonally upon activation during the early stages of an immune response, relatively small numbers of long-lived Tmem cells persist during and after the contraction phase [25]. A defining aspect of Tmem is their capacity to respond, with accelerated kinetics and cytokine release, to a repeat antigen encounter (i.e., a secondary infection). These cells transition abruptly from a state of quiescence requiring minimal energetic demands, to one typified by full-bore effector function and proliferation. The rapid reactivation of Tmem cells must be accounted for metabolically. Within the CD8$^+$ memory lineage, it has been well established that both Tmem generation and persistence, as well as repeat-antigen-induced reactivation, are dependent on the modulation of FAO [17, 26, 27]. Compared to their effector counterparts, Tmem cells possess a greater mitochondrial mass (enhanced mitochondrial biogenesis) and correspondingly express FAO-associated enzymes and carnitine palmitoyltransferase system proteins at a substantially higher level [22, 26, 28] imparting them with a greater spare respiratory capacity (SRC). SRC is defined as the enhanced ability of a cell to generate energy in response to cellular activation or stress [26, 27]. Thus, the development of SRC within Tmem populations effectively preps the cells for reactivation. Additionally, the advantage of augmented SRC is thought to facilitate the survival of Tmem cells under conditions of energetic stress [29].

2.3 Metabolite Acquisition Associated with T Cell Activation

Following TCR ligation and the triggering of costimulation pathways, T cells undergo clonal expansion and transcriptome reprograming to express effector molecules including cytotoxins, cytokines, and cell-surface molecules. This robust increase in cellular growth (accumulation of biomass), proliferation, and protein anabolism demands a matched increase in metabolite uptake to provide components for the generation of cellular machinery and free energy required to catalyze chemical reactions [1, 30, 31]. T cells are incredibly reliant on nutrient absorption for replication and function, and as such, starvation from glucose or amino acids represents one mode of regulation controlling T cell population size and activity [32]. Interestingly, in sites resistant to immune activation (*a.k.a.* immune privileged sites) – e.g., the brain, eye, testicles, and placenta – FAO is the primary source of energy production. Increased FAO supports suppressive immune cell lineages, thereby protecting these vital organs from autoimmunity [33]. T cells control their intake of nutrients through the expression of transporter proteins on the cellular membrane. As will be described, the enhanced expression of the following transporters is essential for the proliferation, differentiation, and effector functions of T cells.

2.3.1 Glucose Uptake

As stated previously, the transition of T cells out of quiescence and into an activated state requires an increased uptake of nutrients to fuel augmented cellular growth and proliferation and meet the anabolic needs of effector molecule production. For Teff subsets, the increased import of glucose is essential for activation-associated functions [34]. A family of glucose transporters (Glut1–14) facilitate the uptake of glucose and related sugars in mammalian cells [35–37]. T cells have been reported to express a select few Glut family transporters, namely Glut1, 2, 3, 6, and 8. Chief among these transporters is the facilitative transporter, Glut1, which is responsible for the basal level of glucose uptake in all resting immune cells [34].

Costimulatory signals received via CD28 ligation and propagated through PI3K/Akt signaling results in the upregulation of Glut1 (and a corresponding downregulation of the carnitine palmitoyltransferase system critical for FAO) [38]. It is the increase in cell-surface expression of Glut1 alone that facilitates the dramatic increase of glucose influx required to manage the cells' heightened glycolytic demands [39, 40]. Interruption of this process has profound effects on T cell function, including their ability to differentiate into proinflammatory lineages. Irrespective of the expression of other glucose transporters, the deletion of Glut1 alone confers a significant deficit to the growth, clonal expansion, and persistence of murine Teff subsets when activated in vivo [34]. Conversely, an overabundance of Glut1 promotes the accumulation of biomass (i.e., cellular growth) in naïve T cells and the acquisition of an activated phenotype [41]. Illustrating the importance of glycolysis in effector molecule anabolism, glucose deficiency diminishes the capacity of cytotoxic $CD8^+$ cells (CTLs) to produce effector cytokines, perforin, and granzymes [41–43]. Tregs, which rely primarily on the oxidation of fatty acids for the generation of cellular energy, appear unaffected by this deletion, persisting and functioning in vivo regardless of Glut1 expression [41].

2.3.2 *Amino Acid Uptake*

T cell activation also results a heightened demand for amino acids, resulting in the upregulation of amino acid transporters following activation. Indeed, as with glucose deficiency, an inadequate supply or uptake of these metabolites can adversely affect the proliferative capacity of activated T cells and can significantly shape their differentiation. Within activated T cells, the amino acid *glutamine* is particularly critical. Glutamine is absorbed and broken down in a process termed *glutaminolysis*. Through glutaminolysis, the amino acid is broken down, and its derivatives can be fed into the TCA cycle, restoring levels of intermediates that are consumed by biosynthetic processes. This allows for the de novo synthesis of lipids and NADPH. The activation-induced enhancement of glutamine absorption is accomplished through several well-characterized transporters. The antiporter, ASC amino acid transporter 2 (ASCT2; also called Slc1a5), is upregulated as a direct result of T cell receptor (TCR) ligation [44]. ASCT2 is surprisingly not necessary for the proliferation of T cells. However, even moderate restrictions to glutamine import can have profound effects on the ability of a T cell to differentiate, particularly into the proinflammatory T helper (Th) cell subsets, Th1 and Th17. In murine models of autoimmunity, the in vivo $CD4^+$ T cell-specific deletion of ASCT2 induces a markedly diminished Th1 and Th17 immune response, alleviating the progression of host-reactive disease in murine EAE (a disease driven by Th1 and Th17 immune responders) [44]. This diminished pathology is the result of deficient induction of proinflammatory Th cell subsets alone, without hampering the induction of Tregs or $CD8^+$ T cells [44]. This fact illustrates the potential heterogeneity of metabolite acquisition needs among T cell lineages.

$CD8^+$ T cell function appears to rely primarily on another amino acid transport pathway, preferentially absorbing the amino acid *leucine* to regulate effector potential. Following exposure to activating signals through TCR ligation and costimulation, both $CD4^+$ and $CD8^+$ T cells upregulate their surface expression of system L neutral amino acid transporter 1 (LAT1; also called Slc7a5). LAT1 is a heterodimeric transporter molecule and is chiefly responsible for the absorption of essential branched-chain (e.g., leucine and isoleucine) and aromatic (e.g., tryptophan and phenylalanine) amino acids while also participating mildly in glutamine transport [45–47]. Specifically, the enhanced leucine influx facilitated by LAT1 is critical for the regulation of $CD8^+$ T cell differentiation and migration [17, 48, 49]. As will be discussed in the sections that follow, concentrations of intracellular amino acids are sensed by the key metabolic regulator, mTOR, with a deficiency in amino acids resulting in the inhibition of anabolic processes and cellular proliferation. Accordingly, LAT1-deficient T cells do not proliferate well and display stunted effector differentiation. Specifically, LAT1-deficiency in $CD4^+$ and $CD8^+$ T cells results in a less pronounced induction of Th1/Th17 cells and CTLs, respectively. These observations were linked to an inability to increase amino acid uptake and a decrease in mTOR activity during T cell activation [45]. Moreover, the diminished uptake of leucine resulted in the loss of c-Myc expression. c-Myc is a key regulator of the glycolytic metabolism essential for activated Teff function, and as such, activation-induced upregulation of Glut1 expression was lacking in LAT1-deficient T cells. Interestingly, this deficit in regulator and transporter expression was only apparent at the protein level; levels of c-Myc and Glut1 mRNA were unaffected [45]. A similar loss of proliferation and effector potential can be accomplished through the system L neutral amino acid transporter inhibitors, BCH (2-aminobicylo-(2,2,1)-heptane-2-carboxylic acid), and brasilicardin A [50, 51]. Thus, it is the enhanced uptake of amino acids facilitated by upregulated expression of amino acid transporters, like ASCT2 and LAT1, which allows for the augmented anabolic metabolism that is essential for adequate T cell activation.

Certain immune cell subtypes, namely, dendritic cells (DCs) and macrophages (Mø), can control the availability of the crucial amino acid *cysteine* to mediate the cellular growth and proliferation of T cells [52]. Due to the typically low extracellular concentrations of the amino acid, DC-/Mø-mediated cysteine release adds an effective level of control to T cell function. Cysteine is processed intracellularly into the tripeptide, *glutathione*, a principal molecule in mechanisms that protect the DNA replication process and, thus, cellular proliferation. Glutathione has been shown to protect T cells from the effector functions of phagocytic cells by functioning as a key antioxidant, allowing for effective T cell targeting of foreign and transformed cells in harsh microenvironments. Both phagocytic cells (e.g., Mø, neutrophils) and activated T cells can initiate respiratory burst, utilizing NADPH to produce and release reactive oxygen species (ROS) to facilitate the destruction of foreign cells [53]. While ROS have been shown to function as important signaling molecules within T cells, promoting, among other things, the production of IL-2, too much intracellular ROS can be damaging. By releasing ROS into the inflammatory milieu, resident T cells may come under oxidative stress, defined by an overabundance of ROS as compared to neutralizing antioxidants. Prolonged oxidative stress can diminish cellular function or induce apoptosis through chemical reactions that alter protein and lipid conformation/function and damage DNA [54–56]. Thus, in order to maintain the functionality of an immune response, mechanisms of protecting T cells against ROS exposure are critical. Therefore, the antioxidant function of glutathione promotes the oxidative balance in conditions of high ROS presence. Interestingly, Tregs too utilize this mechanism to control proinflammatory events. Tregs, while naturally more resistant to oxidative stress-induced apoptosis [57], interact with APCs to facilitate increased cystine uptake by Tregs, thereby competing with local Teff cells for available amino acids and inhibiting the ability of Teff cells to replicate and function [58].

Clearly, an appropriate intracellular concentration of amino acids during T cell activation and expansion is paramount to establish an effective immune response. Indeed, the availability and metabolism of other amino acids such as tryptophan [59, 60] have also been revealed to dictate the effector/regulatory immune axis. The enzyme indoleamine 2,3-dioxygenase (IDO) is responsible for the breakdown of tryptophan, and the metabolites generated by this enzyme as well as the resulting amino acid depletion limit T cell activation and promote suppressive Tregs. IDO is notably produced by tolerogenic dendritic cells and promotes the generation and expansion of Tregs while discouraging the conversion or reprogramming of established Tregs - favoring instead the retention of a traditional, predominantly suppressive phenotype. Importantly abundant IDO levels in the gut and tumor-draining lymph nodes were shown to suppress Treg reprogramming by IL-6 [61–63]. IDO activity is also a potent mechanism for immunosuppression capable of mediating maternal tolerance of the fetus and preventing effective antitumor immunity in the cancer setting [64]. Inhibiting IDO in the cancer setting can be an avenue to overcome tumor-induced immunosuppression [65].

Scarcity of extracellular amino acids can lead to the upregulation of several genes involved in amino acid transport and in the de novo synthesis of nonessential amino acids [47, 59, 66]. Artificial induction of these genes, referred to as the amino acid starvation response (*AAR*), in T cells can suppress the generation of proinflammatory lineages like Th17 cells and reduce effector responses in in vitro and in vivo models of disease [47, 66].

2.4 Key Factors in Metabolic Regulation

An impressive web of receptors, kinases, and transcription factors orchestrate the aforementioned shifting of metabolic lifestyle that is necessary to coordinate the developing immune response (summarized in Fig. 2.3). Several key molecules that govern multiple aspects of cellular metabolism will be described within this chapter. As will be revealed, the precise regulation of these factors and their activities is requisite for appropriate immune function. Their dysregulation, on the other hand, can result in cellular transformation, autoimmunity, or metabolic disorder. How the therapeutic modulation of the expression or function of these metabolic regulators can provide tantalizing avenues for the control of these ailments will also be discussed.

2.4.1 *Cellular Myc (c-Myc)*

c-Myc, the cellular homologue to viral Myc (v-Myc; the avian myelocytomatosis viral oncogene), is a proto-oncogene that can independently transform numerous mammalian cell lineages [67]. The gene was first discovered to be the driving transformative factor in Burkitt's lymphoma, a tumor that arises exclusively from the dysregulation of myc gene expression [68, 69]. In fact, activity-altering mutations or overexpression of c-Myc are such prevalent factors across cancer lineages that

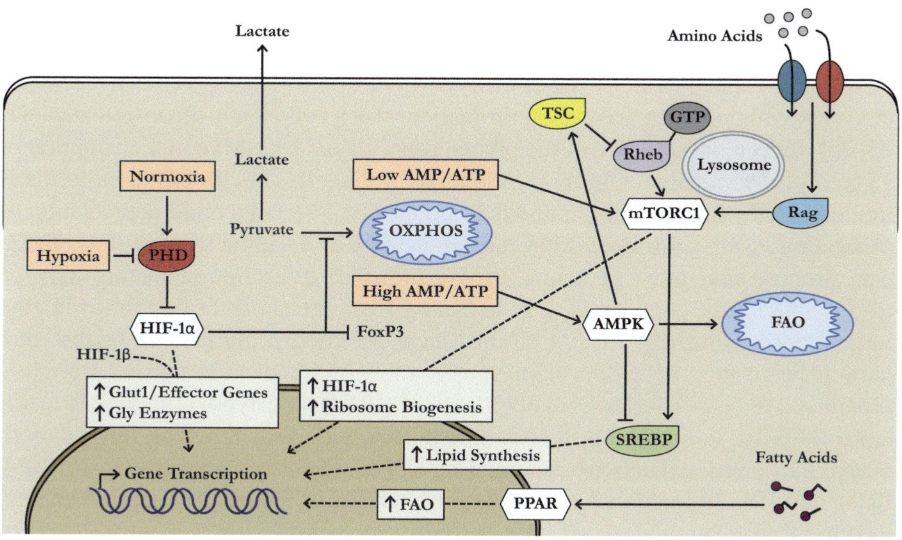

Fig. 2.3 Regulation of metabolic factors and their interplay. Key metabolic factors are highlighted in *white*. Downstream gene regulation is described in terms of the genes upregulated (*up arrow*) upon activation of metabolic factors. Acronyms: *FAO* fatty acid oxidation, *OXPHOS* oxidative phosphorylation

Myc is considered to be a central oncogenic determinant [70]. Since its discovery, the complex regulatory role that Myc plays in cellular metabolism and function has been progressively delineated. c-Myc is categorized as an immediate early gene, meaning that it is rapidly upregulated upon TCR- and costimulatory molecule-triggered signaling cascades. A considerable number of pathways (e.g., NF-κB, MAPK/ERK, PI3Kinase/Akt/mTOR) drive the upregulation of c-Myc expression following cellular activation. Stable expression of c-Myc allows for its heterodimerization with the transcription factor myc-associated factor X (MAX) and subsequent genetic influence over a slew of cellular functions key to effector potential, including the regulation of cell cycle progression, growth and proliferation, and cellular differentiation [70, 71]. Indeed, while global c-Myc deficiency is lethal in in vivo models, homologous deletion of c-Myc in vitro significantly prolongs the time between cellular divisions in affected cells and impairs the dominance of glycolytic metabolism [1, 72]. Specifically, pentose phosphate pathway enzymes are downregulated upon c-Myc deficiency, while other related metabolic pathways remain unaffected. Interestingly, c-Myc also appears to promote the uptake of amino acids, as expression of LAT1 and ASCT2 is correspondingly diminished upon T cell-specific deletion of the molecule (knockout of the receptors also reduces the expression of c-Myc) [1, 44]. This would suggest a multifaceted role of c-Myc in mechanisms that follow T cell activation.

2.4.2 c-Myc Regulates Early Response Pathways Downstream of TCR Ligation

Despite the rapid enhancement of c-Myc expression following TCR engagement, this augmentation does not persist throughout the T cell clonal expansion stage [1, 73–75]. Indeed, irrespective of the pivotal roles played by c-Myc in the adoption of a glycolysis-dominated, rapidly proliferative profile, the activity of the molecule is limited by its relatively short half-life (15 min) [76, 77] and a rapid depreciation of expression ~48 h postinfection in murine models of T cell activation [74]. These data propose that other (potentially Myc-induced) transcription factors may be responsible for the persistence of an activated T cell metabolic profile following the initial reprogramming. Recent studies have revealed a number of factors that participate in this metabolic maintenance, including activating enhancer-binding protein 4 (AP4), interferon regulatory factor-4 (IRF4), and hypoxia-inducible factor alpha (HIF-1α; to be discussed later in detail) [73, 78–80]. Together, through both c-Myc transcriptional activity and persistent TCR signaling, the expression of AP4, IRF4, and HIF-1α increases and resultantly stabilizes the glycolytic profile characteristic of proinflammatory effector lineages. Interestingly, the genetic ablation of these three "second-wave" factors does not impede the initial metabolic reprogramming seen during T cell activation. Rather, deletion of these factors results in a failure of

T cells to sustain aerobic glycolysis and, thus, effector cell function and survival [73, 78–81], as c-Myc levels diminish following T cell activation.

IRF4 has revealed itself to be a rather unique factor in the genetic reprogramming of effector T cells. Several studies have revealed that IRF4 levels in T cells correspond to the relative affinity of TCR-antigen/MHC complex interactions [78, 79, 82]. Specifically, lower-affinity ligands diminished the longevity and intensity of TCR signaling through reduced IRF4, allowing for a fine-tuning of downstream metabolic and effector programs by the strength of T cell activation. Similar to AP4 null experiments, $CD8^+$ T cell-specific deletion of IRF4 results in a reduction of cellular proliferation and effector cytokine production following c-Myc-driven activation [78, 79, 82]. Within $CD4^+$ T cells, IRF4 transcriptional activity appears to coordinate significant pathways in the effector functions of Th lineages and Treg cells. Loss of IRF4 significantly impaired the adoption of effector characteristics upon naïve CD4+ T cell skewing in vitro (revealed by diminished cytokine production and lineage-specific transcription factor expression) and Treg-mediated suppression in vivo [83–87].

c-Myc facilitates an impressive repertoire of mechanisms following T cell activation. Included among these is the upregulation of glucose and amino acid transporters [1] as well as the promotion of glycolysis and glutaminolysis [1, 45]. Indeed, the loss of c-Myc induced a corresponding decrease in amino acid transport expression, inhibiting the central metabolic pathways involved in establishing effector function [1, 44, 45]. Interestingly, LAT1 deletion was revealed to prevent the translation (but not the transcription) of the *c-Myc* gene in a mechanism not involving the amino acid sensor, mTOR. Destabilization of c-Myc expression resulted in the expected reduction in glycolysis (downregulation of Glut1; decreased lactate output) and glutaminolysis (decreased glutamine and arginine uptake) [45]. The mechanism behind the leucine sensitivity of c-Myc has yet to be fully delineated.

2.4.3 Therapeutic Opportunities in Targeting the c-Myc Signaling Pathway

Compounds that target c-Myc in cancer cells may be of particular value therapeutically. Aside from its role in metabolic control [88–90], c-Myc appears to also promote tumor growth through augmented expression of immunomodulatory cell-surface molecules that can facilitate immune evasion [91]. On the other hand, therapeutic downregulation of c-Myc activity within immune cells could allow for the suppression of the activation-/effector-related functions downstream of c-Myc/MAX, thereby dampening unwanted inflammation. The small-molecule inhibitor, 10058-F4, has been shown to obstruct c-Myc/Max heterodimerization, preventing DNA binding/transcriptional activity [92]. Accordingly, 10058-F4 treatment hampers T helper development in in vitro cultures [93].

2.4.4 Mechanistic Target of Rapamycin (mTOR)

The mechanistic target of rapamycin (mTOR; formerly mammalian TOR) is a key regulator of cellular function, integrating intracellular and extracellular signals to coordinate shifting metabolic states with cell growth, proliferation, and longevity. The smooth transition between catabolic and anabolic metabolism in accordance with available extracellular nutrients and intracellular energy storage is paramount to the survival of cells in the dynamic environments that exist within multicellular organisms. mTOR was discovered between 1993 and 1995 to be the molecular target of the cellular proliferation inhibitor, rapamycin [94–97], and it was so named. Constraining mTOR function through rapamycin has substantial effects on the ability of T cells to mount an effective proinflammatory response. As will be discussed in the subsequent sections, the suppression of mTOR activity has profound effects on the metabolic profile, and thus the function, of T cell subsets.

The mTOR protein is a well-conserved serine-threonine kinase within the phosphoinositide 3-kinase (PI3K)-related kinase family and has been found to nucleate in two multi-protein complexes, termed mTOR complexes 1 and 2, respectively (mTORC1 and mTORC2). mTORC1 operates as an environmental sensor, promoting protein synthesis, and cellular growth (mass accumulation) and proliferation when intracellular energy and amino acid concentrations are sufficiently high [98–100]. mTORC2 plays a supporting role by mediating the organization of the actin cytoskeleton and potentially supporting shifting metabolic states [101, 102]. Both mTORC1 and mTORC2 also promote cellular survival (inhibition of autophagy) through phosphorylation of ULK1 and Atg13 by mTORC1 [103–105] or AKT (also called protein kinase B) by mTORC2 [106, 107]. In the following sections, we will discuss the regulation, function, and activity of mTOR complexes in T cell subsets.

2.4.5 Fundamentals of mTORC1 Regulation

mTORC1 is a six-protein complex, consisting of regulatory-associated protein of mTOR (Raptor) [108, 109], mammalian lethal with SEC13 protein 8 (mLST8), proline-rich AKT substrate 40 kDa (PRAS40), DEP-domain-containing mTOR-interacting protein (Deptor) [110], and TELO2-interacting protein 1 homologue and telomere maintenance 2, respectively (Tti1/Tel2) [111]. The distinct functions of each protein have yet to be fully delineated, but significant findings have hinted toward the functional roles of the mTORC1-associated proteins. For example, it has been revealed that Raptor interacts with both mTOR and two downstream substrates, ribosomal protein S6 kinase 1 (S6K1) and eukaryotic initiation factor 4E (eIF4E)-binding protein 1 (4E-BP1), effectively providing the scaffolding to facilitate mTOR kinase activity [108, 109, 112–114]. Knockout studies of Raptor reveal it to be an obligate binding partner for effective mTORC1 activity. Consequences of

S6K1 and 4E-BP1 phosphorylation include enhanced protein synthesis, which facilitates cellular growth and progression through the cell cycle [98, 100, 115, 116]. Tti1/Tel2 constitutively associate with mTOR and appear to coordinate the assembly of the mTORC1 complex. Deletion of either Tti1 or Tel2 results in disassembly of mTORC1 and a loss of mTOR kinase activity [111]. Conversely, PRAS40 and Deptor have been proposed to negatively regulate the complex, as activation of mTORC1 signaling results in the immediate phosphorylation of PRAS40 and Deptor, thereby hindering their ability to bind to the complex and promoting their subsequent degradation. This results in enhanced mTORC1 substrate binding and activity. PRAS40 has been suggested to directly interfere mTORC1 substrate binding [117], while less is known about the exact function of Deptor in mTORC1 inhibition. Indeed, these data and the observation that diminished mTORC1 signaling are associated with PRAS40 and Deptor binding to mTORC1 and support the notion that they act as negative regulators of mTORC1 activity [110, 118, 119].

Pathways downstream of the growth factor receptors, namely, the insulin and Ras signaling pathways, facilitate the activation of mTORC1 activity through the inhibition of the key mTOR regulatory complex, tuberous sclerosis complex (TSC) [120–123]. TSC is a heterodimer of TSC1 (also called hamartin) and TSC2 (also called tuberin) and functions as a GTPase-activating protein (GAP), converting protein-bound GTP to GDP. TSC1 stabilizes its dimeric partner, TSC2, which contains the GAP domain. TSC actively targets the GTP-bound protein, Ras homologous enriched in brain (Rheb). Kinases downstream of growth factor receptors respond to stimulatory signals by phosphorylating TSC2, inhibiting the GAP activity of the protein. Considering that GTP-bound Rheb stimulates the activity of mTORC1 through direct association [118, 124], TSC, therefore, acts to inhibit mTORC1 signaling by dephosphorylating the GTP bound to Rheb, thereby inactivating the protein's effect [120, 125]. Specific mutations within TSC inactivates the complex, leading to uncontrollable cellular growth and proliferation through loss of mTORC1 inhibition and gives rise to the proliferative disease that bears its name, tuberous sclerosis [126]. Ligation of the insulin receptor stimulates a negative feedback loop, where activated S6K1 (downstream of mTORC1) inhibits the stimulatory phosphorylation of the proximal substrate of the insulin signaling pathway, destabilizing insulin receptor signaling and preventing deregulated mTOR activity [127]. Loss of this feedback control through inhibitory mutations or deletions of S6K1 can result in cellular transformation (tumorigenesis) or metabolic disorder (insulin desensitization) [128].

As noted in the "Metabolite Acquisition" section, mTOR functions as a key sensor of intracellular amino acid concentrations [129–132]. While the exact mechanism of amino acid sensing has yet to be determined, it has been well established that the amino acids, glutamine [132–134], leucine [118, 135], and arginine [135] facilitate mTORC1 activation. The lysosome appears to be the staging platform of mTORC1 regulation [136, 137] as stimulatory and inhibitory factors (e.g., Rheb and TSC, respectively) tend to localize in proximity of these organelles [138]. One model suggests that when sufficient amino acid concentrations are sensed, Rag protein heterodimers recruit mTORC1 to the lysosome, allowing for Rheb-dependent activation of mTORC1 and subsequent promotion of cellular growth and proliferation [139].

As described, the activation of the mTORC1 pathway is accomplished through various extracellular and intracellular signals such as the ligation of growth factor receptors or the acquisition of certain nutrients. Additionally, other factors can negatively regulate the downstream signaling of mTORC1, including the energy status of the cell and the oxygen tensions found within the cell's microenvironment. Immune cells carefully coordinate cellular growth and proliferation so that these processes may only occur when sufficient energy stores and biomaterials exist within the cell. mTORC1 is a master regulator of this synchronization, promoting cellular growth only when conditions are ideal. In an example of negative regulation of mTORC1, AMP-activated protein kinase (AMPK) inhibits mTORC1 when cellular energy is lacking (i.e., when AMP/ATP ratios are high); AMPK phosphorylates the Raptor subunit of mTORC1, leading to the recruitment of 14-3-3 protein to Raptor and the inhibition of mTORC1 activity [140]. Additionally, AMPK has been shown to phosphorylate TSC upstream of mTORC1. AMPK phosphorylates TSC2, promoting the activity of TSC and allowing for the control of transcription and cellular growth through mTORC1 inhibition while also protecting against energy-deprivation-induced apoptosis [120]. Low oxygen tension, too, can regulate the activity of mTORC1 [141] by diminishing ATP levels and increasing AMPK activation and by transcriptional activation of regulated in development and DNA damage responses 1 (REDD1), which facilitates TSC2 activation [142, 143].

2.4.6 Fundamentals of mTORC2 Regulation

mTORC2 is a seven-protein complex which shares some protein subunits with mTORC1, including the central protein, mTOR, as well as mLST8, Deptor, and Tti1/Tel2. mTORC2 also includes the proteins rapamycin-insensitive companion of mTOR (Rictor), mammalian stress-activated protein kinase interacting protein (mSIN1), and protein observed with Rictor-1 and Rictor-2, respectively (Protor-1/Protor-2). Like Raptor in mTORC1, Rictor appears to play a central role in stabilizing mTORC2 and is absolutely essential for characteristic mTORC2-mediated phosphorylation of Akt/PKB. Rictor and mSIN1 are stabilized when associated together and maintain the structural integrity of the complex, thus promoting mTORC2 function [144, 145]. Again, Deptor appears to function as a negative regulator of mTORC2 activity, just as it was described with mTORC1 [110]. Uniquely, however, although mLST8 does not appear to be necessary for mTORC1 function, its association with mTORC2 is paramount for the stability and function of the complex. In vitro knockout studies of mLST8 reveal unperturbed phosphorylation of mTORC1 substrates S6K1 and 4E-BP1, while mTORC2-mediated phosphorylation of AKT/PKB was diminished. This inhibition of mTORC2 function yields a phenotype similar to that seen in Rictor knockout studies [146].

Considering that the deletion of mTORC2 activity is lethal in in vivo models and that the complex is insensitive to rapamycin treatment except at prolonged high doses [147], studies in the function of mTORC2 have been met with difficulty. Consequentially, substantially less is known concerning the mTORC2 control mechanisms. What is known is that, like mTORC1, mTORC2 also responds to growth factor signals. Growth factor ligation results in mTORC2-mediated phosphorylation of a unique serine residue on the serine-/threonine-specific kinase, AKT, thereby enhancing the activity of AKT [148, 149]. A second serine residue is phosphorylated by phosphoinositide-dependent kinase-1 (PDK1), coordinating the full activation of AKT. It has been suggested that mSIN1 may coordinate this mTORC2 activity [150]. The complete phosphorylation of AKT results in enhanced proliferation, promotion of a glycolytic metabolism, and cellular migration. Downstream effects of mTORC2 activation will be described in the following section.

2.4.7 Activation of and the Downstream Effects of mTOR Signaling

According to the two-signal model of lymphocyte activation, antigen presentation alone cannot induce an activated state. Instead, the absence of a second costimulatory signal (e.g., CD28 cross-linking or cytokine receptor ligation) results in cellular anergy and an accordingly diminished mTORC1 signaling through S6K1. Similarly, the inhibition of mTOR activity in vivo through rapamycin treatment results in T cell anergy following antigen presentation [151]. It appears that mTOR not only coordinates intracellular nutrient/energy levels and downstream growth factor receptor signaling with cellular activation but that it also senses extracellular environmental cues to determine whether activation or anergy represents the appropriate response.

Increased uptake of amino acids following T cell activation, specifically the absorption of glutamine and leucine, promotes the activity of mTORC1, which in turn, facilitates the adoption of a predominately glycolytic metabolic profile. Genetic deletion of mTOR in $CD4^+$ T cells dampens their proliferative capacity and prevents their differentiation into effector T helper subsets – e.g., Th1, Th2, Th17 – even under the appropriate in vitro skewing conditions. Instead, activating naïve $CD4^+$ T cells in the absence of mTOR activity induces a hypersensitivity to TGF-β signaling and a preferential generation of $Foxp3^+$ Tregs [152]. Interestingly, the inhibition of mTOR activity induces expansion of natural $CD4^+$ Tregs in vivo and the corresponding de novo synthesis of Foxp3 in differentiating naïve CD4+ T cells, supporting the notion that this pathway negatively regulates immune tolerance [153].

Adding complexity to mTOR's role in determining $CD4^+$ T cell fate was the revelation that the two mTOR complexes have distinct effects on this process. Mice lacking Rheb have been used to study the effects of defective mTORC1 activity.

$CD4^+$ T cells from these knockout mice fail to differentiate into proinflammatory Th1 and Th17 effector cells. Treating T cells with rapamycin suppresses rapamycin-sensitive mTORC1 activity and thus inhibits the induction of proinflammatory T helper subsets, Th1 and Th17, but not Th2 [151]. On the other hand, Rictor-deficient $CD4^+$ T cells, which lack mTORC2 activity, display markedly impaired Th2 differentiation [24].

mTOR signaling is also foundational in determining the fate of $CD8^+$ T cells. Upon naïve $CD8^+$ T cell activation, mTOR activity is increased in response to IL-12/STAT4 signaling and costimulation. In these cells, mTOR can upregulate the expression of the transcription factor, T-bet, promoting a cytotoxic $CD8^+$ T cell (CTL) expression profile [48]. While rapamycin treatment of $CD4^+$ T cells leads to cellular anergy and/or the induction of Treg characteristics (including Foxp3 upregulation), $CD8^+$ T cell-specific suppression of mTOR activity results in the generation of memory precursor CD8+ T cells [17, 48, 49]. Indeed, it appears that $CD8^+$ Tmem cells are negatively regulated by mTOR activity. Studies in which TSC1 or TSC2 have been specifically deleted in $CD8^+$ T cells reveal TORC1 to positively enforce effector characteristics in $CD8^+$ cells while inhibiting the formation of Tmem precursors during the early-stage immune response [154–157]. Indeed, TSC2 null mice, which display constitutive mTORC1 activity, boast a $CD8^+$ T cell compartment with enhanced glycolysis and effector function [154]. Inhibition of mTORC2 can similarly enhance the development of Tmem [154].

The downstream effects of mTORC1 and mTORC2 activation are diverse. mTORC1 largely functions to promote pathways involved with anabolic processes. Through the phosphorylation of downstream S6K1 and 4E-BP1, mTORC1 promotes protein synthesis through enhanced ribosome biogenesis and the release of transcriptional inhibition, respectively [100]. mTORC2, too, appears to positively effect protein synthesis through direct association with ribosomal complexes [158]. Lipid and nucleotide synthesis also appear to be under mTOR control. The principal transcriptional regulators of lipogenesis, sterol regulatory element-binding proteins (SREBPs), are activated by both mTORC1 and mTORC2 [159–161]. mTORC1 phosphorylates the SREBP inhibitor, lipin 1, preventing the nuclear translocation of lipin 1 and the subsequent inactivation of SREBP-mediated transcription of nucleotide and lipid synthesis-related genes [161, 162]. mTORC2 has been described to transcriptionally activate SREBP genes through ATK activity, thus also promoting de novo nucleotide and lipid synthesis [160, 163].

2.4.8 *Negative Regulation of mTOR Through AMPK Activity*

The serine/threonine kinase, adenosine monophosphate-activated protein kinase (AMPK), is often cast as an mTOR antagonist. Each molecule directly inhibits the activity of the other [164–167], oscillating the preferential utilization of their respective metabolic pathways; AMKP promotes FAO, while mTOR supports a glycolytic profile. A sensor of energy stress, AMPK is activated, in part, in response to high

AMP/ATP ratios (i.e., low cellular energy), responding by inhibiting anabolic processes and upregulating specific metabolic pathways in order to restore cellular ATP levels. One such avenue for AMPK-mediated ATP production is the promotion of FAO. FAO is promoted through the upregulation of the rate-limiting enzyme for the pathway known as carnitine palmitoyltransferase 1 (CPT-1; a key component of the CTS) [168]. AMPK can also play a role in the maintenance of CPT-1 activity. By inhibiting acetyl-CoA carboxylase, which is itself a suppresser of CPT-1, AMPK further drives the process of FAO [169]. In addition to promoting the transcription of genes that facilitate lipid metabolism, AMPK activation directly inhibits glycolysis and protein synthesis through the inhibition of mTOR activity. Despite the fact that both mTOR and AMPK are induced by TCR stimulation, the latter kinase is a potent negative regulator of the former [166]. AMPK targets the mTORC1 protein, Raptor, disrupting the protein complex and diminishing mTOR kinase activity [140, 170]. Additionally, AMPK activates the TSC by phosphorylating TSC2, further driving the inhibition of mTOR [171]. Reflecting the inverse relationships between AMPK and mTOR activity, T cells lacking AMPK display enhanced mTORC1 signaling and a correspondingly elevated effector cytokine production [172].

AMPK has also been thought of as an enforcer of quiescence. It is easy to fathom that under conditions of metabolic stress (such as glucose scarcity), cell survival may depend on the resetting of the metabolic lifestyle to adapt to austere conditions. In line with this notion, AMPK was recently suggested to be dispensable for the activation and expansion of effector CD8+ T cells but important for the survival of CD8+ T cells during the so-called *contraction phase* of the immune response and the mounting of a recall response upon secondary challenge [173]. $CD8^+$ T cell-specific deletion of AMPK in vivo resulted in a dramatic loss of $CD8^+$ Tmem populations following exposure to parasitic infection [173]. Just as effector cells are dependent on a glycolytic metabolism, Tregs and Tmem rely on FAO. Inhibition of FAO by disruption of AMPK signaling curtails the generation of these key immune populations. Both thymically derived *natural* Tregs and Tregs induced in the periphery display a high degree of AMPK activity. Furthermore, in vivo administration of an AMPK activator (metformin) elevates Treg numbers in a mouse model of asthma while also diminishing cell-surface expression of GLUT1 [23]. Clearly, the reciprocal regulation of mTOR and AMPK represents a principal control mechanism to balance proinflammatory and tolerogenic cues.

2.4.9 Therapeutic Opportunities in Targeting the mTOR Signaling Pathway

Since mTOR is a central player in promoting glycolysis, the metabolic pathway of choice for potently inflammatory effector T cells, inhibiting the kinase has been explored as a means to prevent or down-modulate unwanted immune responses. Indeed, the mTOR agonist, rapamycin, has been administered in the control of the autoimmune disease, systemic lupus erythematosus (SLE) [174, 175]. Rapamycin and sirolimus have been shown to function as immunosuppressants, promoting tolerance in organ and tissue transplants [176].

While mTOR inhibition has the potential to undermine glycolysis and the proinflammatory effector T cells that are heavily dependent on this form of metabolism, somewhat paradoxically, this same strategy can bolster desirable immune responses – particularly through the enhancement of memory T cells. It is important to note that multiple studies have suggested that the memory-like subsets of $CD8^+$ T cells are superior mediators of antitumor immunity [48, 49, 177–179]. In light of this, it stands to reason that modulating the metabolic profile of $CD8^+$ T cells may be an effective strategy to improve their cancer-fighting potential. Sirolimus (rapamycin) administration can enhance the memory CD8+ T cell responses of following vaccination of nonhuman primates [180]. Furthermore, suppressing mTOR activity and glycolysis with transient, high-dose rapamycin treatments can enhance the efficacy of antitumor vaccines as well [181]. An aptamer-conjugated RNAi approach intended to specifically knock down mTORC1 activity has also been shown to suppress tumor progression in vivo [182]. Everolimus, another mTOR inhibitor, is an FDA-approved therapeutic option for the treatment of specific brain tumors (subependymal giant-cell astrocytomas) associated with inactivating TSC mutations. Studies are ongoing in proving the efficacy of everolimus in other TSC-associated diseases [126].

The role played by AMPK in modulating T cell immunity appears to lend itself to pharmaceutical intervention as well. The therapeutic benefit of metformin (N,N-dimethylbiguanide) and 5-aminoimidazole-4-carboxamide-ribonucleoside (AICAR) has been explored in their capacity to correct for insulin resistance by the AMPK-mediated upregulation of the glucose transporter, GLUT4. Both compounds are capable of activating AMPK, effectively facilitating the increased uptake of glucose in the absence of insulin signaling [183, 184]. Metformin and AICAR have both shown efficacy in the treatment of the insulin resistance associated with type-2 diabetes while also efficiently suppressing effector T cell activation and the production of IL-2 [51, 185]. Chemical activation of AMPK reduces the effector functions of murine CTLs [164]. In line with these findings, therapeutic intervention of murine EAE with metformin or AICAR dampens inflammatory cytokine production by self-reactive $CD4^+$ T cells and lessens the severity of disease progression [186]. AICAR treatment also ameliorates disease in murine colitis models as evidenced by stunted body weight loss and subdued colon pathology. Moreover, this treatment inhibited macrophage activation and reduced Th1/Th17 cell frequency and cytokine production [187, 188].

While AMPK-activating agents can negatively impact effector T cell biology, they can have opposite effects on memory T cell subsets. Illustrating this, metformin treatment has been found to markedly enhance the memory $CD8^+$ T cell response to *Listeria monocytogenes* infection in mice [17]. Additionally, others have found that metformin can enhance $CD8^+$ T cell function in tumor sites [189], an effect that likely reflects the more potent antitumor activity of the memory-like $CD8^+$ T cells expected to be bolstered by AMPK activation.

2.4.10 Hypoxia-Inducible Factor 1 (HIF-1)

Great cellular energy demand requires the strict regulation of the byproducts of T cell metabolism. The most efficient method for generating ATP is the catabolism of glucose- or fatty acid-derived pyruvate through oxidative phosphorylation. Hydrogen ions are forced against their concentration and electrochemical gradients into the acidic space between mitochondrial membranes, thereby providing a fuel source to drive the rotary mechanism of ATP synthase [190]. The free electrons generated by this process couple with hydrogen byproducts and free diatomic molecules of oxygen to form water within the mitochondria. This reaction occurs through the action of the terminal enzyme of the electron transport chain, cytochrome c oxidase (COX). COX has been revealed to play a major regulatory role in orchestrating this oxygen-dependent pyruvate metabolism, consuming the majority of the oxygen inspired by the lungs [191]. COX ceases to function under extreme oxygen deprivation (<1 mmHg) [192], a condition termed hypoxia. Specifically, hypoxia refers to tissue oxygen tensions that fall below ~10 mmHg [193]. Low oxygen tensions have a significant impact on immune cell function and proliferation, even when oxygen concentrations have not dropped sufficiently to alter oxidative phosphorylation within the mitochondria. Notably, many tissues that T cells are likely to inhibit – healthy or diseased – can be oxygen scarce. Competition for nutrients (e.g., glucose, fatty acids, amino acids) and oxygen drives characteristic alterations to T cell's metabolism. Consequentially, T cells must be highly adaptive to hypoxic environments and be able to readily alter their metabolic profile in order to function in sites of varying oxygen tensions.

2.4.11 Control of HIF-1 Activity Through the Regulation of the HIF-1α Subunit

The principal sensor of oxygen deprivation is hypoxia-inducible factor alpha (HIF-1α). HIF-1 complexes form as heterodimers of HIF-1α and HIF-1β (also called the aryl hydrocarbon nuclear translator). These factors are expressed, at least at the mRNA level, in every organ of the body [194]. HIF-1β is constitutively expressed at the protein level and functions in a pathway separate from HIF-1α that is associated with the transcriptional activation of enzymes responsible for the catabolism of xenobiotic toxins, such as the environmental pollutant, dioxin [195]. HIF-1α expression, on the other hand, is tightly regulated and is typically stabilized only under hypoxic conditions. The nuclear translocation of HIF-1α and subsequent binding to HIF-1β only occurs following the stabilization of the α-subunit [196]. Within the nucleus, HIF-1α associates with HIB-1β and functions as a transcription factor, binding to specific promotor sequences of hypoxia response elements (HREs) [197, 198] allowing for a robust cellular response to oxygen deprivation. Both subunits possess helix-loop-helix motifs and a Per-ARNT-Sim (PAS) domain that both facilitate DNA binding and α-/β-subunit dimerization [199–202]. Another α-subunit, HIF-3α, was initially reported to propagate an inhibitory function by preventing HIF-1α binding to HIF-1β though the sequestering of the β-subunit, yet recent data has suggested an additional transcriptional regulatory role for the protein [203–205].

The expression of the α-subunit, HIF-1α, is largely reliant on the local oxygen tension. As you breathe, atmospheric oxygen concentrations (159 mmHg) allow for the rapid diffusion of oxygen into the circulatory system which then pushes arterial blood throughout the body, dispensing oxygen where oxygen tensions are low. In tissues, oxygen concentrations are considerably lower (~15–38 mmHg) than in circulating arterial (75–100 mmHg) and venous blood (≥40 mmHg) [206, 207]. Moreover, within the architecture of tissues, oxygen tension can vary dramatically depending on the proximity of a cell to local capillary beds [208, 209]. Although oxygen tensions are lower within tissues than in circulation, the oxygen and nutrient concentrations therein are sufficiently replete due to elaborate networks of capillary beds. Oxygen tensions are particularly low in lymphoid tissues (~8–35 mmHg) where immune cells are derived and mature. The same is generally true of sites of ongoing inflammation or tumor growth, implicating the importance of HIF-1α in immune cell development and function. During an inflammatory response, competition for available oxygen and nutrients increases as immune cells infiltrate and expand in peripheral tissue sites, resulting in nutrient-/oxygen-destitute microenvironments. Within a tumor, this and the prodigious consumption of glucose and oxygen by tumor cells coupled with the inability of angiogenesis to keep pace with the intratumoral demand result in regions of extreme hypoxia [210, 211]. It is the stabilization of HIF-1α expression, largely consequential of hypoxic conditions (> 10 mmHg) alone, that drives much of the cellular adaptations to oxygen deprivation.

HIF-1α expression is largely controlled posttranslationally by prolyl hydroxylases (PHD), which catalyze the formation of L-hydroxyproline by incorporating oxygen to two specific proline residues (P402 and P564) on HIF-1α [212]. Hydroxylation of proline residues increases the affinity of these sites on HIF-1α for the active site of the von Hippel-Lindau/Elongin-C E3 ligase (VHL) complex by nearly 1,000-fold. The bound VHL transfers ubiquitin molecules to L-hydroxyproline residues, marking HIF-1α for proteome-dependent proteolysis [207, 213]. Thus, the repression of HIF-1α hinges on the presence of free oxygen within the cell, and as such, HIF-1α becomes stabilized under hypoxic conditions [214]. Under normoxic conditions (i.e., non-hypoxic; >10 mmHg), HIF-1α is rapidly degraded through the aforementioned process.

However, this oxygen-dependent degradation has been shown to be superseded in $CD4^+$ and $CD8^+$ T cells where high mTORC1 expression is associated with elevated HIF-1α levels even in normoxic conditions [18, 80, 215, 216]. Indeed, TCR ligation and costimulation stabilizes the expression of HIF-1α, possibly through mTORC1-facilitated enhancement HIF-1α protein synthesis [216]. It should be noted that decreased intracellular oxygen tensions resulting from cellular activation may also contribute to the aforementioned HIF-1α stabilization [207].

Proinflammatory cytokines, too, appear to be capable of augmenting HIF-1α expression in T cells. Specifically, interleukin 6 (IL-6) induces HIF-1α stabilization in vitro through the activation of signal transducer and activator of transcription 3 (STAT3)-driven transcriptional regulation [81]. Additionally, numerous studies have elucidated the role of the receptor of activated protein kinase C (RACK1) in the proteasomal degradation of HIF-1α in an oxygen-/PHD-/VHL-independent manner. RACK1 competes with heat-shock protein 90 (HSP90) for binding to the PAS domain of HIF-1α. Upon RACK1/Elongin-C E3 ubiquitin ligase association, RACK1 facilitates the ubiquitination and subsequent proteome-dependent proteolysis of HIF-1α [217–219]. These data suggests various levels of HIF-1α control outside of oxygen tension alone.

2.4.12 Downstream Effects of HIF-1α Stabilization in T Cells

The cellular effects of HIF-1α stabilization are expansive, resulting in the expression of many genes implicated in the control of glucose and pyruvate metabolism in effector T cells. Following TCR ligation and costimulation, T cells undergo clonal expansion and transcriptome reprograming to express effector molecules including cytotoxins, cytokines, and cell-surface molecules. This robust increase in cellular proliferation and protein anabolism demands a matched increase in metabolite uptake to provide components for cellular machinery and free energy to catalyze chemical reactions. HIF-1α expression has been shown to enhance the cell-surface expression of the glucose transporter, Glut1, in $CD4^+$ and $CD8^+$ T cells, supporting the development of proinflammatory T cell lineages [42, 43]. Indeed, within $CD4^+$ T cells, HIF-1 transcriptionally directs the upregulation of the enzymatic machinery required for glycolysis, prioritizing this metabolic pathway which is favored by Teff cells [18]. Interestingly, while c-Myc and mTOR are centrally important for engaging this metabolic reprogramming of recently activated cells, HIF-1 expression appears dispensable at this early juncture. Although HIF-1 controls the transcription of numerous glycolytic genes, HIF-1-deficiency in T cells does not inhibit activation-induced Teff proliferation [80, 81]. It is believed, however, that HIF-1 instead plays an important role in sustaining Teff populations and their glycolytic metabolism following activation.

In $CD4^+$ T cells, HIF-1α reciprocally regulates the key transcriptional regulators of the differentiation programs responsible for Th17 and Treg development, RAR-related orphan receptor gamma (RORγt) and Foxp3, respectively [18, 81]. While Th17 cells are notoriously proinflammatory and functionally share very little with the characteristically immunosuppressive Treg lineage, they nevertheless can be derived in vitro and in peripheral tissues by differentiation pathways that partially overlap, namely, through transforming growth factor beta (TGF-β) receptor signaling. It is the sensing of additional proinflammatory cytokines, like IL-6, that simultaneously promotes the establishment of the Th17 lineage and inhibits the Treg phenotype. Enhanced HIF-1α expression resulting from TCR ligation/IL-6-

mediated STAT3 activation can support Th17 differentiation irrespective of local oxygen tensions. Stabilized HIF-1α subunits directly promote the expression of RORγt in differentiating CD4+ T cells, supporting the adoption of a Th17 fate. HIF-1 further cooperates with RORγt in a transcription factor complex to promote transcription/translation of several Th17-associated genes [81]. Another facet of HIF-1's influence over CD4+ T cell fate decisions appears to involve the downregulation of Foxp3, which can prevent commitment to an induced Treg fate. Indeed, HIF-1 promotes the proteasome-dependent degradation of Foxp3 protein during T cell differentiation. Reflecting these multiple roles of HIF-1, CD4+ T cells from conditional HIF-1α knockout mice are defective in Th17 commitment and are more prone to upregulate Foxp3 in vitro. Moreover, T cell-specific knockout of HIF-1 reduces disease burden in EAE relative to wild type controls. Such protection was seen alongside impaired Th17 immunity and the accumulation of Foxp3$^+$ Tregs when T cells lack HIF-1 expression [18, 81].

The role of HIF-1 in established Tregs appears to be complex. Indeed, HIF-1 deficiency during CD4$^+$ T cell differentiation in vitro results in an accumulation of Foxp3$^+$ T cells [18, 81], suggesting that HIF-1 opposes Treg differentiation. However, other studies suggest that HIF-1α contributes positively to Tregs by promoting transcription at the *Foxp3* and *Ctla4* genes [220]. Moreover, the suppressive capacity of established Tregs in vivo appears to benefit from HIF-1 expression under some conditions as Tregs from HIF-1-deficient mice were found to be less effective suppressors of colitis than wild-type controls [220]. In contrast, other studies have shown that heightened levels of HIF-1 activity can be detrimental to the normal suppressive functions of Tregs. Specific deletion of key components of the HIF-1α degradation machinery (e.g., Deltex, VHL) in Tregs induces robust expression of HIF-1 that is accompanied by unstable expression of Foxp3 and compromised suppressive function [221, 222]. In one study, these ineffective Tregs were found to have uncharacteristic, Th1 effector-like attributes that resulted from HIF-1-mediated upregulation of Teff cell genes (e.g., *Ifng*) and metabolic genes responsible for promoting glycolysis [23, 222]. These findings suggest that a precise balance of HIF-1 expression may be required for optimal Treg function.

HIF-1α also influences the development of CD8$^+$ cytotoxic T lymphocytes (CTLs). The cellular activities and functions of activated CD8$^+$ T cells, like those of their CD4$^+$ counterparts, are fueled primarily by a glycolysis-dominated metabolism [18, 81, 223]. Newly activated CD8+ T cells upregulate HIF-1 through the action of mTORC1. This activation begins with TCR signaling/CD28 costimulation, which induces c-Myc expression and subsequent activation of c-Myc-controlled genes that are crucial for efficient cellular activation and expansion. A second wave of augmented gene expression, coordinated by HIF-1 and other transcriptional regulators (e.g., AP4, IRF4), continues to enforce a glycolytic metabolic profile through the upregulation of enzymes involved in glycolysis and glutaminolysis [224].

Additionally, HIF-1α stabilization appears to promote the establishment of effector functions within CTLs. HIF-1 drives the expression of key effector molecules, such as perforin and granzymes, which partially account for the cytotoxic capacity of these cells [80]. Genetic deletion of VHL (the complex responsible for oxygen-dependent HIF-1α degradation) in vivo results in elevated expression of granzyme B and effector cytokines such as IFN-γ and TNF-α. These enhancements, which result from elevated HIF-1 activity, translate into improved CTL-mediated immunity in models of melanoma and viral infection [223]. Boosting HIF-1 levels by exposing cells to hypoxic culture conditions can also enhance the expression of Glut1 and effector molecules [80], further supporting the role of HIF-1 activity in promoting glycolysis and CTL function. Correspondingly, HIF-1 deficiency impairs the acquisition of these effector molecules.

In keeping with the theme of contrasting metabolic lifestyle and requirements for Teff and Tmem lineages, HIF-1 appears to negatively impact memory $CD8^+$ T cells. Chemical inhibition of glycolysis (and HIF-1 expression) promotes a memory CD8+ phenotype instead of a short-lived, effector fate. Indeed, these conditions result in a more memory-like phenotype as evidenced by low levels of KLRG1 and an abundance of surface CD62L. CD62L, also called L-selectin, is expressed most notably in naïve and memory T cells, facilitating the trafficking of these cells to secondary lymphoid organs. CD62L appears a key checkpoint of T cell function as its expression has been shown to control lytic activity of tumor reactive T cells and the memory response to previously encountered antigen [225, 226]. Conversely, forced utilization of glycolysis in $CD8^+$ T cells restricts the generation of memory cells. Interestingly, memory $CD8^+$ T cells that accumulate during inhibition of glycolysis and HIF-1 downregulation appear to be more potent mediators of antitumor immunity than their short-lived Teff counterparts [20].

2.4.13 Therapeutic Opportunities in Targeting the HIF-1 Signaling Pathway

Interest in generating potent HIF inhibitors has been understandably whetted by the possibility of simultaneously disrupting its multiple tumor-abetting functions (e.g., upregulation of glycolytic enzymes, promotion of angiogenesis, facilitation of metastasis). Indeed, HIF-1 is thought to facilitate tumor growth in its characteristically oxygen-deprived microenvironments, and, in keeping, HIF-1 expression trends upward as certain brain tumors increase in grade [227]. Currently, several HIF-1α inhibitors are undergoing clinical or preclinical trials. The small-molecule inhibitor, topotecan, is a well-established anti-neoplastic agent that inhibits HIF-1α expression through a topoisomerase I-mediated mechanism [228, 229]. A daily low-dose regimen of topotecan was revealed to destabilize HIF-1α expression and thus diminish tumor growth and angiogenesis in murine tumor models [228, 230]. Another molecule being explored as a molecular HIF antagonist is PX-478, which showed early promise in treating human tumor xenografts via diminishing HIF-1α and Glut1 expression [231–234].

In keeping with the notion that HIF-1 stabilization promotes cellular survival, activation of HIF-1α within $CD8^+$ T cells has been shown to prevent immune cell exhaustion, a consequence of chronic viral infection or persistent tumor growth. Loss of VHL derails HIF regulation and thereby enhances effector T cell potential and longevity [223]. Within $CD4^+$ T cells, too, HIF-1 has revealed itself to be targetable. Digoxin, a cardiac glycoside, has been shown to inhibit HIF-1α expression and disrupt HIF-mediated transcriptional activation [235]. Administration of digoxin or its derivatives to naïve $CD4^+$ T cell cultures effectively hinders the skewing of the proinflammatory Th17 subset. Indeed, digoxin treatment in in vivo models of EAE markedly suppressed disease progression [236]. Although digoxin has been shown to directly inhibit the Th17-promoting transcription factor RORγt, digoxin-mediated inhibition of HIF-1α mechanisms, namely, the adoption of a glycolytic profile, is likely another contributing factor to the ineffectual generation of Th17 cells [18, 81].

2.4.14 Peroxisome Proliferator-Activated Receptor (PPAR)

Peroxisome proliferator-activated receptors (PPARs) represent a group of nuclear receptor proteins that are activated by fatty acids and initiate gene transcription to coordinate both lipid and carbohydrate metabolism with cellular proliferation and differentiation. Activated PPARs associate with retinoid x receptors (RXRs) to transcriptionally regulate genes with PPAR response element (PPRE) promotor regions [237, 238]. Three PPAR isoforms exist – α, δ (occasionally defined as β or β/δ), and γ – which possess specific tissue expression patterns and differ in their physiological consequences following ligand binding. Ligation of PPARs by fatty acids can promote the dissociation of corepressors and association of co-activators, facilitating the direct binding of PPAR to DNA and the activation of gene transcription. Furthermore, co-activator complexes have been shown to direct chromatin remodeling through histone acetyltransferase activity, further promoting gene transcription [239–244].

While little is known about the direct mechanistic role of PPARδ (also called PPARβ or PPARβ/δ), the functions of the α and γ isoforms have been described at length in both immune and nonimmune cells. PPARα expression is constrained to areas of elevated metabolic activity (e.g., the liver, brown adipose, cardiac and skeletal muscle fibers, intestinal mucosa) and its ligand-dependent activation is associated with augmented fatty acid metabolism. PPARα activation induces the transcriptional upregulation of genes that facilitate intracellular fatty acid transport into sites of β-oxidation as well as the enzymes that are associated with this catabolic pathway [245, 246]. PPARγ, on the other hand, principally promotes anabolic processes, participating in adipogenesis (γ expression remains elevated in both brown and white adipose), lipid biosynthesis, and insulin sensitivity [247]. PPARδ appears to augment the accumulation of lipids within a cell through the repression of metabolic mechanisms and the increased expression of influx-specific transport molecules [248]. Importantly, while other isoforms require corepressor release to facilitate DNA binding, PPARδ has been shown to associate with PPRE promotor regions irrespective of corepressor binding [249]. This unique ability of PPARδ could represent a specific regulatory function, regulating the transcriptional activity of lipid metabolism-associated PPARα and γ.

PPARs are expressed in a variety of immune cell subtypes – e.g., T and B cells, macrophages, and dendritic cells – and have been reported to primarily support anti-inflammatory and tolerogenic activity [239, 246, 247, 250–253]. Indeed, PPARs have been revealed to transcriptionally downregulate several genes associated with immune cell activation and T cell effector function in inflammation. Notably, PPARs control the activity of T cell activation- and proliferation-related transcription factors that include activator protein-1 (AP-1), nuclear factor of activated T cells (NFAT), nuclear factor-κB (NF-κB), and signal transducers and activators of transcription (STATs) [251, 254–256], as well as expression of the stimulatory cytokine, IL-2 [257].

2.4.15 Downstream Effects of PPAR Ligation in Effector T Cells

A consequence of the PPAR-mediated promotion of lipid metabolism (at the expense of glycolytic metabolism) is the downregulation of effector responses, supporting the notion that PPARs function indirectly as immunosuppressors. Indeed, Th1-/Th17-mediated EAE disease is exacerbated in mice with diminished PPARγ (PPAR$\gamma^{+/-}$) as compared to their wild-type littermates. This was revealed by enhanced neuron demyelination and ex vivo IFNγ-production from CNS-infiltrating Th1 cells in PPAR$\gamma^{+/-}$ mice. Correspondingly, PPARγ antagonists also enhanced the clinical scores of EAE symptoms as well as the induction of cellular activation and IFNγ-production, suggesting an augmented proinflammatory response [258, 259]. PPARγ *agonists*, on the other hand, ameliorated EAE severity as evidenced by a reduction of leukocyte infiltration into the CNS, a diminishing of proinflammatory cytokine release following in vitro reactivation, and an overall reduced disease duration [260, 261]. Considering the significant Th17 component in EAE disease progression, PPARγ activity was hypothesized to interfere with Th17 differentiation. Interestingly, PPARγ agonists specifically interfered with the induction of RORγt expression in vitro, without directly altering the expression of other pro- and anti-inflammatory transcription factors – e.g., T-bet (Th1), GATA3 (Th2), or Foxp3 (Treg) [262]. While these data suggest that PPARγ does not directly interfere with Th1 skewing in vitro, significant studies have illustrated the PPAR-mediated control of IL-12 production and signaling in vivo, which in turn would effectively diminish Th1 responses in a physiological setting [263, 264]. Indeed, the previous reports of augmented Th1 effector responses upon diminished PPARγ expression [258, 265] have been well supported [263, 266, 267].

Interestingly, while the transcriptional regulation of *RORc* (the gene encoding RORγt) falls under the control of PPARγ, expression of T-bet (the Th1-associated transcription factor) appears to be mediated by PPARδ. Agonists of the δ isoform result in similar amelioration of CNS inflammation and demyelination in EAE, although not through the reduction of antigen-specific Teff cellular proliferation. Instead, PPARδ agonists hindered the production of effector cytokines, IL-12 and IL-23, resulting in diminished Th1/Th17 differentiation and thus control of T helper-specific cytokine production – e.g., IFNγ, IL-17 – and the expression of the Th1-favoring T-bet [256, 266]. Notably, PPARδ agonists did not altogether down-regulate cytokine expression, but propagated the production of anti-inflammatory cytokines, IL-4 and IL-10, further suppressing the effector functions of some T helper subsets. PPARδ knockout results in the converse effect, enhancing Th1/Th17 responses, prolonging the induction phase of EAE, and worsening the disease burden [266].

While PPARδ and γ agonists directly regulate the effector functions of Th17 cells, the loss of PPARα does not appear to alter Th17 effector cytokine (IL-17) production, suggesting that there are differing methods of proinflammatory regulation between PPAR isoforms [256, 262, 266]. Ligand-dependent activation of PPARα effectively protects against autoimmunity, but apparently through the control of a different immune axis. Indeed, PPARα activation prevents the development of severe EAE in mice by reducing the generation of potentially pathologic Th1 cells in favor of an augmented Th2 cell population. This is accomplished through the direct regulation of the Th1- and Th2-specific transcription factors, T-bet and GATA3, respectively, resulting in diminished IFNγ (Th1) and heightened IL-4 and IL5 (Th2) [268, 269]. Also, similar to PPARγ, activation of PPARα also suppresses the production of proinflammatory cytokines by innate cells within the CNS – e.g., microglia and astrocytes [270, 271].

Recently, a role for PPARγ was revealed in a unique subset of Treg cells. Visceral adipose tissues (VATs) house a population of Treg cells that suppress inflammation specifically within this niche. Interestingly, their presence and function are linked to a reversal of insulin resistance. VAT Tregs were found to uniquely express PPARγ, and they display a gene expression profile significantly distinct from their lymphoid tissue-dwelling counterparts. Mice lacking PPARγ specifically within the Treg compartment (*Foxp3*-cre/*Pparg*$^{\text{flox/flox}}$) have lower VAT Treg frequencies and are susceptible to organismal metabolic dysfunction [272]. It may be possible that PPAR activation contributes to the heightened lipid-dominated metabolic profile that imparts VAT Tregs with the survival advantage optimal for their unique microenvironment.

2.4.16 Therapeutic Opportunities in Targeting the PPAR Signaling Pathway

Although the complete mechanistic picture of PPAR-mediated immunosuppression remains to be resolved, it is clear that manipulating the activity of any PPAR isoforms alters a cell's metabolic ability, resulting in support of either pro- or anti-inflammatory conditions. PPAR activation drives enhanced lipid anabolism and curtails the glycolytic profile required for complete Teff function. Inhibition of the PPAR pathway results in the antithesis, namely, a prolonged and more intense inflammatory response. While the involvement of PPAR in lipid metabolism-favoring Treg populations was not directly addressed in many of the aforementioned studies, there is reason to believe that the immune-dampening effects of PPAR activity may also involve these tolerance-promoting cells. For instance, the suppression-inducing effects of the PPARγ agonists 15d-PGJ2 and Ciglitazone can be heightened by the co-administration of retinoic acid, a known promoter of Treg-mediated function [273]. Ciglitazone was revealed to promote the conversion of Teff cells to induced Tregs in vitro through the enhanced secretion of retinoic acid from DCs and a corresponding decrease in pro-effector IL-12 production. Moreover, high-dose Ciglitazone was sufficient to induce this conversion in a DC-independent manner [273]. This synthetic ligand also was found to offer a protective effect in a model of graft versus host disease by enhancing Tregs, further supporting the notion that PPAR activation induces tolerogenic conditions [274]. Moreover, PPARα or γ ligation in human $CD4^+$ T cells treated with TGF-β can induce the stabilized expression of Foxp3 and a Treg phenotype [275]. It is also noteworthy that PPAR agonists can directly activate AMPK [169], a metabolic regulator important for inhibiting the adoption of the glycolytic metabolism that can negatively affect the generation of Tregs.

PPAR agonists are currently being clinically vetted for their ability to suppress unwanted inflammation. Experimental ligation of PPAR has been shown to successfully ameliorate immunopathology in several animal models of autoimmunity [261, 276–279], supporting the notion that PPAR activation can inhibit proinflammatory immune responses. The PPAR ligands, rosiglitazone and pioglitazone, were found to diminish Th2- and Th17-specific cytokine production and afford protection from asthma. Considering that exogenous IL-17 negates the protective effect of these drugs [280], rosiglitazone and pioglitazone may have the potential to specifically combat Th17-driven pathologies. In one multiple sclerosis trial, pioglitazone (in combination with IFNβ-1α) yielded promising MRI results (i.e., less gray matter atrophy and trends toward decreased lesions) in a 1-year follow-up [281]. Also, encouragingly, in a randomized, placebo-controlled study, rosiglitazone was found to provide a promising strategy to ameliorate ulcerative colitis in patients suffering from the disease [282].

2.5 Concluding Statements

The mechanisms that control the balance of proinflammatory and tolerogenic T cell lineages are closely tied to the activity and the maintenance of specific metabolic pathways. Activating stimuli necessitate increased absorption and utilization of metabolites to facilitate costly cellular proliferation, effector function, and control of apoptosis in many cell types. Yet the particular nutrients absorbed and metabolic enzymes expressed to meet these demands can vary between subsets of T cells. Teff cells, responding to TCR and costimulatory cascades, will principally metabolize glucose through aerobic glycolysis, while shunting off glycolytic intermediates for anabolic processes and maintaining some meager levels of OXPHOS. Tmem and Treg prioritize lipid metabolism, procuring fatty acids for the majority of their energetic needs and facilitating their longevity even in sites of inflammation and tumor growth.

The key regulators of the metabolic programming described here control the expression of genes related to nutrient absorption and metabolism. As such, the manipulating the function of these proteins has proven to be an effective level of control over immune responses in disease settings. In brief, c-Myc expression is dramatically upregulated shortly after cellular activation, promoting a glucose absorption and glycolytic metabolic profile and inducing the expression of transcription factors that maintain this glycolysis-dominant profile (e.g., mTOR, HIF-1α, AP4, IRF4, Bim). Both mTOR and AMPK are upregulated upon TCR and growth hormone receptor ligation and the interplay of their kinase activity promotes either anabolic machinery, cellular growth, and proliferation (mTOR) or the regeneration of energy storages through enhanced catabolism (AMPK). Lipid-mediated ligation of PPARs activate the transcription of genes that directly upregulate FAO, facilitating the distinct functions and longevity associated with Tmem and Treg.

Intense study of the mechanisms that control T cell immunity have revealed pathways that can be manipulated for effective control of immune disease and disorders. Although great bounds have been made in delineating these systems, questions remain concerning how they might be more effectively manipulated to provide therapeutic benefit. As laboratory technology continues to advance, a more intimate understanding of the interplay between c-Myc, mTOR, AMPK, HIF-1α, PPARs, etc. in relation to T cell function will no doubt soon come to light.

Acknowledgment Our research is supported by the grants from the Bloomberg-Kimmel Institute, the Melanoma Research Alliance (Established Investigator Award), the National Institutes of Health (RO1AI099300 and RO1AI089830), Department of Defense (PC130767), "Kelly's Dream" Foundation, the Janey Fund, the Seraph Foundation, gifts from Bill and Betty Topecer and Dorothy Needle, and the Roswell Park Alliance Foundation. FP is a Stewart Trust Scholar.

References

1. Wang R et al (2011) The transcription factor Myc controls metabolic reprogramming upon T lymphocyte activation. Immunity 35(6):871–882
2. Zhu J, Yamane H, Paul WE (2010) Differentiation of effector CD4 T cell populations (*). Annu Rev Immunol 28:445–489
3. Littman DR, Rudensky AY (2010) Th17 and regulatory T cells in mediating and restraining inflammation. Cell 140(6):845–858
4. Rudensky AY (2011) Regulatory T cells and Foxp3. Immunol Rev 241(1):260–268
5. Vignali DA, Collison LW, Workman CJ (2008) How regulatory T cells work. Nat Rev Immunol 8(7):523–532

6. Nishikawa H, Sakaguchi S (2010) Regulatory T cells in tumor immunity. Int J Cancer 127(4):759–767
7. Josefowicz SZ, Lu LF, Rudensky AY (2012) Regulatory T cells: mechanisms of differentiation and function. Annu Rev Immunol 30:531–564
8. Schell JC, Rutter J (2013) The long and winding road to the mitochondrial pyruvate carrier. Cancer Metab 1(1):6
9. Warburg O (1956) On the origin of cancer cells. Science 123(3191):309–314
10. Fox CJ, Hammerman PS, Thompson CB (2005) Fuel feeds function: energy metabolism and the T-cell response. Nat Rev Immunol 5(11):844–852
11. Vander Heiden MG, Cantley LC, Thompson CB (2009) Understanding the Warburg effect: the metabolic requirements of cell proliferation. Science 324(5930):1029–1033
12. Pearce EL, Pearce EJ (2013) Metabolic pathways in immune cell activation and quiescence. Immunity 38(4):633–643
13. Pearce EL et al (2013) Fueling immunity: insights into metabolism and lymphocyte function. Science 342(6155):1242454
14. Segal AW (2008) The function of the NADPH oxidase of phagocytes and its relationship to other NOXs in plants, invertebrates, and mammals. Int J Biochem Cell Biol 40(4):604–618
15. Wahl DR et al (2012) Distinct metabolic programs in activated T cells: opportunities for selective immunomodulation. Immunol Rev 249(1):104–115
16. De Boer RJ, Homann D, Perelson AS (2003) Different dynamics of CD4+ and CD8+ T cell responses during and after acute lymphocytic choriomeningitis virus infection. J Immunol 171(8):3928–3935
17. Pearce EL et al (2009) Enhancing CD8 T-cell memory by modulating fatty acid metabolism. Nature 460(7251):103–107
18. Shi LZ et al (2011) HIF1alpha-dependent glycolytic pathway orchestrates a metabolic checkpoint for the differentiation of TH17 and Treg cells. J Exp Med 208(7):1367–1376
19. Dardalhon V et al (2008) Role of Th1 and Th17 cells in organ-specific autoimmunity. J Autoimmun 31(3):252–256
20. Sukumar M et al (2013) Inhibiting glycolytic metabolism enhances CD8+ T cell memory and antitumor function. J Clin Invest 123(10):4479–4488
21. Schwenk RW et al (2010) Fatty acid transport across the cell membrane: regulation by fatty acid transporters. Prostaglandins Leukot Essent Fat Acids 82(4–6):149–154
22. Lochner M, Berod L, Sparwasser T (2015) Fatty acid metabolism in the regulation of T cell function. Trends Immunol 36(2):81–91
23. Michalek RD et al (2011) Cutting edge: distinct glycolytic and lipid oxidative metabolic programs are essential for effector and regulatory CD4+ T cell subsets. J Immunol 186(6):3299–3303
24. Delgoffe GM et al (2011) The kinase mTOR regulates the differentiation of helper T cells through the selective activation of signaling by mTORC1 and mTORC2. Nat Immunol 12(4):295–303
25. Kaech SM, Cui W (2012) Transcriptional control of effector and memory CD8+ T cell differentiation. Nat Rev Immunol 12(11):749–761
26. van der Windt GJ et al (2012) Mitochondrial respiratory capacity is a critical regulator of CD8+ T cell memory development. Immunity 36(1):68–78
27. van der Windt GJ et al (2013) CD8 memory T cells have a bioenergetic advantage that underlies their rapid recall ability. Proc Natl Acad Sci U S A 110(35):14336–14341
28. Fraser KA et al (2013) Preexisting high frequencies of memory CD8+ T cells favor rapid memory differentiation and preservation of proliferative potential upon boosting. Immunity 39(1):171–183
29. Nicholls DG (2009) Spare respiratory capacity, oxidative stress and excitotoxicity. Biochem Soc Trans 37(Pt 6):1385–1388
30. Newsholme EA, Crabtree B, Ardawi MS (1985) The role of high rates of glycolysis and glutamine utilization in rapidly dividing cells. Biosci Rep 5(5):393–400

31. van Stipdonk MJ et al (2003) Dynamic programming of CD8+ T lymphocyte responses. Nat Immunol 4(4):361–365
32. Rathmell JC et al (2000) In the absence of extrinsic signals, nutrient utilization by lymphocytes is insufficient to maintain either cell size or viability. Mol Cell 6(3):683–692
33. Newell MK et al (2006) Cellular metabolism as a basis for immune privilege. J Immune Based Ther Vaccines 4:1
34. Macintyre AN et al (2014) The glucose transporter Glut1 is selectively essential for CD4 T cell activation and effector function. Cell Metab 20(1):61–72
35. Mueckler M, Thorens B (2013) The SLC2 (GLUT) family of membrane transporters. Mol Asp Med 34(2–3):121–138
36. Scheepers A, Joost HG, Schurmann A (2004) The glucose transporter families SGLT and GLUT: molecular basis of normal and aberrant function. JPEN J Parenter Enteral Nutr 28(5):364–371
37. Wood IS, Trayhurn P (2003) Glucose transporters (GLUT and SGLT): expanded families of sugar transport proteins. Br J Nutr 89(1):3–9
38. Qu Q et al (2016) Fatty acid oxidation and carnitine palmitoyltransferase I: emerging therapeutic targets in cancer. Cell Death Dis 7:e2226
39. Chakrabarti R et al (1994) Changes in glucose transport and transporter isoforms during the activation of human peripheral blood lymphocytes by phytohemagglutinin. J Immunol 152(6):2660–2668
40. Frauwirth KA et al (2002) The CD28 signaling pathway regulates glucose metabolism. Immunity 16(6):769–777
41. Jacobs SR et al (2008) Glucose uptake is limiting in T cell activation and requires CD28-mediated Akt-dependent and independent pathways. J Immunol 180(7):4476–4486
42. Cham CM et al (2008) Glucose deprivation inhibits multiple key gene expression events and effector functions in CD8+ T cells. Eur J Immunol 38(9):2438–2450
43. Cham CM, Gajewski TF (2005) Glucose availability regulates IFN-gamma production and p70S6 kinase activation in CD8+ effector T cells. J Immunol 174(8):4670–4677
44. Nakaya M et al (2014) Inflammatory T cell responses rely on amino acid transporter ASCT2 facilitation of glutamine uptake and mTORC1 kinase activation. Immunity 40(5):692–705
45. Sinclair LV et al (2013) Control of amino-acid transport by antigen receptors coordinates the metabolic reprogramming essential for T cell differentiation. Nat Immunol 14(5):500–508
46. Verrey F et al (2004) CATs and HATs: the SLC7 family of amino acid transporters. Pflugers Arch 447(5):532–542
47. Hayashi K et al (2013) LAT1 is a critical transporter of essential amino acids for immune reactions in activated human T cells. J Immunol 191(8):4080–4085
48. Rao RR et al (2010) The mTOR kinase determines effector versus memory CD8+ T cell fate by regulating the expression of transcription factors T-bet and Eomesodermin. Immunity 32(1):67–78
49. Araki K et al (2009) mTOR regulates memory CD8 T-cell differentiation. Nature 460(7251):108–112
50. Usui T et al (2006) Brasilicardin a, a natural immunosuppressant, targets amino acid transport system L. Chem Biol 13(11):1153–1160
51. Zheng Y et al (2009) Anergic T cells are metabolically anergic. J Immunol 183(10):6095–6101
52. Yan Z, Banerjee R (2010) Redox remodeling as an immunoregulatory strategy. Biochemistry 49(6):1059–1066
53. Dahlgren C, Karlsson A (1999) Respiratory burst in human neutrophils. J Immunol Methods 232(1–2):3–14
54. Sies H (1997) Oxidative stress: oxidants and antioxidants. Exp Physiol 82(2):291–295
55. Thoren FB et al (2007) Cutting edge: Antioxidative properties of myeloid dendritic cells: protection of T cells and NK cells from oxygen radical-induced inactivation and apoptosis. J Immunol 179(1):21–25

56. Cemerski S, van Meerwijk JP, Romagnoli P (2003) Oxidative-stress-induced T lymphocyte hyporesponsiveness is caused by structural modification rather than proteasomal degradation of crucial TCR signaling molecules. Eur J Immunol 33(8):2178–2185
57. Mougiakakos D, Johansson CC, Kiessling R (2009) Naturally occurring regulatory T cells show reduced sensitivity toward oxidative stress-induced cell death. Blood 113(15):3542–3545
58. Yan Z, Garg SK, Banerjee R (2010) Regulatory T cells interfere with glutathione metabolism in dendritic cells and T cells. J Biol Chem 285(53):41525–41532
59. Munn DH, Mellor AL (2013) Indoleamine 2, 3 dioxygenase and metabolic control of immune responses. Trends Immunol 34(3):137–143
60. Prodinger J et al (2016) The tryptophan metabolite picolinic acid suppresses proliferation and metabolic activity of CD4+ T cells and inhibits c-Myc activation. J Leukoc Biol 99(4):583–594
61. Baban B et al (2009) IDO activates regulatory T cells and blocks their conversion into Th17-like T cells. J Immunol 183(4):2475–2483
62. Sharma MD et al (2007) Plasmacytoid dendritic cells from mouse tumor-draining lymph nodes directly activate mature Tregs via indoleamine 2,3-dioxygenase. J Clin Invest 117(9):2570–2582
63. Sharma MD et al (2013) An inherently bifunctional subset of Foxp3+ T helper cells is controlled by the transcription factor eos. Immunity 38(5):998–1012
64. Mellor AL, Munn DH (2004) IDO expression by dendritic cells: tolerance and tryptophan catabolism. Nat Rev Immunol 4(10):762–774
65. Qian F et al (2009) Inhibition of tumor cell growth, invasion, and metastasis by EXEL-2880 (XL880, GSK1363089), a novel inhibitor of HGF and VEGF receptor tyrosine kinases. Cancer Res 69(20):8009–8016
66. Carlson TJ et al (2014) Halofuginone-induced amino acid starvation regulates Stat3-dependent Th17 effector function and reduces established autoimmune inflammation. J Immunol 192(5):2167–2176
67. Lee CM, Reddy EP (1999) The v-myc oncogene. Oncogene 18(19):2997–3003
68. Boxer LM, Dang CV (2001) Translocations involving c-myc and c-myc function. Oncogene 20(40):5595–5610
69. Erikson J et al (1983) Transcriptional activation of the translocated c-myc oncogene in burkitt lymphoma. Proc Natl Acad Sci U S A 80(3):820–824
70. Dang CV (1999) C-Myc target genes involved in cell growth, apoptosis, and metabolism. Mol Cell Biol 19(1):1–11
71. Douglas NC et al (2001) Defining the specific physiological requirements for c-Myc in T cell development. Nat Immunol 2(4):307–315
72. Mateyak MK et al (1997) Phenotypes of c-Myc-deficient rat fibroblasts isolated by targeted homologous recombination. Cell Growth Differ 8(10):1039–1048
73. Chou C et al (2014) C-Myc-induced transcription factor AP4 is required for host protection mediated by CD8+ T cells. Nat Immunol 15(9):884–893
74. Best JA et al (2013) Transcriptional insights into the CD8(+) T cell response to infection and memory T cell formation. Nat Immunol 14(4):404–412
75. Nie Z et al (2012) C-Myc is a universal amplifier of expressed genes in lymphocytes and embryonic stem cells. Cell 151(1):68–79
76. Hann SR, Eisenman RN (1984) Proteins encoded by the human c-myc oncogene: differential expression in neoplastic cells. Mol Cell Biol 4(11):2486–2497
77. Hann SR et al (1983) Proteins encoded by v-myc and c-myc oncogenes: identification and localization in acute leukemia virus transformants and bursal lymphoma cell lines. Cell 34(3):789–798
78. Man K et al (2013) The transcription factor IRF4 is essential for TCR affinity-mediated metabolic programming and clonal expansion of T cells. Nat Immunol 14(11):1155–1165
79. Yao S et al (2013) Interferon regulatory factor 4 sustains CD8(+) T cell expansion and effector differentiation. Immunity 39(5):833–845

80. Finlay DK et al (2012) PDK1 regulation of mTOR and hypoxia-inducible factor 1 integrate metabolism and migration of CD8+ T cells. J Exp Med 209(13):2441–2453
81. Dang EV et al (2011) Control of T(H)17/T(reg) balance by hypoxia-inducible factor 1. Cell 146(5):772–784
82. Nayar R et al (2014) Graded levels of IRF4 regulate CD8+ T cell differentiation and expansion, but not attrition, in response to acute virus infection. J Immunol 192(12):5881–5893
83. Brustle A et al (2007) The development of inflammatory T(H)-17 cells requires interferon-regulatory factor 4. Nat Immunol 8(9):958–966
84. Cretney E et al (2011) The transcription factors blimp-1 and IRF4 jointly control the differentiation and function of effector regulatory T cells. Nat Immunol 12(4):304–311
85. Lohoff M et al (2002) Dysregulated T helper cell differentiation in the absence of interferon regulatory factor 4. Proc Natl Acad Sci U S A 99(18):11808–11812
86. Mittrucker HW et al (1997) Requirement for the transcription factor LSIRF/IRF4 for mature B and T lymphocyte function. Science 275(5299):540–543
87. Zheng Y et al (2009) Regulatory T-cell suppressor program co-opts transcription factor IRF4 to control T(H)2 responses. Nature 458(7236):351–356
88. Dang CV et al (2008) The interplay between MYC and HIF in cancer. Nat Rev Cancer 8(1):51–56
89. Miller DM et al (2012) C-Myc and cancer metabolism. Clin Cancer Res 18(20):5546–5553
90. Cairns RA, Harris IS, Mak TW (2011) Regulation of cancer cell metabolism. Nat Rev Cancer 11(2):85–95
91. Casey SC et al (2016) MYC regulates the antitumor immune response through CD47 and PD-L1. Science 352(6282):227–231
92. Yin X et al (2003) Low molecular weight inhibitors of Myc-max interaction and function. Oncogene 22(40):6151–6159
93. Bandukwala HS et al (2012) Selective inhibition of CD4+ T-cell cytokine production and autoimmunity by BET protein and c-Myc inhibitors. Proc Natl Acad Sci U S A 109(36):14532–14537
94. Cafferkey R et al (1993) Dominant missense mutations in a novel yeast protein related to mammalian phosphatidylinositol 3-kinase and VPS34 abrogate rapamycin cytotoxicity. Mol Cell Biol 13(10):6012–6023
95. Brown EJ et al (1994) A mammalian protein targeted by G1-arresting rapamycin-receptor complex. Nature 369(6483):756–758
96. Sabatini DM et al (1994) RAFT1: a mammalian protein that binds to FKBP12 in a rapamycin-dependent fashion and is homologous to yeast TORs. Cell 78(1):35–43
97. Sabers CJ et al (1995) Isolation of a protein target of the FKBP12-rapamycin complex in mammalian cells. J Biol Chem 270(2):815–822
98. Fingar DC, Blenis J (2004) Target of rapamycin (TOR): an integrator of nutrient and growth factor signals and coordinator of cell growth and cell cycle progression. Oncogene 23(18):3151–3171
99. Wullschleger S, Loewith R, Hall MN (2006) TOR signaling in growth and metabolism. Cell 124(3):471–484
100. Ma XM, Blenis J (2009) Molecular mechanisms of mTOR-mediated translational control. Nat Rev Mol Cell Biol 10(5):307–318
101. Sarbassov DD et al (2004) Rictor, a novel binding partner of mTOR, defines a rapamycin-insensitive and raptor-independent pathway that regulates the cytoskeleton. Curr Biol 14(14):1296–1302
102. Jacinto E et al (2004) Mammalian TOR complex 2 controls the actin cytoskeleton and is rapamycin insensitive. Nat Cell Biol 6(11):1122–1128
103. Hosokawa N et al (2009) Nutrient-dependent mTORC1 association with the ULK1-Atg 13-FIP200 complex required for autophagy. Mol Biol Cell 20(7):1981–1991
104. Jung CH et al (2009) ULK-Atg 13-FIP200 complexes mediate mTOR signaling to the autophagy machinery. Mol Biol Cell 20(7):1992–2003

105. Ganley IG et al (2009) ULK1.ATG13.FIP200 complex mediates mTOR signaling and is essential for autophagy. J Biol Chem 284(18):12297–12305
106. Hresko RC, Mueckler M (2005) mTOR.RICTOR is the Ser 473 kinase for Akt/protein kinase B in 3T3-L1 adipocytes. J Biol Chem 280(49):40406–40416
107. Sarbassov DD et al (2005) Phosphorylation and regulation of Akt/PKB by the rictor-mTOR complex. Science 307(5712):1098–1101
108. Hara K et al (2002) Raptor, a binding partner of target of rapamycin (TOR), mediates TOR action. Cell 110(2):177–189
109. Kim DH et al (2002) mTOR interacts with raptor to form a nutrient-sensitive complex that signals to the cell growth machinery. Cell 110(2):163–175
110. Peterson TR et al (2009) DEPTOR is an mTOR inhibitor frequently overexpressed in multiple myeloma cells and required for their survival. Cell 137(5):873–886
111. Kaizuka T et al (2010) Tti 1 and Tel 2 are critical factors in mammalian target of rapamycin complex assembly. J Biol Chem 285(26):20109–20116
112. Nojima H et al (2003) The mammalian target of rapamycin (mTOR) partner, raptor, binds the mTOR substrates p 70 S6 kinase and 4E-BP1 through their TOR signaling (TOS) motif. J Biol Chem 278(18):15461–15464
113. Loewith R et al (2002) Two TOR complexes, only one of which is rapamycin sensitive, have distinct roles in cell growth control. Mol Cell 10(3):457–468
114. Foster KG et al (2010) Regulation of mTOR complex 1 (mTORC1) by raptor Ser 863 and multisite phosphorylation. J Biol Chem 285(1):80–94
115. Fingar DC et al (2004) mTOR controls cell cycle progression through its cell growth effectors S6K1 and 4E-BP1/eukaryotic translation initiation factor 4E. Mol Cell Biol 24(1):200–216
116. Schalm SS et al (2003) TOS motif-mediated raptor binding regulates 4E-BP1 multisite phosphorylation and function. Curr Biol 13(10):797–806
117. Wang L et al (2007) PRAS40 regulates mTORC1 kinase activity by functioning as a direct inhibitor of substrate binding. J Biol Chem 282(27):20036–20044
118. Sancak Y et al (2008) The rag GTPases bind raptor and mediate amino acid signaling to mTORC1. Science 320(5882):1496–1501
119. Vander Haar E et al (2007) Insulin signalling to mTOR mediated by the Akt/PKB substrate PRAS40. Nat Cell Biol 9(3):316–323
120. Inoki K, Zhu T, Guan KL (2003) TSC2 mediates cellular energy response to control cell growth and survival. Cell 115(5):577–590
121. Potter CJ, Pedraza LG, Xu T (2002) Akt regulates growth by directly phosphorylating Tsc 2. Nat Cell Biol 4(9):658–665
122. Ma L et al (2005) Phosphorylation and functional inactivation of TSC2 by Erk implications for tuberous sclerosis and cancer pathogenesis. Cell 121(2):179–193
123. Roux PP et al (2004) Tumor-promoting phorbol esters and activated Ras inactivate the tuberous sclerosis tumor suppressor complex via p 90 ribosomal S6 kinase. Proc Natl Acad Sci U S A 101(37):13489–13494
124. Long X et al (2005) Rheb binds and regulates the mTOR kinase. Curr Biol 15(8):702–713
125. Tee AR et al (2003) Tuberous sclerosis complex gene products, Tuberin and Hamartin, control mTOR signaling by acting as a GTPase-activating protein complex toward Rheb. Curr Biol 13(15):1259–1268
126. Curatolo P, Moavero R (2012) mTOR inhibitors in tuberous sclerosis complex. Curr Neuropharmacol 10(4):404–415
127. Harrington LE et al (2005) Interleukin 17-producing CD4+ effector T cells develop via a lineage distinct from the T helper type 1 and 2 lineages. Nat Immunol 6(11):1123–1132
128. Manning BD (2004) Balancing Akt with S6K: implications for both metabolic diseases and tumorigenesis. J Cell Biol 167(3):399–403
129. Powell JD, Delgoffe GM (2010) The mammalian target of rapamycin: linking T cell differentiation, function, and metabolism. Immunity 33(3):301–311

130. Beugnet A et al (2003) Regulation of targets of mTOR (mammalian target of rapamycin) signalling by intracellular amino acid availability. Biochem J 372(Pt 2):555–566
131. Schriever SC et al (2013) Cellular signaling of amino acids towards mTORC1 activation in impaired human leucine catabolism. J Nutr Biochem 24(5):824–831
132. Duran RV et al (2012) Glutaminolysis activates rag-mTORC1 signaling. Mol Cell 47(3):349–358
133. Nicklin P et al (2009) Bidirectional transport of amino acids regulates mTOR and autophagy. Cell 136(3):521–534
134. Kim SG et al (2013) Metabolic stress controls mTORC1 lysosomal localization and dimerization by regulating the TTT-RUVBL1/2 complex. Mol Cell 49(1):172–185
135. Hara K et al (1998) Amino acid sufficiency and mTOR regulate p 70 S6 kinase and eIF-4E BP1 through a common effector mechanism. J Biol Chem 273(23):14484–14494
136. Jewell JL, Guan KL (2013) Nutrient signaling to mTOR and cell growth. Trends Biochem Sci 38(5):233–242
137. Jewell JL et al (2015) Metabolism. Differential regulation of mTORC1 by leucine and glutamine. Science 347(6218):194–198
138. Zoncu R et al (2011) mTORC1 senses lysosomal amino acids through an inside-out mechanism that requires the vacuolar H(+)-ATPase. Science 334(6056):678–683
139. Sancak Y, Sabatini DM (2009) Rag proteins regulate amino-acid-induced mTORC1 signalling. Biochem Soc Trans 37(Pt 1):289–290
140. Gwinn DM et al (2008) AMPK phosphorylation of raptor mediates a metabolic checkpoint. Mol Cell 30(2):214–226
141. Wouters BG, Koritzinsky M (2008) Hypoxia signalling through mTOR and the unfolded protein response in cancer. Nat Rev Cancer 8(11):851–864
142. DeYoung MP et al (2008) Hypoxia regulates TSC1/2-mTOR signaling and tumor suppression through REDD1-mediated 14-3-3 shuttling. Genes Dev 22(2):239–251
143. Katiyar S et al (2009) REDD1, an inhibitor of mTOR signalling, is regulated by the CUL4A-DDB1 ubiquitin ligase. EMBO Rep 10(8):866–872
144. Frias MA et al (2006) mSin1 is necessary for Akt/PKB phosphorylation, and its isoforms define three distinct mTORC2s. Curr Biol 16(18):1865–1870
145. Jacinto E et al (2006) SIN1/MIP1 maintains rictor-mTOR complex integrity and regulates Akt phosphorylation and substrate specificity. Cell 127(1):125–137
146. Guertin DA et al (2006) Ablation in mice of the mTORC components raptor, rictor, or mLST8 reveals that mTORC2 is required for signaling to Akt-FOXO and PKCalpha, but not S6K1. Dev Cell 11(6):859–871
147. Sarbassov DD et al (2006) Prolonged rapamycin treatment inhibits mTORC2 assembly and Akt/PKB. Mol Cell 22(2):159–168
148. Guertin DA, Sabatini DM (2007) Defining the role of mTOR in cancer. Cancer Cell 12(1):9–22
149. Populo H, Lopes JM, Soares P (2012) The mTOR signalling pathway in human cancer. Int J Mol Sci 13(2):1886–1918
150. Lawlor MA, Alessi DR (2001) PKB/Akt: a key mediator of cell proliferation, survival and insulin responses? J Cell Sci 114(Pt 16):2903–2910
151. Zheng Y et al (2007) A role for mammalian target of rapamycin in regulating T cell activation versus anergy. J Immunol 178(4):2163–2170
152. Delgoffe GM et al (2009) The mTOR kinase differentially regulates effector and regulatory T cell lineage commitment. Immunity 30(6):832–844
153. Chi H (2012) Regulation and function of mTOR signalling in T cell fate decisions. Nat Rev Immunol 12(5):325–338
154. Pollizzi KN et al (2015) mTORC1 and mTORC2 selectively regulate CD8(+) T cell differentiation. J Clin Invest 125(5):2090–2108
155. Kaech SM, Ahmed R (2003) Immunology. CD8 T cells remember with a little help. Science 300(5617):263–265
156. Kaech SM et al (2003) Selective expression of the interleukin 7 receptor identifies effector CD8 T cells that give rise to long-lived memory cells. Nat Immunol 4(12):1191–1198

157. Shrestha S et al (2014) Tsc 1 promotes the differentiation of memory CD8+ T cells via orchestrating the transcriptional and metabolic programs. Proc Natl Acad Sci U S A 111(41):14858–14863
158. Oh WJ et al (2010) mTORC2 can associate with ribosomes to promote cotranslational phosphorylation and stability of nascent Akt polypeptide. EMBO J 29(23):3939–3951
159. Porstmann T et al (2008) SREBP activity is regulated by mTORC1 and contributes to Akt-dependent cell growth. Cell Metab 8(3):224–236
160. Hagiwara A et al (2012) Hepatic mTORC2 activates glycolysis and lipogenesis through Akt, glucokinase, and SREBP1c. Cell Metab 15(5):725–738
161. Duvel K et al (2010) Activation of a metabolic gene regulatory network downstream of mTOR complex 1. Mol Cell 39(2):171–183
162. Peterson TR et al (2011) mTOR complex 1 regulates lipin 1 localization to control the SREBP pathway. Cell 146(3):408–420
163. Yuan M et al (2012) Identification of Akt-independent regulation of hepatic lipogenesis by mammalian target of rapamycin (mTOR) complex 2. J Biol Chem 287(35):29579–29588
164. Blagih J et al (2015) The energy sensor AMPK regulates T cell metabolic adaptation and effector responses in vivo. Immunity 42(1):41–54
165. Faubert B et al (2013) AMPK is a negative regulator of the Warburg effect and suppresses tumor growth in vivo. Cell Metab 17(1):113–124
166. Tamas P et al (2006) Regulation of the energy sensor AMP-activated protein kinase by antigen receptor and Ca2+ in T lymphocytes. J Exp Med 203(7):1665–1670
167. MacIver NJ et al (2011) The liver kinase B1 is a central regulator of T cell development, activation, and metabolism. J Immunol 187(8):4187–4198
168. Deberardinis RJ, Lum JJ, Thompson CB (2006) Phosphatidylinositol 3-kinase-dependent modulation of carnitine palmitoyltransferase 1A expression regulates lipid metabolism during hematopoietic cell growth. J Biol Chem 281(49):37372–37380
169. Lee WH, Kim SG (2010) AMPK-dependent metabolic regulation by PPAR agonists. PPAR Res 2010
170. Suzuki T et al (2013) Inhibition of AMPK catabolic action by GSK3. Mol Cell 50(3):407–419
171. Inoki K, Kim J, Guan KL (2012) AMPK and mTOR in cellular energy homeostasis and drug targets. Annu Rev Pharmacol Toxicol 52:381–400
172. Carretero J et al (2007) Dysfunctional AMPK activity, signalling through mTOR and survival in response to energetic stress in LKB1-deficient lung cancer. Oncogene 26(11):1616–1625
173. Rolf J et al (2013) AMPKalpha1: a glucose sensor that controls CD8 T-cell memory. Eur J Immunol 43(4):889–896
174. Fernandez D et al (2006) Rapamycin reduces disease activity and normalizes T cell activation-induced calcium fluxing in patients with systemic lupus erythematosus. Arthritis Rheum 54(9):2983–2988
175. Warner LM, Adams LM, Sehgal SN (1994) Rapamycin prolongs survival and arrests pathophysiologic changes in murine systemic lupus erythematosus. Arthritis Rheum 37(2):289–297
176. Augustine JJ, Bodziak KA, Hricik DE (2007) Use of sirolimus in solid organ transplantation. Drugs 67(3):369–391
177. Gao Y, Whitaker-Dowling P, Bergman I (2015) Memory antitumor T-cells resist inhibition by immune suppressor cells. Anticancer Res 35(9):4593–4597
178. Klebanoff CA, Gattinoni L, Restifo NP (2006) CD8+ T-cell memory in tumor immunology and immunotherapy. Immunol Rev 211:214–224
179. Klebanoff CA et al (2005) Central memory self/tumor-reactive CD8+ T cells confer superior antitumor immunity compared with effector memory T cells. Proc Natl Acad Sci U S A 102(27):9571–9576
180. Turner AP et al (2011) Sirolimus enhances the magnitude and quality of viral-specific CD8+ T-cell responses to vaccinia virus vaccination in rhesus macaques. Am J Transplant 11(3):613–618
181. Li Q et al (2012) Regulating mammalian target of rapamycin to tune vaccination-induced CD8(+) T cell responses for tumor immunity. J Immunol 188(7):3080–3087

182. Berezhnoy A et al (2014) Aptamer-targeted inhibition of mTOR in T cells enhances antitumor immunity. J Clin Invest 124(1):188–197
183. Foretz M et al (2014) Metformin: from mechanisms of action to therapies. Cell Metab 20(6):953–966
184. Sajan MP et al (2010) AICAR and metformin, but not exercise, increase muscle glucose transport through AMPK-, ERK-, and PDK1-dependent activation of atypical PKC. Am J Physiol Endocrinol Metab 298(2):E179–E192
185. Jhun BS et al (2005) AICAR suppresses IL-2 expression through inhibition of GSK-3 phosphorylation and NF-AT activation in Jurkat T cells. Biochem Biophys Res Commun 332(2):339–346
186. Nath N et al (2009) Metformin attenuated the autoimmune disease of the central nervous system in animal models of multiple sclerosis. J Immunol 182(12):8005–8014
187. Bai A et al (2010) AMPK agonist downregulates innate and adaptive immune responses in TNBS-induced murine acute and relapsing colitis. Biochem Pharmacol 80(11):1708–1717
188. Bai A et al (2010) Novel anti-inflammatory action of 5-aminoimidazole-4-carboxamide ribonucleoside with protective effect in dextran sulfate sodium-induced acute and chronic colitis. J Pharmacol Exp Ther 333(3):717–725
189. Eikawa S et al (2015) Immune-mediated antitumor effect by type 2 diabetes drug, metformin. Proc Natl Acad Sci U S A 112(6):1809–1814
190. Walker JE (2013) The ATP synthase: the understood, the uncertain and the unknown. Biochem Soc Trans 41(1):1–16
191. Li Y et al (2006) Cytochrome c oxidase subunit IV is essential for assembly and respiratory function of the enzyme complex. J Bioenerg Biomembr 38(5–6):283–291
192. Jimenez-Gutierrez LR et al (2014) The cytochrome c oxidase and its mitochondrial function in the whiteleg shrimp *Litopenaeus vannamei* during hypoxia. J Bioenerg Biomembr 46(3):189–196
193. Hockel M, Vaupel P (2001) Tumor hypoxia: definitions and current clinical, biologic, and molecular aspects. J Natl Cancer Inst 93(4):266–276
194. Semenza GL et al (1997) Structural and functional analysis of hypoxia-inducible factor 1. Kidney Int 51(2):553–555
195. Eguchi H et al (1997) A nuclear localization signal of human aryl hydrocarbon receptor nuclear translocator/hypoxia-inducible factor 1beta is a novel bipartite type recognized by the two components of nuclear pore-targeting complex. J Biol Chem 272(28):17640–17647
196. Freeburg PB, Abrahamson DR (2004) Divergent expression patterns for hypoxia-inducible factor-1beta and aryl hydrocarbon receptor nuclear transporter-2 in developing kidney. J Am Soc Nephrol 15(10):2569–2578
197. Schodel J et al (2011) High-resolution genome-wide mapping of HIF-binding sites by ChIP-seq. Blood 117(23):e207–e217
198. Wenger RH, Stiehl DP, Camenisch G (2005) Integration of oxygen signaling at the consensus HRE. Sci STKE 2005(306):re12
199. Wang GL et al (1995) Hypoxia-inducible factor 1 is a basic-helix-loop-helix-PAS heterodimer regulated by cellular O2 tension. Proc Natl Acad Sci U S A 92(12):5510–5514
200. Chapman-Smith A, Lutwyche JK, Whitelaw ML (2004) Contribution of the per/Arnt/Sim (PAS) domains to DNA binding by the basic helix-loop-helix PAS transcriptional regulators. J Biol Chem 279(7):5353–5362
201. Yang J et al (2005) Functions of the per/ARNT/Sim domains of the hypoxia-inducible factor. J Biol Chem 280(43):36047–36054
202. Ziello JE, Jovin IS, Huang Y (2007) Hypoxia-inducible factor (HIF)-1 regulatory pathway and its potential for therapeutic intervention in malignancy and ischemia. Yale J Biol Med 80(2):51–60
203. Makino Y et al (2001) Inhibitory PAS domain protein is a negative regulator of hypoxia-inducible gene expression. Nature 414(6863):550–554

204. Semenza GL (2010) Defining the role of hypoxia-inducible factor 1 in cancer biology and therapeutics. Oncogene 29(5):625–634
205. Zhang P et al (2014) Hypoxia-inducible factor 3 is an oxygen-dependent transcription activator and regulates a distinct transcriptional response to hypoxia. Cell Rep 6(6):1110–1121
206. Jiang BH et al (1996) Hypoxia-inducible factor 1 levels vary exponentially over a physiologically relevant range of O2 tension. Am J Phys 271(4 Pt 1):C1172–C1180
207. McNamee EN et al (2013) Hypoxia and hypoxia-inducible factors as regulators of T cell development, differentiation, and function. Immunol Res 55(1–3):58–70
208. Vaupel P, Harrison L (2004) Tumor hypoxia: causative factors, compensatory mechanisms, and cellular response. Oncologist 9(Suppl 5):4–9
209. Tsai AG, Johnson PC, Intaglietta M (2003) Oxygen gradients in the microcirculation. Physiol Rev 83(3):933–963
210. Chang CH et al (2015) Metabolic competition in the tumor microenvironment is a driver of cancer progression. Cell 162(6):1229–1241
211. Gilkes DM, Semenza GL, Wirtz D (2014) Hypoxia and the extracellular matrix: drivers of tumour metastasis. Nat Rev Cancer 14(6):430–439
212. Gorres KL, Raines RT (2010) Prolyl 4-hydroxylase. Crit Rev Biochem Mol Biol 45(2):106–124
213. Ohh M et al (2000) Ubiquitination of hypoxia-inducible factor requires direct binding to the beta-domain of the von Hippel-Lindau protein. Nat Cell Biol 2(7):423–427
214. Schofield CJ, Ratcliffe PJ (2004) Oxygen sensing by HIF hydroxylases. Nat Rev Mol Cell Biol 5(5):343–354
215. Hudson CC et al (2002) Regulation of hypoxia-inducible factor 1alpha expression and function by the mammalian target of rapamycin. Mol Cell Biol 22(20):7004–7014
216. Nakamura H et al (2005) TCR engagement increases hypoxia-inducible factor-1 alpha protein synthesis via rapamycin-sensitive pathway under hypoxic conditions in human peripheral T cells. J Immunol 174(12):7592–7599
217. Baek JH et al (2007) Spermidine/spermine N(1)-acetyltransferase-1 binds to hypoxia-inducible factor-1alpha (HIF-1alpha) and RACK1 and promotes ubiquitination and degradation of HIF-1alpha. J Biol Chem 282(46):33358–33366
218. Liu YV et al (2007) RACK1 competes with HSP90 for binding to HIF-1alpha and is required for O(2)-independent and HSP90 inhibitor-induced degradation of HIF-1alpha. Mol Cell 25(2):207–217
219. Liu YV et al (2007) Calcineurin promotes hypoxia-inducible factor 1alpha expression by dephosphorylating RACK1 and blocking RACK1 dimerization. J Biol Chem 282(51):37064–37073
220. Clambey ET et al (2012) Hypoxia-inducible factor-1 alpha-dependent induction of FoxP3 drives regulatory T-cell abundance and function during inflammatory hypoxia of the mucosa. Proc Natl Acad Sci U S A 109(41):E2784–E2793
221. Hsiao HW et al (2015) Deltex1 antagonizes HIF-1alpha and sustains the stability of regulatory T cells in vivo. Nat Commun 6:6353
222. Lee JH et al (2015) E3 ubiquitin ligase VHL regulates hypoxia-inducible factor-1alpha to maintain regulatory T cell stability and suppressive capacity. Immunity 42(6):1062–1074
223. Doedens AL et al (2013) Hypoxia-inducible factors enhance the effector responses of CD8(+) T cells to persistent antigen. Nat Immunol 14(11):1173–1182
224. Chisolm DA, Weinmann AS (2015) TCR-signaling events in cellular metabolism and specialization. Front Immunol 6:292
225. Hengel RL et al (2003) Cutting edge: L-selectin (CD62L) expression distinguishes small resting memory CD4+ T cells that preferentially respond to recall antigen. J Immunol 170(1):28–32
226. Yang S et al (2011) The shedding of CD62L (L-selectin) regulates the acquisition of lytic activity in human tumor reactive T lymphocytes. PLoS One 6(7):e22560

227. Zagzag D et al (2000) Expression of hypoxia-inducible factor 1alpha in brain tumors: association with angiogenesis, invasion, and progression. Cancer 88(11):2606–2618
228. Rapisarda A et al (2004) Schedule-dependent inhibition of hypoxia-inducible factor-1alpha protein accumulation, angiogenesis, and tumor growth by topotecan in U251-HRE glioblastoma xenografts. Cancer Res 64(19):6845–6848
229. Rapisarda A et al (2002) Identification of small molecule inhibitors of hypoxia-inducible factor 1 transcriptional activation pathway. Cancer Res 62(15):4316–4324
230. Rapisarda A et al (2009) Increased antitumor activity of bevacizumab in combination with hypoxia inducible factor-1 inhibition. Mol Cancer Ther 8(7):1867–1877
231. Koh MY et al (2008) Molecular mechanisms for the activity of PX-478, an antitumor inhibitor of the hypoxia-inducible factor-1alpha. Mol Cancer Ther 7(1):90–100
232. Welsh S et al (2004) Antitumor activity and pharmacodynamic properties of PX-478, an inhibitor of hypoxia-inducible factor-1alpha. Mol Cancer Ther 3(3):233–244
233. Lang M et al (2016) Arsenic trioxide plus PX-478 achieves effective treatment in pancreatic ductal adenocarcinoma. Cancer Lett 378(2):87–96
234. Zhao T et al (2015) Inhibition of HIF-1alpha by PX-478 enhances the anti-tumor effect of gemcitabine by inducing immunogenic cell death in pancreatic ductal adenocarcinoma. Oncotarget 6(4):2250–2262
235. Zhang H et al (2008) Mitochondrial autophagy is an HIF-1-dependent adaptive metabolic response to hypoxia. J Biol Chem 283(16):10892–10903
236. Huh JR et al (2011) Digoxin and its derivatives suppress TH17 cell differentiation by antagonizing RORgammat activity. Nature 472(7344):486–490
237. Osada S et al (1997) Identification of an extended half-site motif required for the function of peroxisome proliferator-activated receptor alpha. Genes Cells 2(5):315–327
238. Krey G et al (1993) Xenopus peroxisome proliferator activated receptors: genomic organization, response element recognition, heterodimer formation with retinoid X receptor and activation by fatty acids. J Steroid Biochem Mol Biol 47(1–6):65–73
239. Alleva DG et al (2002) Regulation of murine macrophage proinflammatory and anti-inflammatory cytokines by ligands for peroxisome proliferator-activated receptor-gamma: counter-regulatory activity by IFN-gamma. J Leukoc Biol 71(4):677–685
240. DiRenzo J et al (1997) Peroxisome proliferator-activated receptors and retinoic acid receptors differentially control the interactions of retinoid X receptor heterodimers with ligands, coactivators, and corepressors. Mol Cell Biol 17(4):2166–2176
241. Dowell P et al (1997) p300 functions as a coactivator for the peroxisome proliferator-activated receptor alpha. J Biol Chem 272(52):33435–33443
242. Yuan CX et al (1998) The TRAP220 component of a thyroid hormone receptor- associated protein (TRAP) coactivator complex interacts directly with nuclear receptors in a ligand-dependent fashion. Proc Natl Acad Sci U S A 95(14):7939–7944
243. Zhu Y et al (1996) Cloning and identification of mouse steroid receptor coactivator-1 (mSRC-1), as a coactivator of peroxisome proliferator-activated receptor gamma. Gene Expr 6(3):185–195
244. Zhu Y et al (1997) Isolation and characterization of PBP, a protein that interacts with peroxisome proliferator-activated receptor. J Biol Chem 272(41):25500–25506
245. Desvergne B, Wahli W (1999) Peroxisome proliferator-activated receptors: nuclear control of metabolism. Endocr Rev 20(5):649–688
246. Jones DC, Ding X, Daynes RA (2002) Nuclear receptor peroxisome proliferator-activated receptor alpha (PPARalpha) is expressed in resting murine lymphocytes. The PPARalpha in T and B lymphocytes is both transactivation and transrepression competent. J Biol Chem 277(9):6838–6845
247. Harris SG, Phipps RP (2001) The nuclear receptor PPAR gamma is expressed by mouse T lymphocytes and PPAR gamma agonists induce apoptosis. Eur J Immunol 31(4):1098–1105
248. Vosper H et al (2001) The peroxisome proliferator-activated receptor delta promotes lipid accumulation in human macrophages. J Biol Chem 276(47):44258–44265

249. Shi Y, Hon M, Evans RM (2002) The peroxisome proliferator-activated receptor delta, an integrator of transcriptional repression and nuclear receptor signaling. Proc Natl Acad Sci U S A 99(5):2613–2618
250. Chinetti G et al (1998) Activation of proliferator-activated receptors alpha and gamma induces apoptosis of human monocyte-derived macrophages. J Biol Chem 273(40):25573–25580
251. Delerive P et al (1999) Peroxisome proliferator-activated receptor activators inhibit thrombin-induced endothelin-1 production in human vascular endothelial cells by inhibiting the activator protein-1 signaling pathway. Circ Res 85(5):394–402
252. Gosset P et al (2001) Peroxisome proliferator-activated receptor gamma activators affect the maturation of human monocyte-derived dendritic cells. Eur J Immunol 31(10):2857–2865
253. Padilla J et al (2000) Peroxisome proliferator activator receptor-gamma agonists and 15-deoxy-Delta(12,14)(12,14)-PGJ(2) induce apoptosis in normal and malignant B-lineage cells. J Immunol 165(12):6941–6948
254. Delerive P et al (1999) Peroxisome proliferator-activated receptor alpha negatively regulates the vascular inflammatory gene response by negative cross-talk with transcription factors NF-kappaB and AP-1. J Biol Chem 274(45):32048–32054
255. Delerive P et al (2000) Oxidized phospholipids activate PPARalpha in a phospholipase A2-dependent manner. FEBS Lett 471(1):34–38
256. Peters JM, Hollingshead HE, Gonzalez FJ (2008) Role of peroxisome-proliferator-activated receptor beta/delta (PPARbeta/delta) in gastrointestinal tract function and disease. Clin Sci (Lond) 115(4):107–127
257. Yang XY et al (2000) Activation of human T lymphocytes is inhibited by peroxisome proliferator-activated receptor gamma (PPARgamma) agonists. PPARgamma co-association with transcription factor NFAT. J Biol Chem 275(7):4541–4544
258. Raikwar HP et al (2005) PPARgamma antagonists exacerbate neural antigen-specific Th1 response and experimental allergic encephalomyelitis. J Neuroimmunol 167(1–2):99–107
259. Raikwar HP et al (2006) PPARgamma antagonists reverse the inhibition of neural antigen-specific Th1 response and experimental allergic encephalomyelitis by Ciglitazone and 15-deoxy-Delta12,14-prostaglandin J2. J Neuroimmunol 178(1–2):76–86
260. Niino M et al (2001) Amelioration of experimental autoimmune encephalomyelitis in C57BL/6 mice by an agonist of peroxisome proliferator-activated receptor-gamma. J Neuroimmunol 116(1):40–48
261. Feinstein DL et al (2002) Peroxisome proliferator-activated receptor-gamma agonists prevent experimental autoimmune encephalomyelitis. Ann Neurol 51(6):694–702
262. Klotz L et al (2009) The nuclear receptor PPAR gamma selectively inhibits Th17 differentiation in a T cell-intrinsic fashion and suppresses CNS autoimmunity. J Exp Med 206(10):2079–2089
263. Natarajan C, Bright JJ (2002) Peroxisome proliferator-activated receptor-gamma agonists inhibit experimental allergic encephalomyelitis by blocking IL-12 production, IL-12 signaling and Th1 differentiation. Genes Immun 3(2):59–70
264. Jakobsen MA et al (2006) Peroxisome proliferator-activated receptor alpha, delta, gamma1 and gamma2 expressions are present in human monocyte-derived dendritic cells and modulate dendritic cell maturation by addition of subtype-specific ligands. Scand J Immunol 63(5):330–337
265. Storer PD et al (2005) Peroxisome proliferator-activated receptor-gamma agonists inhibit the activation of microglia and astrocytes: implications for multiple sclerosis. J Neuroimmunol 161(1–2):113–122
266. Kanakasabai S et al (2010) Peroxisome proliferator-activated receptor delta agonists inhibit T helper type 1 (Th1) and Th17 responses in experimental allergic encephalomyelitis. Immunology 130(4):572–588
267. Natarajan C et al (2003) Peroxisome proliferator-activated receptor-gamma-deficient heterozygous mice develop an exacerbated neural antigen-induced Th1 response and experimental allergic encephalomyelitis. J Immunol 171(11):5743–5750

268. Gocke AR et al (2009) Transcriptional modulation of the immune response by peroxisome proliferator-activated receptor-{alpha} agonists in autoimmune disease. J Immunol 182(7):4479–4487
269. Cunard R et al (2002) Regulation of cytokine expression by ligands of peroxisome proliferator activated receptors. J Immunol 168(6):2795–2802
270. Xu J, Racke MK, Drew PD (2007) Peroxisome proliferator-activated receptor-alpha agonist fenofibrate regulates IL-12 family cytokine expression in the CNS: relevance to multiple sclerosis. J Neurochem 103(5):1801–1810
271. Kapadia R, Yi JH, Vemuganti R (2008) Mechanisms of anti-inflammatory and neuroprotective actions of PPAR-gamma agonists. Front Biosci 13:1813–1826
272. Cipolletta D et al (2012) PPAR-gamma is a major driver of the accumulation and phenotype of adipose tissue Treg cells. Nature 486(7404):549–553
273. Housley WJ et al (2009) PPARgamma regulates retinoic acid-mediated DC induction of Tregs. J Leukoc Biol 86(2):293–301
274. Wohlfert EA et al (2007) Peroxisome proliferator-activated receptor gamma (PPARgamma) and immunoregulation: enhancement of regulatory T cells through PPARgamma-dependent and -independent mechanisms. J Immunol 178(7):4129–4135
275. Lei J et al (2010) Peroxisome proliferator-activated receptor alpha and gamma agonists together with TGF-beta convert human CD4+CD25- T cells into functional Foxp3+ regulatory T cells. J Immunol 185(12):7186–7198
276. Bassaganya-Riera J et al (2004) Activation of PPAR gamma and delta by conjugated linoleic acid mediates protection from experimental inflammatory bowel disease. Gastroenterology 127(3):777–791
277. Diab A et al (2002) Peroxisome proliferator-activated receptor-gamma agonist 15-deoxy-Delta(12,14)-prostaglandin J(2) ameliorates experimental autoimmune encephalomyelitis. J Immunol 168(5):2508–2515
278. Mueller C et al (2003) Peroxisome proliferator-activated receptor gamma ligands attenuate immunological symptoms of experimental allergic asthma. Arch Biochem Biophys 418(2):186–196
279. Beales PE et al (1998) Troglitazone prevents insulin dependent diabetes in the non-obese diabetic mouse. Eur J Pharmacol 357(2–3):221–225
280. Park SJ et al (2009) Peroxisome proliferator-activated receptor gamma agonist down-regulates IL-17 expression in a murine model of allergic airway inflammation. J Immunol 183(5):3259–3267
281. Kaiser CC et al (2009) A pilot test of pioglitazone as an add-on in patients with relapsing remitting multiple sclerosis. J Neuroimmunol 211(1–2):124–130
282. Lewis JD et al (2008) Rosiglitazone for active ulcerative colitis: a randomized placebo-controlled trial. Gastroenterology 134(3):688–695

Chapter 3
Transcriptional Regulation of T Cell Metabolism Reprograming

Jinxia Zhang, Guoyu Bi, Yu Xia, Pingfei Li, Xiaofei Deng, Zhengping Wei, and Xiang-Ping Yang

Abstract T cell activation, differentiation, and function are tightly regulated by a complex network of transcription factors, epigenetic modifications, and signaling pathways of both TCR and cytokines. Over the past decade, it is increasingly clear that T cell immune responses are also regulated by their associated metabolic reprograming. Compared with relatively well-understood transcriptional regulation of T cell activation, differentiation, and function, less is known about the transcriptional regulation of T cell metabolic reprograming during T cell immune responses. In this review, we first describe how signaling pathways of TCR and cytokines regulate metabolic reprograming and then focus on transcription factors that control metabolic pathways and outcomes of T cell immune responses. A better understanding of T cell metabolic regulation will provide new strategies and targets for the treatment of T cell-related diseases.

Keywords T cell differentiation • T cell metabolism • Transcription factor • Glycolysis

3.1 Introduction

T cells are activated upon engagement of their receptors with p-MHC complexes presented by antigen-presenting cells. Orchestrated by local tissue cytokine milieu, activated T cells proliferate and differentiate into multiple effector subsets that are critical for controlling miscellaneous pathogens [1]. After elimination of invaded microbes, the majority of T effector cells undergo apoptosis, and a small amount of them become memory T cells that can respond rapidly against reinfection [2, 3]. In addition, the differentiation of T effector cells and formation of T memory cells are

J. Zhang • G. Bi • Y. Xia • P. Li • X. Deng • Z. Wei • X.-P. Yang (✉)
Department of Immunology, School of Basic Medicine, Tongji Medical School, Huazhong University of Science and Technology, 13 Hangkong Road, Wuhan, China, 430030
e-mail: yangxp@hust.edu.cn

B. Li, F. Pan (eds.), *Immune Metabolism in Health and Tumor*, Advances in Experimental Medicine and Biology 1011, DOI 10.1007/978-94-024-1170-6_3

closely associated with metabolic reprograming. While the transcriptional regulation of T cell differentiation and $CD8^+$ memory T cell formation have been intensively studied over the last two decades [4–9], the topic of transcriptional regulation of T cell metabolism and associated T cell functions has recently emerged [10, 11].

Numerous studies have shown that different T cell subsets adopt distinct metabolic pathways to meet their bioenergetic and biosynthetic demands; quiescent and less proliferative T cells, including naïve T cells, memory T cells, and regulatory T cells (Treg), often use fatty acid oxidation (FAO) as their main metabolic resource [12–15]. In contrast, activated effector T cells undergo a drastic metabolic switch from FAO to aerobic glycolysis. The phenomenon that cancer cells exhibit a high rate of glycolysis even in the presence of normal oxygen levels (aerobic glycolysis) was discovered by Otto Warburg and was termed the Warburg effect [16]. Although glycolysis is not efficient in the sense of ATP production when compared with TCA [17, 18], this procedure produces large amounts of metabolic intermediates for biosynthesis that support robust proliferation after T cell activation [14, 19]. T cell metabolic reprograming during T cell immune responses, i.e., TCR activation, T effector cell generation, and formation of T memory cells, is orchestrated by a series of transcription factors [10, 11] (Fig. 3.1). Intervention of T cell metabolism can alter T cell specification and functions and offers potential novel strategies for the treatment of T cell-related diseases.

In this chapter, we first discussed signaling pathways, including TCR signaling pathway and cytokine signaling pathways, i.e., IL-2, IL-7, and IL-15, that are involved in T cell metabolic reprograming and then focused on key transcription factors that regulate T cell metabolism; afterward, we discussed the potential implication of targeting the metabolic nature of T cells in diseases.

3.2 TCR Signaling and T Cell Metabolism

Engagement of TCR with specific antigens presented by MHC on antigen-presenting cells activates multiple signaling pathways [20–22]. Both tyrosine kinases and serine kinases are crucial enzymes for the full activation of T cells [23]. Detailed signaling pathways activated by TCR ligation had been thoroughly reviewed [21, 24]. Recently, it has been shown that TCR-driven intracellular signaling pathways are specifically regulated by the multiplicity of immuno-receptor tyrosine-based activation motifs (ITAMs) in the TCR-CD3 complex [25]. Guy et al. found that a lower multiplicity of ITAMs induces cytokine production, whereas a higher multiplicity of ITAMs induces Notch-1 pathway, followed by c-Myc expression [25]. This ultimately leads to T cell proliferation, suggesting that initial cytokine production and proliferation are separable pathways [25].

Currently, all transcription factors known to mediate or sustain the glycolysis pathway of T effector cells, including Myc, HIF-1α, IRF4, SREBPs, and AP4, can be induced by TCR activation [26–30]. It is known that single-cell level TCR strength contributes to the preferred specialization program it initiates (i.e., diverse

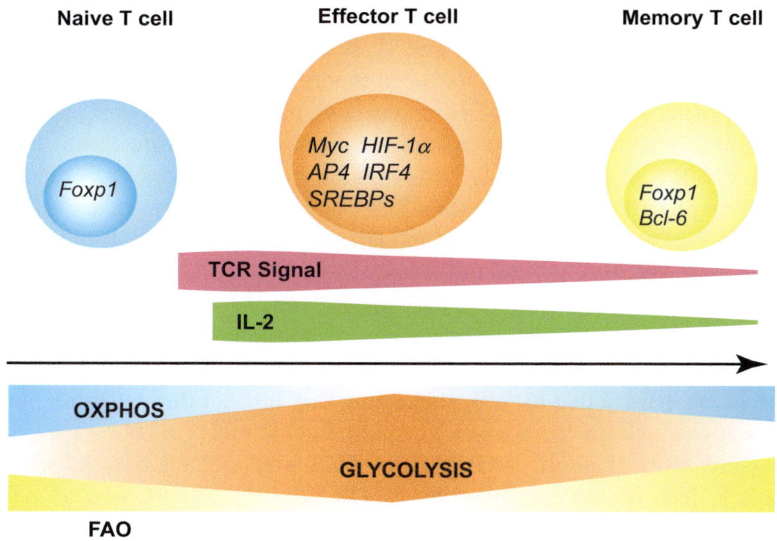

Fig. 3.1 The metabolic reprograming of T cells during T cell activation is driven by TCR and IL-2 signaling pathways and associated transcriptional factors. Foxp1 is critical for the maintenance of the quiescent status of naïve T cells that relies primarily on oxidative phosphorylation (OXPHOS) for their metabolic needs. Myc is an early transcriptional factor induced by TCR stimulation and plays non-redundant roles in initiating glycolysis and glutaminolysis reprograming and promoting T effector cell specification. Other key transcriptional factors, including HIF-1α, AP4, IRF4, and SREBPs, are also induced by TCR and IL-2 signaling, and these transcriptional factors are required for maintaining the glycolysis, glutaminolysis, and lipid biosynthesis pathways. During the later phase of T cell immune responses, IL-2 signaling diminishes and Bcl-6 expression is upregulated. Bcl-6 is essential for the metabolic reprograming from glycolytic pathways in T effector cells toward fatty acid oxidation (FAO) in memory T cells

$CD4^+$ T cell differentiation; effector versus memory) [31]. Similarly, the induction levels of Myc, IRF4, and AP4 upon TCR activation are correlated with TCR signal strength [26, 28, 30].

In addition to the induction of key transcription factors, TCR stimulation also activates critical signaling pathways for T cell metabolism reprograming, including mechanistic target of rapamycin (mTOR) pathway [32–34]. The notable function of mTOR activation is to promote cellular growth and catabolic processes by sensing and integrating diverse signals of growth factors, nutrients availability, and energy levels [35]. There are two cellular mTOR multi-protein complexes: mTORC1 and mTORC2 [35]. mTORC1 consists of mTOR, Raptor, GβL (mammalian lethal with SEC13 protein 8) and domain-containing mTOR-interacting protein (DEPTOR). mTORC1 activation potentiates general cellular translation of mRNA by phosphorylating downstream targets such as 4E-BP1 and p70 S6 kinase, induces ribosome biogenesis, and enhances mitochondrial activity or adipogenesis. mTORC2, which consists of mTOR, Rictor, GβL, Sin1, PRR5/Protor-1, and DEPTOR, promotes cell survival through the activation of AKT [36]. mTORC2 also regulates cytoskeletal dynamics, ion transport, and growth by activating PKCα and phosphorylating SGK1, respectively [37–39]. TCR and IL-2 rapidly activate both mTORC1 and mTORC2 pathways. Recently, it has been reported that in $CD8^+$ T cell, PI3K and AKT are not required for TCR-induced mTOR pathways; however, inhibition or deficiency of PDK1, an essential regulator of glucose metabolism, diminishes TCR-induced mTOR activation [40, 41] . This suggests that other serine/threonine kinases downstream of PDK1 might activate mTOR signaling pathways in activated $CD8^+$ T cells. The period of stimulation and antigen avidity affect the magnitude of mTOR activation [42, 43]. Although mTOR is not required for TCR-induced activation and IL-2 production, it is indispensable for Th1, Th2, and Th17 cell differentiation [44, 45]. Deletion of mTOR leads to spontaneous generation of $Foxp3^+$ Treg cells [44]; however, deletion of Rheb or Rictor does not recapitulate the mTOR-deficient phenotype on Treg cells [46, 47], suggesting that distinct downstream complexes might be involved in the convergence of Foxp3 induction.

TCR-induced metabolic switch from TCA and OXPHOS to glycolysis in activated T cells is also controlled by mTOR [26, 41]. Suppression of mTORC1 activation by rapamycin results in lower expression of c-Myc and reduced rate of glycolysis; in addition, *c-Myc*$^{-/-}$ T cells have reduced mTORC1 activation in response to TCR signals, suggesting that a possible mutual regulation mechanism exists between mTOR and Myc [26, 41]. In CTLs, however, rapamycin does not affect Myc expression; instead, it inhibits HIF-1α expression and sequential glycolysis. In contrast, *Tsc2*$^{-/-}$ T cells with constitutive mTOR activation are highly glycolytic in response to TCR [48].

mTOR activity can be further regulated by co-stimulatory or co-inhibitory signals: both CD28 and OX40 enhance TCR-induced PI3K-AKT signaling and further upregulate mTOR activity induced by TCR; this facilitates sustained T cell activation [49–52]. In contrast, PD1 negatively regulates mTOR activation to mediate immune tolerance [53].

3.3 IL-2, IL-7, and IL-15 Signaling Involved in the Regulation of Metabolism

The common cytokine receptor γ-chain (γ_c) (IL-2Rγ) is indispensable for the signaling pathways initiated by IL-2, IL-4, IL-5, IL-7, IL-15, IL-21, and TSLP [54]. These cytokines play crucial roles in the development, proliferation, survival, and differentiation of multiple innate and adaptive cell lineages [54]. Mutations of this gene result in loss of T cells and NK cells and a disease called X-linked severe combined immunodeficiency [55]. Thus, it is not surprising that several common γ-cytokines, especially IL-2, IL-7, and IL-15, regulate T cell metabolism.

Interleukin-2 is a pleiotropic cytokine produced upon TCR activation, which plays pivotal and complex roles in host immune responses [56]. Although initially identified as a T cell growth factor, IL-2 regulates almost every T cell subset differentiation, promoting Th1, Th2, and Treg differentiation [57–59] while inhibiting Th17 and Tfh cell differentiation [60, 61]. Moreover, IL-2 is essential for the generation of both $CD8^+$ T effector cells and memory cells [62].

IL-2 activates both JAK/STAT pathway and PI3K/AKT/mTOR pathway in T cells [56]. Recently, Ray et al. showed that Tfh cells, when compared with Th1 cells, have reduced mTOR activity, accompanied with less proliferation, glycolysis, and mitochondrial respiration in an acute virus infection model [63]. In this model, IL-2 signaling is critical for repressing the differentiation of Tfh cells and promoting Th1 cell differentiation, consistent with previous reports [61]. The activation of the AKT kinase and mTORC1 signaling were both necessary and sufficient for the transition of Tfh toward Th1 cells [63].

IL-2 signaling also regulates the expression of several transcription factors, including Myc, AP4, HIF-1α, and Bcl-6 that control metabolic gene expression programs in T cells [26, 30, 41]. Myc, AP4, and HIF-1α are reported to mediate glycolysis, and Bcl-6 suppresses glycolysis and promotes oxidative phosphorylation (OXPHOS), which is required in the formation of memory T cells [26, 30, 41, 64].

IL-7 is another common γ-chain cytokine that plays a critical role in T cell development, homeostasis, and quiescence [54]. IL-7 signaling is required for the survival of naïve T cells by increasing the expression of the antiapoptotic proteins Bcl-2 and Mcl-1 [65, 66]. In addition, IL-7 signaling is also essential in maintaining the basal levels of glycolysis and preventing atrophy in resting T cells [67]. The tonic and sustained activation of PI3K/AKT/mTOR activation induced by IL-7 is crucial for maintaining the quiescence state of naïve T cells, as deletion of TSC1 results in enhanced mTOR activation and the loss of quiescent phenotype of naïve T cells [68]. IL-7 also takes part in the development of memory T cells, which requires complicated metabolic reprogramming [69–71]. In the formation process of memory T cells, T cells start to dampen the metabolic gene expression programs majorly glycolysis established during effector T cell differentiation; however, the expression of genes encoding components of the catabolic pathways is promoted to meet the needs required for the quiescent, long-lived phenotype of memory cells. In particular, metabolic pathways, such as the fatty acid oxidation (FAO) and lipolysis pathways, are upregulated to promote memory cell formation [69–71]. IL-7 enhances the FAO pathway in part through regulating the expression of the glycerol channel receptor aquaporin 9 (Aqp9) in $CD8^+$ T cells [71]. Aqp9-deficient $CD8^+$ T cells expressed less amounts of numerous components in the FAO pathway and could not establish robust long-term memory, although the underlying mechanism remains unclear [71].

Similar to IL-7, IL-15 also promotes the survival of naïve and memory $CD8^+$ T cells and is critical for optimal T memory responses [72, 73] . IL-15 regulates $CD8^+$ memory cell spare respiratory capacity (SRC), which represents the ability of cells to produce extra energy in response to increased stress and is often associated with cell survival. In addition, IL-15 promotes mitochondrial oxidative metabolism and biogenesis by increasing the expression of carnitine palmitoyltransferase (CPT1a), a rate-limiting enzyme that controls mitochondrial FAO [69]. T memory cells need substantial amounts of fatty acids to fuel the mitochondrial OXPHOS that is acquired intracellularly by active lipolysis [70, 74]. IL-15 enhances the expression of lysosomal acid lipase and is essential for the intrinsic generation of fatty acids in memory T cells [70]. Knowledge of transcription factors, especially how STATs mediate the induction of key proteins involved in FAO and lipolysis, would indeed be informative.

Despite similar signaling networks triggered by IL-2, IL-7, and IL-15, they have differential physiological effects in T cell biology. Distinct phenotypes associated with IL-2, IL-7, and IL-15 regulated T cell biology, including metabolic reprograming, might arise from the differential expression of receptors, the availability of cytokines, and distinct cytokine receptor intracellular trafficking kinetics [75].

3.4 Transcription Factors Involved in T Cell Metabolism Reprograming

As discussed previously, the differentiation of naïve T cells into diverse effector T cells requires cellular metabolic switch from majorly OXPHOS to glycolysis and glutaminolysis; whereas, during the process of the formation of memory T cells, metabolism pathways were shifted from glycolysis, glutaminolysis, and lipid biosynthesis pathways to FAO and OXPHOS [7, 12, 19]. This metabolic reprograming in different T cell subsets is regulated by several key transcription factors and multiple signaling pathways (Fig. 3.2).

3.4.1 Myc

The myelocytomatosis oncogene (Myc) belongs to the family of helix-loop-helix-leucine zipper transcription factors that form a heterodimer with Myc-associated protein X (MAX) to bind specific DNA sites termed E-box [76]. Previously found and known as an important oncogene, Myc plays critical roles in the regulation of cell growth, proliferation, apoptosis, and pluripotency [76]. Though Myc has been intensively studied over the last three decades, the mechanisms of how Myc functions and regulates transcription remain obscure. With the advent of CHIP-Seq technology, genome-wide binding sites of Myc have been mapped out in different cell types [77, 78]. Unexpectedly, Myc binds to widespread DNA regions that account for 10–15% of genome [77, 78]. This has led to speculations that instead of specifically regulating gene expression, Myc functions largely as a universal amplifier of RNAs [79, 80].

Myc expression in cells is subjected to tight regulation at both transcriptional and posttranscriptional levels [76]. Multiple mitogenic growth factors induce Myc expression, and both RNA and protein of Myc are very short-lived [81]. The phosphorylation at serine 62 by ERK kinase stabilizes its protein, whereas phosphorylation at threonine 58 of Myc by GSK3β targets it to proteasome-mediated degradation [81].

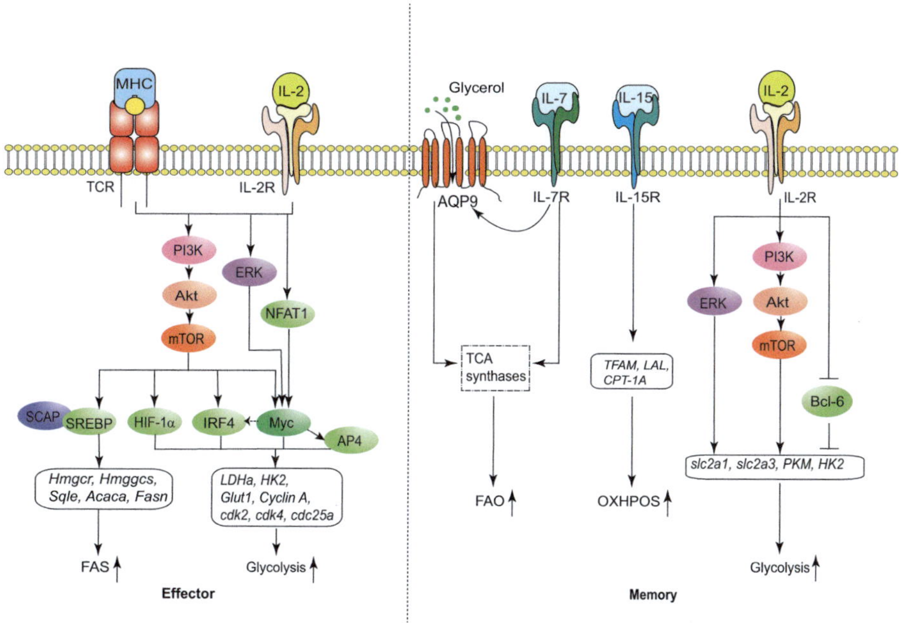

Fig. 3.2 Signaling pathways and transcriptional factors mediate the metabolic reprograming in T effector and memory cells. In activated T effector cells, both TCR and IL-2 induce PI3K/AKT/mTOR activation that is required for the induction of Myc, HIF-1α, IRF4, and SREBPs. Myc, HIF-1α, and IRF4 induce expression of a variety of glycolytic genes, including *ldha*, *hk2*, *glut1*, *cyclin A*, *cdk2*, *cdk4*, and *cdc25a*; SREBPs induce the expression of key lipid biosynthesis genes, including *hmgcr*, *hmggcs*, *sqle*, *acaca*, and *fasn*. ERK and NF-AT1 also contribute to the induction of Myc, whereas Myc is essential for the further induction of AP4, which is required for the maintenance of the glycolytic pathways in cytolytic T cells. In memory T cells, the metabolism is reverted back to catabolic metabolism, resulting in the suppression of glycolytic pathways and the upregulation of mitochondrial biogenesis and OXPHOS and FAO pathways. IL-7 induces the expression of AQP9 that mediates glycerol transport and triacylglycerol (TAG) synthesis, which are essential for the survival of memory T cells. IL-15 induces the expression of mitochondrial transcription factor A (TFAM), lysosomal acid lipase (LAL), and carnitine palmitoyltransferase 1A (CPT1A) that promote mitochondrial biogenesis, fatty acid generation, and mitochondrial spare respiratory capacity of memory T cells respectively. Diminished IL-2 signaling in the later phase of T cell immune responses leads to increased expression of Bcl-6 that suppresses the expression of key glycolytic genes such as *slc2a1*, *slc2a2*, *PKM*, and *HK2*

Myc is an essential transcription factor in regulating the activation-induced glycolysis switch in T cells [26]. TCR activation and IL-2 induce rapid Myc expression with detectable mRNA and Myc protein as early as 2 h after stimulation, although the mechanisms are different [82]. TCR induces Myc expression in a digital fashion, and the strength of TCR determines the on/off switch of Myc expression, whereas IL-2 induces the expression of Myc in an analogous fashion [82]. ERK-, mTOR-, AKT-mediated signaling pathways and transcription factor NF-AT1 are attributed to the maximum induction of Myc upon T cell activation [26, 83]. IL-2 also enhances the expression of slc7a5, an amino acid transporter, to increase the translation of

Myc mRNA [82]. IL-2 sustains the level of Myc protein in activated T cells that can be blocked by tofacitinib, an inhibitor of JAKs, suggesting that the JAK/STAT pathway is involved in the induction of Myc [82].

In early studies with fibroblast and lymphocyte cells, Myc was shown to bind and regulate almost all glycolytic enzyme genes and genes involved in mitochondrial biogenesis [84]. In T cells, Myc also regulates a variety of genes that are involved in cell cycle progression, glycolysis, glutaminolysis, and lipid synthesis [26]. The maximum expressions of glucose transporter *slc2a1*, *Cyclin A*, and cyclin-dependent kinases such as *cdk2*, *cdk4*, and *cdc25a* require Myc [26]. The role of Myc in mediating the TCR-induced metabolism switch is non-redundant: deletion of Myc in T cells results in reduced glucose intake, reduced glycolysis, and glutaminolysis but has no obvious effect on pyruvate metabolic influx, FAO influx, and oxidation consumption rate [26]. In contrast, deletion of HIF-1α does not affect the glycolysis and cell proliferation for at least 24 h post TCR stimulation, suggesting that Myc might act as an upstream factor of HIF-1α in driving the metabolic switch to glycolysis [26]. In $CD8^+$ T effector cells, Myc can further induce AP4 expression [30]. Although Myc and AP4 share a significant amount of target genes genome-wide, AP4 is essential for mediating the glycolysis reprogramming and functions of $CD8^+$ T cells in the later phases of immune responses in several infection models [30], indicating a critical role of the temporal induction of Myc and AP4 in the regulation of metabolic reprograming and functions of $CD8^+$ T cell in vivo.

3.4.2 HIF-1α

Hypoxia-inducible factor 1α (HIF-1α) is a member of a transcription factor family that senses the shortage of oxygen and coordinates cellular responses to hypoxia to ensure optimal functional and metabolic adaption in a lower oxygen condition [85]. Deregulation of the HIF pathways occurs often in inflammation, immunity, and cancer [85]. HIF-1 is a heterodimeric basic loop-helix-loop protein containing α- and β-units and regulates expression of genes whose promoters contain the hypoxia response consensus sequences (HREs). Although both subunits are constitutively synthesized, the stability of subunit α is regulated by an oxygen-dependent post-translational regulation. Under normal oxygen conditions, the iron-dependent enzyme prolyl hydroxylases (PHDs) hydroxylates the subunit α, and the hydroxylated α-subunit can be recognized by the von Hippel-Lindau protein (VHL) and targeted by proteasome-mediated degradation. In lower oxygen conditions, the PHD is inactive and the α-subunit is stabilized.

In T cells, the stabilization and the induction of HIF-1α can be regulated via an oxygen-independent manner [27]. TCR ligation induces detectable and substantial accumulations of HIF-1α mRNA and protein as early as 2 h under normal oxygen conditions [86]. The combination of Th17 differentiation cytokines, IL-6, and TGF-β significantly enhances the HIF-1α induction in a STAT3-depedent manner [27, 87]. Other T cell subset differentiation cytokines have little effect on the further induction of HIF-1α. Activated $CD8^+$ T cells also elevate the expression of HIF-1α and HIF-2α in normoxic conditions [41, 45].

Although it has been suggested that HIF-1α is not required for the initial glycolysis at least 24 h after TCR stimulation, deletion of HIF-1α significantly reduced Th17 differentiation and the expression of key glycolytic enzymes LDHα and HK2 [27, 86]. More importantly, inhibition of glycolysis also results in reduced Th17 differentiation, indicating an intimate relationship between metabolism and Th17 differentiation [27, 87]. In addition, HIF-1α directly induces the transcription of RORγt, the master transcription factor of Th17 cells, and promotes Foxp3 protein degradation [87].

Previous reports suggested that TCR-induced HIF-1α is not due to the stabilization of mRNA but to the newly synthesized protein in an mTOR-dependent, rapamycin-sensitive manner [41, 88]. In $CD8^+$ T cells, the mTOR activation is essential for HIF-1α induction, glucose uptake, and glycolysis. HIF-1α is not required for the initial proliferation and activation of $CD8^+$ T cells but is required for the sustained glycolysis and T cell trafficking. Interestingly, the mTOR activation is independent of PI3K and AKT pathways but dependent on PDK activation in $CD8^+$ T cells [41]. Further investigation is needed in order to determine which serine/threonine kinase downstream of PDK mediates the mTOR activation. During the activation of T cells, HIF-1α induces miR-210 that can negatively regulate HIF-1α, which forms a feedback inhibitory loop of HIF-1α activity [89].

3.4.3 IRF4

IRF4 is a member of the family of transcription factors of interferon regulatory factor (IRF), which consists of nine members, IRF1 through IRF9, in mice and humans [90]. IRFs have important and diverse functions in the regulation of innate and adaptive immune responses, and IRF4 is the most studied transcription factor in T cells [91]. The N-terminal DNA-binding domain of IRF4 is highly conserved and binds to a 5′-GAAA-3′ motif, a part of the canonical IFN-stimulated response element (ISRE, A/GNGAAANNGAAACT) [91]. The C-terminal regulatory domain of IRF4 mediates homo- and heteromeric interactions with other transcription factors, including IRFs such as IRF8; an autoinhibitory region residing in the last 30 amino acids of the C-terminal regulatory domain of IRF4 can prevent the binding activities of IRF4 to its targeted sequences [91].

IRF4 plays crucial roles in T cell differentiation. IRF4 promotes differentiation of naïve $CD4^+$ T cells into Th2, Th9, Th17, or Tfh cells and is required for the function of effector regulatory T cells [92–96]. Moreover, IRF4 is essential for the sustained differentiation of cytotoxic effector $CD8^+$ T cells and for $CD8^+$ T cell memory formation [28, 97, 98].

IRF4 often synergizes with other transcription factors to mediate its biological functional activities because its binding affinity to the consensus motif is relatively low. IRF4 can function as either transcriptional activator or repressor [91]. A variety of transcription factors including PU.1, SPI-B, JUN/BATF heterodimer, STAT3, NF-AT, homeobox protein, BCL-6, FOXP3, ROR-γt, and SMADs have been shown to interact with IRF4 in T cells [91]. It has been suggested that during Th17 differentiation, IRF4/BATF complex promotes chromatin accessibility to other transcription factors, including STAT3 and ROR-γt [99–101].

Similar to HIF-1α and Myc, IRF4 is strongly induced within a few hours upon TCR stimulation and expressed in almost all T cell subsets [28]. The levels of IRF4 expression correlate with the strength of the TCR stimulation and determine the capacity of T cells to sustain proliferation [28]. Both mTOR and ITK activations are required for the TCR-mediated IRF4 expression [97, 102]. C-REL and NF-AT also contribute to the TCR-mediated induction of IRF4 [103].

In activated $CD8^+$ T cells, IRF4 is not required for initial proliferation but is indispensable to sustain the clonal T cell expansion [28, 104]. T cells lacking IRF4 could not expand properly and give appropriate immune responses [28, 97, 104]. *IRF4*$^{-/-}$ mice have significantly reduced antigen-specific $CD8^+$ T effector cells and are more susceptible to infections in different infectious models [28, 97, 104]. IRF4 is required for the glycolytic metabolism of $CD8^+$ T effector cells by upregulating key genes associated with the glycolysis pathway, i.e., *HIF-1α*, *Foxo1*, *Slc2a1*, *Slc2a3*, and the hexokinase (*HK2*) [28]. Myc and IRF4 have overlapping functions in the regulation of glycolysis and proliferation of T effector cells; however, they may operate in a temporal manner, and while the induction of Myc is rapid and transient, the expression of IRF4 peaks during the phase of the clonal expansion [28]. In addition to regulating the metabolism, IRF4 could also promote T effector cell differentiation at the expense of T memory development by inducing key transcription factors in T effector differentiation, including *prdm-1* (which encodes Blimp-1), *Runx3*, and *Tcf3* [28].

3.4.4 Bcl-6

Bcl-6 was initially identified as a proto-oncogene in non-Hodgkin's lymphoma with the origin of germinal center [105]. It is a zinc finger transcription factor expressed in both T cells and B cells [106]. In general, Bcl-6 acts as a sequence-specific repressor of transcription via interactions with several corepressor complexes including BCoR, N-CoR, SMRT, CtBP, BAZF, PLZF, MIZ1, and others [105]. The repressive actions of Bcl-6 on transcription can be antagonized by Blimp-1 that is encoded by the gene *Prdm1*. Bcl-6 is required for the formation of germinal center, as *Bcl-6*-deficient mice lack germinal center B cells and have an absence of affinity maturation. In addition, Bcl-6 is the master transcription factor for Tfh cell differentiation.

The regulation of Bcl-6 expression is complex in T cells. Many cytokines induce Bcl-6 expression in T cells, for example, activation of STAT1 and STAT3 by IL-6, activation of STAT4 by IL-12, activation of STAT3 by IL-21, and activation of STAT1 by IFN-γ [105]. The formation of memory T cells is accompanied with the metabolic switch from glycolysis to fatty acid oxidation, and inhibiting glycolysis promotes the transition between effector T cells and memory T cells [107, 108]. The molecular events and transcription factors that induce glycolysis in effector T cells have been intensively studied [10]. However, less is known about the transcription factors that repress glycolysis and promote FAO and OXPHOS in memory T cells. Recently, Bcl-6 has been shown to be essential in inhibiting glycolysis [64]. Bcl-6 directly binds and represses genes that encode key glycolytic proteins, such as glucose transporters *slc2a1* and *slc2a3*, and enzymes *PKM* and *HK2* to effectively dampen glycolysis that is associated with effector T cell differentiation [64]. The effect of Bcl-6 on the inhibition of glycolysis can be antagonized by T-bet, which is consistent with the opposing roles of Bcl-6 and T-bet in Tfh and Th1 differentiation [109].

The Bcl-6 expression is reversely correlated with IL-2 expression during immune responses [61, 64, 110, 111]. IL-2 is critical for the T effector cell expansion, and expression of IL-2 is tightly regulated during immune responses. During the early phases of immune responses, IL-2 is highly induced by the expanding T effector cells, whereas in the latter phases of immune responses, IL-2 starts to be limited, and most T effector cells undergo apoptosis with a small fraction of the T effector cells becoming memory T cells. Bcl-6 expression is repressed by high amounts of IL-2 in effector T cells, and the repression is relieved once IL-2 expression decreases, which coincides with the timing of beginning transition to memory potential [7, 112]. Currently, it is unclear whether Blimp-1 acts in an opposite manner as Bcl-6 does in the regulation of T cell metabolism and memory cell generation.

3.4.5 SREBPA

Sterol regulatory element-binding proteins (SREBPs) belong to the basic helix-loop-helix-leucine zipper (bHLH-Zip) family of transcription factors that are essential for lipid homeostasis [113]. They are synthesized as inactive precursors bound to the endoplasmic reticulum [113]. The lower sterol levels can activate the proteases S2P and S1P to cleave the SREBP precursors and release the N-terminal domain of SREBPS containing the bHLH-Zip region for binding DNA in the nucleus. Activated SREBPs induce the expression of more than 30 genes dedicated to the synthesis and uptake of cholesterol, fatty acids, triglycerides, and phospholipids, as well as the NADPH cofactor required to synthesize these molecules. There are three isoforms in mammalian genome, designated as SREBP-1a, SREBP-1c, and SREBP-2.

In addition to reprograming the T cell metabolism to glycolysis and glutaminolysis after T cell activation, biosynthesis pathways including lipid synthesis are also upregulated to meet the macromolecule demands for rapid and robust cell divisions [15]. TCR activation induces *Srebf1* and *Srebf2*, the genes that encode SREBP-1 and SREBP-2, respectively [29]. In T cells, upregulations of SREBP-1 and SREBP-2 lead to the induction of key components of the lipid biosynthesis pathways, such as

Hmgcr, *Hmggcs*, *Sqle*, *Acaca*, and *Fasn* [29]. In *Scap*$^{-/-}$ mice where the SCAP protein, a protein associated with SREBPs and required for SREBP cleavage and activation, was deleted, CD8^{+} T cells failed to blast and expand upon viral infection, although SCAP is not required for CD8^{+} T cell homeostatic proliferation [29] . Deletion of *Scap* in T cells does not change the proximal TCR signaling but results in compromised lipid homeostasis. The defect of blastogenesis and proliferation in *Scap*$^{-/-}$ T cells can be reversed by adding exogenous cholesterol. In line with this, inhibiting ACAT1 activity in T cells potentiates CD8^{+} antitumor activities by increasing cholesterol levels in the plasma membrane and enhancing T cell signaling [114].

3.4.6 AP4

Activator protein 4, encoded by *Tfap4*, is a helix-loop-helix transcriptional factor that can bind to enhancers through homo-dimerization and regulate gene expression. AP4 is required for the repression of the *cd4* expression in immature DN thymocytes and CD8^{+} cells via interaction with Runx1 and binding to the proximal *cd4* enhancer. Both *Tfap4*-deficient mice and mice harboring point mutations in the *cd4* silencer for AP4 binding have increased frequencies of CD4-expressing effector/memory CD8^{+} T cells [115].

Although Myc is essential for the metabolic switch to glycolysis and glutaminolysis, the expression of Myc is very transient, suggesting that other molecules downstream of Myc might exist to mediate the glycolytic pathway initiated by TCR and IL-2. Recently, it has been shown that IL-2 induces AP4 at both transcriptional and posttranscriptional levels in activated CD8^{+} T cells, and TCR stimulation sustains its expression [30]. Myc is essential for the induction of AP4 and binds directly to the *Tfap4* locus. *Tfap4*-deficient mice are more susceptible to West Nile virus infection and have significantly fewer KLRG^{+}CD8^{+} effector cells, due to impaired clonal expansion [30]. This defect is CD8^{+} T cell intrinsic. Interestingly, the numbers of antigen-specific CD8^{+} T cells at day 3 were comparable between WT and *Tfap4*-deficient mice after virus infection, suggesting the initial T cell proliferation was intact. Ex vivo recovered CD8^{+} T cells following virus infection from AP4-deficeint mice expressed lesser amounts of genes encoding key glycolytic proteins, including PKM2 and HK2 compared with the CD8^{+} T cells recovered from the WT mice [30]. AP4 and Myc shared a large population of binding sites genome-wide, including many genes involved in metabolism pathways. However, ectopic expression of stabilized form of Myc in the *Tfap4*$^{-/-}$ could not fully restore the expansion and metabolic defect, suggesting that the temporal switch from Myc to AP4 is necessary for sustained T cell metabolism and functions.

3.4.7 Foxp1

Foxp1 is one of the "winged-helix" transcription factors coded by the forkhead (Fox) gene family and plays an essential role in maintaining naïve T cell quiescence [116, 117]. In the absence of Foxp1, naïve $CD8^+$ T cells have an effector phenotype and function and can proliferate directly in response to homeostatic cytokine IL-7 [117]. Foxp1 antagonizes the function of Foxo1 in regulating the expression of IL-7Rα [117]. *Foxp1*-deficient $CD8^+$ T cells have enhanced activation of the PI3K/Akt/mTOR signaling pathway and expression of glycolytic genes in response to IL-7 [118]. Mechanistically, Foxp1 induces the expression of PI3K interacting protein 1, a negative regulator of PI3K, and suppresses the expression of E2fs, the critical components for cell cycle progression and proliferation [118]. Foxp1 maintains naive $CD8^+$ T cell quiescence by simultaneously repressing both cellular metabolism and cell cycle.

3.5 Targeting Metabolism and Transcription Factors in Diseases

Transcription factors are key cellular components that regulate almost every aspect of T cell biology, including development, differentiation, and memory formation [1]. These processes are accompanied with T cell metabolism reprograming. The transcriptional regulation of T cell development and differentiation has been intensively studied over the last few decades and is now relatively well understood [4]; however, less is known about the transcriptional regulation of T cell metabolism reprograming [10, 11]. Understanding this in great detail would shed new insights on how to beneficially manipulate T cell immune responses in a variety of diseases.

It is known that targeting NF-AT pathways during T cell activation has beneficial effects in transplantation. As calcineurin inhibitors, both Cyclosporine A and FK506 are widely used in transplantation to inhibit NF-AT activity and T cell responses. However, blocking NF-AT activation in all T cells could also affect Treg cell functions [119]. Thus, it is reasonable to speculate that selectively targeting T cell metabolism reprograming and related T cell subsets might represent a better strategy for immune modulation.

Digoxin and its derivatives are potent inhibitors of Th17 differentiation that counteract RORγt activity [120]. It has been shown that digoxin and other RORγt-specific inhibitors can ameliorate Th17 cell-mediated autoimmune diseases in mice [120, 121]. Interestingly, a recent drug screen found digoxin inhibits HIF-1α activity as well [122], which is required for the high glycolysis rate during Th17 differentiation [27]. It is possible that digoxin- and RORγt-specific inhibitors not only inhibit RORγt activity but also inhibit HIF-1α activity and associated glycolysis during Th17 differentiation.

Myc is essential for T cell proliferation and activation-induced glycolytic reprograming [26]. Targeting Myc could be a potent immunosuppressive strategy. Bromodomain-containing proteins (BRD2, BRD3, and BRD4) bind acetylated histones and can recruit additional transcriptional coactivators to potentiate transcription [123]. In cancers, a BRD4 BET inhibitor JQ1 inhibits Myc expression, Myc-mediated transcriptome, and tumor growth [123, 124]. Another BET inhibitor, I-BET-762, suppresses the production of pro-inflammatory cytokine productions in macrophages and ameliorates acute inflammation in mice [125]. I-BET-76 inhibits not only pro-inflammatory cytokine production in T cells, including GM-CSF and IL-17, but also the pathogenic ability of differentiated Th1 cells [125]. Interestingly, Myc expression is also inhibited by I-BET-76 and 10058-F4, a characterized c-Myc/Max inhibitor. In addition, 10058-F4 has similar effects on T cell functions and can also disrupt the N-Myc/Max interaction [126]. In a mouse model, 10058-F4 inhibits neuronal differentiation in N-Myc-amplified neuroblastoma cells and extends survival of N-Myc transgenic mice by inducing cell cycle arrest and apoptosis of neuroblastoma cells [126]. Furthermore, Myc inhibition leads to metabolic changes with accumulation of lipid droplets in tumor cells [126]. The efficacy of BET inhibitors on T cell-mediated diseases and whether they act on T cell metabolism need further investigation.

mTOR regulates T effector differentiation and plays essential roles in mediating the glycolytic reprograming. Targeting mTOR is effective in suppressing T cell immune responses. The mTORC1 inhibitor rapamycin has been used to prevent transplant rejection [127], initially attributed to its ability to inhibit T cell proliferation [128]. However, later studies have shown that rapamycin is an inefficient proliferation inhibitor and is more effective in regulation of T cell metabolism, effector T cell responses, and Treg cell differentiation promotion [33]. Lower doses of rapamycin only inhibit mTORC1 activity; however, higher doses of rapamycin particularly under conditions of prolonged exposure can inhibit mTORC2 activity [129]. This might explain the ability of rapamycin to promote Treg cell generation. In addition, rapamycin can also promote memory T cell generation, which is beneficial in enhancing vaccine responses in a nonhuman primate model of vaccinia virus vaccination [130]. While rapamycin sterically disrupts the formation of the mTOR complex, efforts have been made for the development of mTOR kinase inhibitors that would inhibit the activity of both mTORC1 and mTORC2 [131, 132].

Modulation of the T cell metabolism can affect T cell functions especially for $CD8^+$ cell-mediated antitumor activity. Although glycolysis is required for T effector cell functions, inhibition of glycolysis in activated $CD8^+$ T cells enhanced the generation of memory cells and antitumor functionality [107]. This could be due to the effect that activated $CD8^+$ T cells with higher glycolysis rates were less prone to become T memory cells [107]. Indeed, enhancing glycolytic metabolism by overexpressing the glycolytic enzyme phosphoglycerate mutase-1 severely impaired the ability of $CD8^+$ T cells to form long-term memory T cells [107]. Recently, another study showed that inhibiting cholesterol esterification in T cells by genetic ablation or pharmacological inhibition of ACAT1, a key cholesterol esterification enzyme, also led to enhanced $CD8^+$ antitumor activities [114]. Deficiency or inhibition of

ACAT1 increases cholesterol level in the plasma membrane of $CD8^+$ T cells and results in more efficient formation of the immunological synapse and enhanced T cell signaling [114]. Tumor microenvironment including lower concentrations of oxygen and glucose that limit aerobic glycolysis can suppress the functions of tumor-infiltrating T cells [133]. The glycolytic metabolite phosphoenolpyruvate (PEP) can sustain T cell receptor-mediated Ca^{2+}-NF-AT signaling and effector functions by repressing sarco/ER Ca^{2+}-ATPase (SERCA) activity [134]. Reprograming tumor-reactive metabolisms by overexpression of phosphoenolpyruvate carboxykinase 1 (PCK1) to increase PEP production bolstered antitumor T cell responses [134].

3.6 Conclusion Marks

Cellular metabolic pathways provide not only energy but also a variety of metabolic intermediates that can regulate T cell differentiation and functions. The regulation of metabolic reprogramming during T cell immune responses involves a complex network of cytokines, enzymes, membrane transporters, and transcription factors. Although the roles of some transcription factors mediating the metabolic glycolysis switch associated with T cell activation and differentiation have been elucidated, our knowledge regarding transcriptional regulation of lipid metabolism associated with T effector cells and T memory cells is still very limited. Dysregulation of T cell metabolisms is involved in the development of autoimmune diseases, infectious diseases, and tumorigenesis. Understanding mechanistic details pertaining to transcriptional regulation of T cell metabolism will provide opportunities to discover novel prevention and therapeutic drugs for diseases associated with T cell malfunctions.

Acknowledgments We apologize to our colleagues whose work was not cited, due to space limitation. Supported by the National Scientific Foundation of China (C3140851), 973 Program (2013CB530505), and Integrated Innovative Team for Major Human Diseases Program of Tongji Medical College, HUST to X.P.Y.

References

1. Zhu J, Yamane H, Paul WE (2010) Differentiation of effector CD4 T cell populations (*). Annu Rev Immunol 28:445–489
2. Ahmed R, Gray D (1996) Immunological memory and protective immunity: understanding their relation. Science 272(5258):54–60
3. Harrington LE, Janowski KM, Oliver JR, Zajac AJ, Weaver CT (2008) Memory CD4 T cells emerge from effector T-cell progenitors. Nature 452(7185):356–360
4. Zhu J, Paul WE (2010) Peripheral CD4+ T-cell differentiation regulated by networks of cytokines and transcription factors. Immunol Rev 238(1):247–262
5. Williams MA, Bevan MJ (2007) Effector and memory CTL differentiation. Annu Rev Immunol 25:171–192

6. Joshi NS, Kaech SM (2008) Effector CD8 T cell development: a balancing act between memory cell potential and terminal differentiation. J Immunol 180(3):1309–1315
7. Kaech SM, Cui W (2012) Transcriptional control of effector and memory CD8+ T cell differentiation. Nat Rev Immunol 12(11):749–761
8. Jameson SC, Masopust D (2009) Diversity in T cell memory: an embarrassment of riches. Immunity 31(6):859–871
9. Kaech SM, Wherry EJ (2007) Heterogeneity and cell-fate decisions in effector and memory CD8+ T cell differentiation during viral infection. Immunity 27(3):393–405
10. Man K, Kallies A (2015) Synchronizing transcriptional control of T cell metabolism and function. Nat Rev Immunol 15(9):574–584
11. Hough KP, Chisolm DA, Weinmann AS (2015) Transcriptional regulation of T cell metabolism. Mol Immunol 68(2 Pt C):520–526
12. Pearce EL (2010) Metabolism in T cell activation and differentiation. Curr Opin Immunol 22(3):314–320
13. Pearce EL, Pearce EJ (2013) Metabolic pathways in immune cell activation and quiescence. Immunity 38(4):633–643
14. Pearce EL, Poffenberger MC, Chang CH, Jones RG (2013) Fueling immunity: insights into metabolism and lymphocyte function. Science 342(6155):1242454
15. Lochner M, Berod L, Sparwasser T (2015) Fatty acid metabolism in the regulation of T cell function. Trends Immunol 36(2):81–91
16. Warburg O (1956) On the origin of cancer cells. Science 123(3191):309–314
17. Vander Heiden MG, Cantley LC, Thompson CB (2009) Understanding the Warburg effect: the metabolic requirements of cell proliferation. Science 324(5930):1029–1033
18. Lunt SY, Vander Heiden MG (2011) Aerobic glycolysis: meeting the metabolic requirements of cell proliferation. Annu Rev Cell Dev Biol 27:441–464
19. MacIver NJ, Michalek RD, Rathmell JC (2013) Metabolic regulation of T lymphocytes. Annu Rev Immunol 31:259–283
20. Rudolph MG, Stanfield RL, Wilson IA (2006) How TCRs bind MHCs, peptides, and coreceptors. Annu Rev Immunol 24:419–466
21. Smith-Garvin JE, Koretzky GA, Jordan MS (2009) T cell activation. Annu Rev Immunol 27:591–619
22. Rossjohn J, Gras S, Miles JJ, Turner SJ, Godfrey DI, McCluskey J (2015) T cell antigen receptor recognition of antigen-presenting molecules. Annu Rev Immunol 33:169–200
23. Navarro MN, Cantrell DA (2014) Serine-threonine kinases in TCR signaling. Nat Immunol 15(9):808–814
24. Malissen B, Bongrand P (2015) Early T cell activation: integrating biochemical, structural, and biophysical cues. Annu Rev Immunol 33:539–561
25. Guy CS, Vignali KM, Temirov J, Bettini ML, Overacre AE, Smeltzer M et al (2013) Distinct TCR signaling pathways drive proliferation and cytokine production in T cells. Nat Immunol 14(3):262–270
26. Wang R, Dillon CP, Shi LZ, Milasta S, Carter R, Finkelstein D et al (2011) The transcription factor Myc controls metabolic reprogramming upon T lymphocyte activation. Immunity 35(6):871–882
27. Shi LZ, Wang R, Huang G, Vogel P, Neale G, Green DR et al (2011) HIF1alpha-dependent glycolytic pathway orchestrates a metabolic checkpoint for the differentiation of TH17 and Treg cells. J Exp Med 208(7):1367–1376
28. Man K, Miasari M, Shi W, Xin A, Henstridge DC, Preston S et al (2013) The transcription factor IRF4 is essential for TCR affinity-mediated metabolic programming and clonal expansion of T cells. Nat Immunol 14(11):1155–1165
29. Kidani Y, Elsaesser H, Hock MB, Vergnes L, Williams KJ, Argus JP et al (2013) Sterol regulatory element-binding proteins are essential for the metabolic programming of effector T cells and adaptive immunity. Nat Immunol 14(5):489–499

30. Chou C, Pinto AK, Curtis JD, Persaud SP, Cella M, Lin CC et al (2014) c-Myc-induced transcription factor AP4 is required for host protection mediated by CD8+ T cells. Nat Immunol 15(9):884–893
31. Buchholz VR, Schumacher TN, Busch DH (2015) T cell fate at the single-cell level. Annu Rev Immunol 34:65
32. Chi H (2012) Regulation and function of mTOR signalling in T cell fate decisions. Nat Rev Immunol 12(5):325–338
33. Waickman AT, Powell JD (2012) mTOR, metabolism, and the regulation of T-cell differentiation and function. Immunol Rev 249(1):43–58
34. Pollizzi KN, Powell JD (2014) Integrating canonical and metabolic signalling programmes in the regulation of T cell responses. Nat Rev Immunol 14(7):435–446
35. Laplante M, Sabatini DM (2012) mTOR signaling in growth control and disease. Cell 149(2):274–293
36. Jacinto E, Facchinetti V, Liu D, Soto N, Wei S, Jung SY et al (2006) SIN1/MIP1 maintains rictor-mTOR complex integrity and regulates Akt phosphorylation and substrate specificity. Cell 127(1):125–137
37. Ikenoue T, Inoki K, Yang Q, Zhou X, Guan KL (2008) Essential function of TORC2 in PKC and Akt turn motif phosphorylation, maturation and signalling. EMBO J 27(14):1919–1931
38. Yan L, Mieulet V, Lamb RF (2008) mTORC2 is the hydrophobic motif kinase for SGK1. Biochem J 416(3):e19–e21
39. Garcia-Martinez JM, Alessi DR (2008) mTOR complex 2 (mTORC2) controls hydrophobic motif phosphorylation and activation of serum- and glucocorticoid-induced protein kinase 1 (SGK1). Biochem J 416(3):375–385
40. Macintyre AN, Finlay D, Preston G, Sinclair LV, Waugh CM, Tamas P et al (2011) Protein kinase B controls transcriptional programs that direct cytotoxic T cell fate but is dispensable for T cell metabolism. Immunity 34(2):224–236
41. Finlay DK, Rosenzweig E, Sinclair LV, Feijoo-Carnero C, Hukelmann JL, Rolf J et al (2012) PDK1 regulation of mTOR and hypoxia-inducible factor 1 integrate metabolism and migration of CD8+ T cells. J Exp Med 209(13):2441–2453
42. Turner MS, Kane LP, Morel PA (2009) Dominant role of antigen dose in CD4+Foxp3+ regulatory T cell induction and expansion. J Immunol 183(8):4895–4903
43. Katzman SD, O'Gorman WE, Villarino AV, Gallo E, Friedman RS, Krummel MF et al (2010) Duration of antigen receptor signaling determines T-cell tolerance or activation. Proc Natl Acad Sci U S A 107(42):18085–18090
44. Delgoffe GM, Kole TP, Zheng Y, Zarek PE, Matthews KL, Xiao B et al (2009) The mTOR kinase differentially regulates effector and regulatory T cell lineage commitment. Immunity 30(6):832–844
45. Doedens AL, Phan AT, Stradner MH, Fujimoto JK, Nguyen JV, Yang E et al (2013) Hypoxia-inducible factors enhance the effector responses of CD8(+) T cells to persistent antigen. Nat Immunol 14(11):1173–1182
46. Delgoffe GM, Pollizzi KN, Waickman AT, Heikamp E, Meyers DJ, Horton MR et al (2011) The kinase mTOR regulates the differentiation of helper T cells through the selective activation of signaling by mTORC1 and mTORC2. Nat Immunol 12(4):295–303
47. Lee K, Gudapati P, Dragovic S, Spencer C, Joyce S, Killeen N et al (2010) Mammalian target of rapamycin protein complex 2 regulates differentiation of Th1 and Th2 cell subsets via distinct signaling pathways. Immunity 32(6):743–753
48. Michalek RD, Gerriets VA, Nichols AG, Inoue M, Kazmin D, Chang CY et al (2011) Estrogen-related receptor-alpha is a metabolic regulator of effector T-cell activation and differentiation. Proc Natl Acad Sci U S A 108(45):18348–18353
49. Colombetti S, Basso V, Mueller DL, Mondino A (2006) Prolonged TCR/CD28 engagement drives IL-2-independent T cell clonal expansion through signaling mediated by the mammalian target of rapamycin. J Immunol 176(5):2730–2738

50. Zheng Y, Collins SL, Lutz MA, Allen AN, Kole TP, Zarek PE et al (2007) A role for mammalian target of rapamycin in regulating T cell activation versus anergy. J Immunol 178(4):2163–2170
51. Zheng Y, Delgoffe GM, Meyer CF, Chan W, Powell JD (2009) Anergic T cells are metabolically anergic. J Immunol 183(10):6095–6101
52. So T, Choi H, Croft M (2011) OX40 complexes with phosphoinositide 3-kinase and protein kinase B (PKB) to augment TCR-dependent PKB signaling. J Immunol 186(6):3547–3555
53. Francisco LM, Salinas VH, Brown KE, Vanguri VK, Freeman GJ, Kuchroo VK et al (2009) PD-L1 regulates the development, maintenance, and function of induced regulatory T cells. J Exp Med 206(13):3015–3029
54. Rochman Y, Spolski R, Leonard WJ (2009) New insights into the regulation of T cells by gamma(c) family cytokines. Nat Rev Immunol 9(7):480–490
55. Noguchi M, Yi H, Rosenblatt HM, Filipovich AH, Adelstein S, Modi WS et al (1993) Interleukin-2 receptor gamma chain mutation results in X-linked severe combined immunodeficiency in humans. Cell 73(1):147–157
56. Liao W, Lin JX, Leonard WJ (2013) Interleukin-2 at the crossroads of effector responses, tolerance, and immunotherapy. Immunity 38(1):13–25
57. Zhou W, Zhang F, Aune TM (2003) Either IL-2 or IL-12 is sufficient to direct Th1 differentiation by nonobese diabetic T cells. J Immunol 170(2):735–740
58. Cote-Sierra J, Foucras G, Guo L, Chiodetti L, Young HA, Hu-Li J et al (2004) Interleukin 2 plays a central role in Th2 differentiation. Proc Natl Acad Sci U S A 101(11):3880–3885
59. de la Rosa M, Rutz S, Dorninger H, Scheffold A (2004) Interleukin-2 is essential for CD4+CD25+ regulatory T cell function. Eur J Immunol 34(9):2480–2488
60. Yang XP, Ghoreschi K, Steward-Tharp SM, Rodriguez-Canales J, Zhu J, Grainger JR et al (2011) Opposing regulation of the locus encoding IL-17 through direct, reciprocal actions of STAT3 and STAT5. Nat Immunol 12(3):247–254
61. Ballesteros-Tato A, Leon B, Graf BA, Moquin A, Adams PS, Lund FE et al (2012) Interleukin-2 inhibits germinal center formation by limiting T follicular helper cell differentiation. Immunity 36(5):847–856
62. Boyman O, Sprent J (2012) The role of interleukin-2 during homeostasis and activation of the immune system. Nat Rev Immunol 12(3):180–190
63. Ray JP, Staron MM, Shyer JA, Ho PC, Marshall HD, Gray SM et al (2015) The interleukin-2-mTORc1 kinase Axis defines the signaling, differentiation, and metabolism of T helper 1 and follicular B helper T cells. Immunity 43(4):690–702
64. Oestreich KJ, Read KA, Gilbertson SE, Hough KP, McDonald PW, Krishnamoorthy V et al (2014) Bcl-6 directly represses the gene program of the glycolysis pathway. Nat Immunol 15(10):957–964
65. von Freeden-Jeffry U, Solvason N, Howard M, Murray R (1997) The earliest T lineage-committed cells depend on IL-7 for Bcl-2 expression and normal cell cycle progression. Immunity 7(1):147–154
66. Dunkle A, Dzhagalov I, He YW (2011) Cytokine-dependent and cytokine-independent roles for Mcl-1: genetic evidence for multiple mechanisms by which Mcl-1 promotes survival in primary T lymphocytes. Cell Death Dis 2:e214
67. Jacobs SR, Michalek RD, Rathmell JC (2010) IL-7 is essential for homeostatic control of T cell metabolism in vivo. J Immunol 184(7):3461–3469
68. Yang K, Neale G, Green DR, He W, Chi H (2011) The tumor suppressor Tsc1 enforces quiescence of naive T cells to promote immune homeostasis and function. Nat Immunol 12(9):888–897
69. van der Windt GJ, Everts B, Chang CH, Curtis JD, Freitas TC, Amiel E et al (2012) Mitochondrial respiratory capacity is a critical regulator of CD8+ T cell memory development. Immunity 36(1):68–78

70. O'Sullivan D, van der Windt GJ, Huang SC, Curtis JD, Chang CH, Buck MD et al (2014) Memory CD8(+) T cells use cell-intrinsic lipolysis to support the metabolic programming necessary for development. Immunity 41(1):75–88
71. Cui G, Staron MM, Gray SM, Ho PC, Amezquita RA, Wu J et al (2015) IL-7-induced glycerol transport and TAG synthesis promotes memory CD8+ T cell longevity. Cell 161(4):750–761
72. Berard M, Brandt K, Bulfone-Paus S, Tough DF (2003) IL-15 promotes the survival of naive and memory phenotype CD8+ T cells. J Immunol 170(10):5018–5026
73. Richer MJ, Pewe LL, Hancox LS, Hartwig SM, Varga SM, Harty JT (2015) Inflammatory IL-15 is required for optimal memory T cell responses. J Clin Invest 125(9):3477–3490
74. Pearce EL, Walsh MC, Cejas PJ, Harms GM, Shen H, Wang LS et al (2009) Enhancing CD8 T-cell memory by modulating fatty acid metabolism. Nature 460(7251):103–107
75. Arneja A, Johnson H, Gabrovsek L, Lauffenburger DA, White FM (2014) Qualitatively different T cell phenotypic responses to IL-2 versus IL-15 are unified by identical dependences on receptor signal strength and duration. J Immunol 192(1):123–135
76. Dang CV (2012) MYC on the path to cancer. Cell 149(1):22–35
77. Zeller KI, Zhao X, Lee CW, Chiu KP, Yao F, Yustein JT et al (2006) Global mapping of c-Myc binding sites and target gene networks in human B cells. Proc Natl Acad Sci U S A 103(47):17834–17839
78. Chen X, Xu H, Yuan P, Fang F, Huss M, Vega VB et al (2008) Integration of external signaling pathways with the core transcriptional network in embryonic stem cells. Cell 133(6):1106–1117
79. Nie Z, Hu G, Wei G, Cui K, Yamane A, Resch W et al (2012) c-Myc is a universal amplifier of expressed genes in lymphocytes and embryonic stem cells. Cell 151(1):68–79
80. Lin CY, Loven J, Rahl PB, Paranal RM, Burge CB, Bradner JE et al (2012) Transcriptional amplification in tumor cells with elevated c-Myc. Cell 151(1):56–67
81. Thomas LR, Tansey WP (2011) Proteolytic control of the oncoprotein transcription factor Myc. Adv Cancer Res 110:77–106
82. Preston GC, Sinclair LV, Kaskar A, Hukelmann JL, Navarro MN, Ferrero I et al (2015) Single cell tuning of Myc expression by antigen receptor signal strength and interleukin-2 in T lymphocytes. EMBO J 34(15):2008–2024
83. Mognol GP, de Araujo-Souza PS, Robbs BK, Teixeira LK, Viola JP (2012) Transcriptional regulation of the c-Myc promoter by NFAT1 involves negative and positive NFAT-responsive elements. Cell Cycle 11(5):1014–1028
84. Li F, Wang Y, Zeller KI, Potter JJ, Wonsey DR, O'Donnell KA et al (2005) Myc stimulates nuclearly encoded mitochondrial genes and mitochondrial biogenesis. Mol Cell Biol 25(14):6225–6234
85. Palazon A, Goldrath AW, Nizet V, Johnson RS (2014) HIF transcription factors, inflammation, and immunity. Immunity 41(4):518–528
86. Wang R, Green DR (2012) Metabolic checkpoints in activated T cells. Nat Immunol 13(10):907–915
87. Dang EV, Barbi J, Yang HY, Jinasena D, Yu H, Zheng Y et al (2011) Control of T(H)17/T(reg) balance by hypoxia-inducible factor 1. Cell 146(5):772–784
88. Nakamura H, Makino Y, Okamoto K, Poellinger L, Ohnuma K, Morimoto C et al (2005) TCR engagement increases hypoxia-inducible factor-1 alpha protein synthesis via rapamycin-sensitive pathway under hypoxic conditions in human peripheral T cells. J Immunol 174(12):7592–7599
89. Wang H, Flach H, Onizawa M, Wei L, McManus MT, Weiss A (2014) Negative regulation of Hif1a expression and TH17 differentiation by the hypoxia-regulated microRNA miR-210. Nat Immunol 15(4):393–401
90. Taniguchi T, Ogasawara K, Takaoka A, Tanaka N (2001) IRF family of transcription factors as regulators of host defense. Annu Rev Immunol 19:623–655
91. Huber M, Lohoff M (2014) IRF4 at the crossroads of effector T-cell fate decision. Eur J Immunol 44(7):1886–1895

92. Lohoff M, Mittrucker HW, Prechtl S, Bischof S, Sommer F, Kock S et al (2002) Dysregulated T helper cell differentiation in the absence of interferon regulatory factor 4. Proc Natl Acad Sci U S A 99(18):11808–11812
93. Staudt V, Bothur E, Klein M, Lingnau K, Reuter S, Grebe N et al (2010) Interferon-regulatory factor 4 is essential for the developmental program of T helper 9 cells. Immunity 33(2):192–202
94. Brustle A, Heink S, Huber M, Rosenplanter C, Stadelmann C, Yu P et al (2007) The development of inflammatory T(H)-17 cells requires interferon-regulatory factor 4. Nat Immunol 8(9):958–966
95. Kwon H, Thierry-Mieg D, Thierry-Mieg J, Kim HP, Oh J, Tunyaplin C et al (2009) Analysis of interleukin-21-induced Prdm1 gene regulation reveals functional cooperation of STAT3 and IRF4 transcription factors. Immunity 31(6):941–952
96. Cretney E, Xin A, Shi W, Minnich M, Masson F, Miasari M et al (2011) The transcription factors Blimp-1 and IRF4 jointly control the differentiation and function of effector regulatory T cells. Nat Immunol 12(4):304–311
97. Yao S, Buzo BF, Pham D, Jiang L, Taparowsky EJ, Kaplan MH et al (2013) Interferon regulatory factor 4 sustains CD8(+) T cell expansion and effector differentiation. Immunity 39(5):833–845
98. Grusdat M, McIlwain DR, Xu HC, Pozdeev VI, Knievel J, Crome SQ et al (2014) IRF4 and BATF are critical for CD8(+) T-cell function following infection with LCMV. Cell Death Differ 21(7):1050–1060
99. Li P, Spolski R, Liao W, Wang L, Murphy TL, Murphy KM et al (2012) BATF-JUN is critical for IRF4-mediated transcription in T cells. Nature 490(7421):543–546
100. Glasmacher E, Agrawal S, Chang AB, Murphy TL, Zeng W, Vander Lugt B et al (2012) A genomic regulatory element that directs assembly and function of immune-specific AP-1-IRF complexes. Science 338(6109):975–980
101. Ciofani M, Madar A, Galan C, Sellars M, Mace K, Pauli F et al (2012) A validated regulatory network for Th17 cell specification. Cell 151(2):289–303
102. Nayar R, Enos M, Prince A, Shin H, Hemmers S, Jiang JK et al (2012) TCR signaling via Tec kinase ITK and interferon regulatory factor 4 (IRF4) regulates CD8+ T-cell differentiation. Proc Natl Acad Sci U S A 109(41):E2794–E2802
103. Grumont RJ, Gerondakis S (2000) Rel induces interferon regulatory factor 4 (IRF-4) expression in lymphocytes: modulation of interferon-regulated gene expression by rel/nuclear factor kappaB. J Exp Med 191(8):1281–1292
104. Nayar R, Schutten E, Bautista B, Daniels K, Prince AL, Enos M et al (2014) Graded levels of IRF4 regulate CD8+ T cell differentiation and expansion, but not attrition, in response to acute virus infection. J Immunol 192(12):5881–5893
105. Crotty S, Johnston RJ, Schoenberger SP (2010) Effectors and memories: Bcl-6 and Blimp-1 in T and B lymphocyte differentiation. Nat Immunol 11(2):114–120
106. Beaulieu AM, Sant'Angelo DB (2011) The BTB-ZF family of transcription factors: key regulators of lineage commitment and effector function development in the immune system. J Immunol 187(6):2841–2847
107. Sukumar M, Liu J, Ji Y, Subramanian M, Crompton JG, Yu Z et al (2013) Inhibiting glycolytic metabolism enhances CD8+ T cell memory and antitumor function. J Clin Invest 123(10):4479–4488
108. Man K, Kallies A (2014) Bcl-6 gets T cells off the sugar. Nat Immunol 15(10):904–905
109. Yu D, Rao S, Tsai LM, Lee SK, He Y, Sutcliffe EL et al (2009) The transcriptional repressor Bcl-6 directs T follicular helper cell lineage commitment. Immunity 31(3):457–468
110. Pipkin ME, Sacks JA, Cruz-Guilloty F, Lichtenheld MG, Bevan MJ, Rao A (2010) Interleukin-2 and inflammation induce distinct transcriptional programs that promote the differentiation of effector cytolytic T cells. Immunity 32(1):79–90

111. Oestreich KJ, Mohn SE, Weinmann AS (2012) Molecular mechanisms that control the expression and activity of Bcl-6 in TH1 cells to regulate flexibility with a TFH-like gene profile. Nat Immunol 13(4):405–411
112. Chang JT, Wherry EJ, Goldrath AW (2014) Molecular regulation of effector and memory T cell differentiation. Nat Immunol 15(12):1104–1115
113. Horton JD, Goldstein JL, Brown MS (2002) SREBPs: activators of the complete program of cholesterol and fatty acid synthesis in the liver. J Clin Invest 109(9):1125–1131
114. Yang W, Bai Y, Xiong Y, Zhang J, Chen S, Zheng X et al (2016) Potentiating the antitumour response of CD8(+) T cells by modulating cholesterol metabolism. Nature 531(7596):651–655
115. Egawa T, Littman DR (2011) Transcription factor AP4 modulates reversible and epigenetic silencing of the Cd4 gene. Proc Natl Acad Sci U S A 108(36):14873–14878
116. Jonsson H, Peng SL (2005) Forkhead transcription factors in immunology. Cell Mol Life Sci 62(4):397–409
117. Feng X, Wang H, Takata H, Day TJ, Willen J, Hu H (2011) Transcription factor Foxp1 exerts essential cell-intrinsic regulation of the quiescence of naive T cells. Nat Immunol 12(6):544–550
118. Wei H, Geng J, Shi B, Liu Z, Wang YH, Stevens AC et al (2016) Cutting edge: Foxp1 controls naive CD8+ T cell quiescence by simultaneously repressing key pathways in cellular metabolism and cell cycle progression. J Immunol 196:3537–3541
119. Bopp T, Palmetshofer A, Serfling E, Heib V, Schmitt S, Richter C et al (2005) NFATc2 and NFATc3 transcription factors play a crucial role in suppression of CD4+ T lymphocytes by CD4+ CD25+ regulatory T cells. J Exp Med 201(2):181–187
120. Huh JR, Leung MW, Huang P, Ryan DA, Krout MR, Malapaka RR et al (2011) Digoxin and its derivatives suppress TH17 cell differentiation by antagonizing RORgammat activity. Nature 472(7344):486–490
121. Xiao S, Yosef N, Yang J, Wang Y, Zhou L, Zhu C et al (2014) Small-molecule RORgammat antagonists inhibit T helper 17 cell transcriptional network by divergent mechanisms. Immunity 40(4):477–489
122. Zhang H, Qian DZ, Tan YS, Lee K, Gao P, Ren YR et al (2008) Digoxin and other cardiac glycosides inhibit HIF-1alpha synthesis and block tumor growth. Proc Natl Acad Sci U S A 105(50):19579–19586
123. Delmore JE, Issa GC, Lemieux ME, Rahl PB, Shi J, Jacobs HM et al (2011) BET bromodomain inhibition as a therapeutic strategy to target c-Myc. Cell 146(6):904–917
124. Mertz JA, Conery AR, Bryant BM, Sandy P, Balasubramanian S, Mele DA et al (2011) Targeting MYC dependence in cancer by inhibiting BET bromodomains. Proc Natl Acad Sci U S A 108(40):16669–16674
125. Bandukwala HS, Gagnon J, Togher S, Greenbaum JA, Lamperti ED, Parr NJ et al (2012) Selective inhibition of CD4+ T-cell cytokine production and autoimmunity by BET protein and c-Myc inhibitors. Proc Natl Acad Sci U S A 109(36):14532–14537
126. Zirath H, Frenzel A, Oliynyk G, Segerstrom L, Westermark UK, Larsson K et al (2013) MYC inhibition induces metabolic changes leading to accumulation of lipid droplets in tumor cells. Proc Natl Acad Sci U S A 110(25):10258–10263
127. Touzot M, Soulillou JP, Dantal J (2012) Mechanistic target of rapamycin inhibitors in solid organ transplantation: from benchside to clinical use. Curr Opin Organ Transplant 17(6):626–633
128. Powell JD, Lerner CG, Schwartz RH (1999) Inhibition of cell cycle progression by rapamycin induces T cell clonal anergy even in the presence of costimulation. J Immunol 162(5):2775–2784
129. Sarbassov DD, Ali SM, Sengupta S, Sheen JH, Hsu PP, Bagley AF et al (2006) Prolonged rapamycin treatment inhibits mTORC2 assembly and Akt/PKB. Mol Cell 22(2):159–168
130. Turner AP, Shaffer VO, Araki K, Martens C, Turner PL, Gangappa S et al (2011) Sirolimus enhances the magnitude and quality of viral-specific CD8+ T-cell responses to vaccinia virus vaccination in rhesus macaques. Am J Transplant 11(3):613–618

131. Thoreen CC, Kang SA, Chang JW, Liu Q, Zhang J, Gao Y et al (2009) An ATP-competitive mammalian target of rapamycin inhibitor reveals rapamycin-resistant functions of mTORC1. J Biol Chem 284(12):8023–8032
132. Feldman ME, Apsel B, Uotila A, Loewith R, Knight ZA, Ruggero D et al (2009) Active-site inhibitors of mTOR target rapamycin-resistant outputs of mTORC1 and mTORC2. PLoS Biol 7(2):e38
133. Chang CH, Qiu J, O'Sullivan D, Buck MD, Noguchi T, Curtis JD et al (2015) Metabolic competition in the tumor microenvironment is a driver of cancer progression. Cell 162(6):1229–1241
134. Ho PC, Bihuniak JD, Macintyre AN, Staron M, Liu X, Amezquita R et al (2015) Phosphoenolpyruvate is a metabolic checkpoint of anti-tumor T cell responses. Cell 162(6):1217–1228

Chapter 4
Adipose Tissue-Resident Regulatory T Cells

Fuxiang Zhu, Aiting Wang, Yangyang Li, Rui Liang, Dan Li, and Bin Li

Abstract Tissue-resident immune cells play critical roles in regulating tissue function and homeostasis. Obesity-associated visceral adipose tissue inflammation is attributed to the accumulation of M1 macrophages which produce inflammatory cytokines like TNF-α, IL-6, and expansion of effector T cells like Th1 cells, $CD8^+$ cytotoxic T cells which produce interferon-γ to further add to the severity of inflammation in the visceral adipose tissue. Regulatory T cells have been reported to exert key roles in suppressing inflammation, thus maintaining the homeostasis of immune responses, and visceral adipose Tregs exert critical roles in defending against obesity-associated metabolic disorders. They inhibit the infiltration of effector T cells and facilitate the reconstitution of adipose tissue macrophages from M1 to M2

F. Zhu
Key Laboratory of Molecular Virology & Immunology, Institut Pasteur of Shanghai, Shanghai Institutes for Biological Sciences, University of Chinese Academy of Sciences, Shanghai 200031, China

A. Wang • R. Liang
CAS Center for Excellence in Molecular Cell Science, Unit of Molecular Immunology, Institut Pasteur of Shanghai, Shanghai Institutes for Biological Sciences, University of Chinese Academy of Sciences, Chinese Academy of Sciences, Shanghai 200025, China

Shanghai Institute of Immunology & Department of Immunology and Microbiology, Shanghai Jiao Tong University School of Medicine, Shanghai 200025, China

Y. Li • D. Li
Shanghai Institute of Immunology, Shanghai JiaoTong University School of Medicine, Shanghai 200025, China

Department of Immunology and Microbiology, Shanghai Jiao Tong University School of Medicine, Shanghai 200025, China

B. Li (✉)
Shanghai Institute of Immunology, Department of Immunology and Microbiology, Shanghai Jiao Tong University, Shanghai 200025, China

School of Medicine, Shanghai Jiao Tong University, Shanghai 200025, China

CAS Center for Excellence in Molecular Cell Science, CAS Key Laboratory of Molecular Virology and Immunology, Unit of Molecular Immunology, Institut Pasteur of Shanghai, Shanghai Institutes for Biological Sciences, University of the Chinese Academy of Sciences, Chinese Academy of Sciences, Shanghai 200025, China
e-mail: binli@shsmu.edu.cn

B. Li, F. Pan (eds.), *Immune Metabolism in Health and Tumor*, Advances in Experimental Medicine and Biology 1011, DOI 10.1007/978-94-024-1170-6_4

phenotype. What is more, they can take up lipids from the adipocytes through CD36 which is driven by PPARγ. Here we review the recent progress in adipose tissue-resident regulatory T cells (Tregs), a subpopulation of $CD4^+$ T cells which suppress adipose tissue inflammation.

Keywords VAT Tregs • Obesity-associated metabolic disorders

4.1 Introduction

In recent years obesity has become one of the major health problems across the world, and the impact of obesity-associated metabolic disorders like insulin-resistant type 2 diabetes increases every day [8]. Obesity-associated visceral adipose tissue inflammation is attributed to the accumulation of M1 macrophages, which produce inflammatory cytokines like TNF-α, IL-6, and expansion of effector T cells. Th1 $CD4^+$ and $CD8^+$ cytotoxic T cells (which produce interferon-γ) further add to the severity of inflammation in the visceral adipose tissue [15]. Regulatory T cells, on the other hand, suppress inflammation, thus maintaining the homeostasis of immune responses [23]. The presence of visceral adipose tissue (VAT) Tregs causes persistence of chronic inflammation in the adipose tissue [17]. Chipolletta et al. reported that in slim individuals, the number of VAT Tregs increases with age, but in obese individuals, it decreases with age. These findings suggest that Tregs contribute to obesity and metabolism throughout the body.

4.2 VAT Tregs Protect Against Obesity-Associated Metabolic Disorders

$CD4^+CD25^+$ Tregs were discovered by Sakaguchi et al. in 1995. These cells play an indispensable role in the maintenance of immunologic self-tolerance and immune homeostasis [23]. It is widely accepted that “natural” Tregs originate in the thymus and migrate to peripheral tissues. Tregs can also be induced from peripheral naïve $CD4^+$ T cells. Tregs are distinguished from other $CD4^+$ T cells in their high constitutive expression of CD25, transcription factor forkhead P3(FOXP3), glucocorticoid induced tumor necrosis factor (GITR), and cytotoxic T lymphocyte antigen (CTLA-4) [23]. FOXP3 is the master regulator that mediates the development and function of Tregs. However, the mechanisms by which Tregs suppress immune responses are still not well understood. It is proposed that CTLA-4 on the surface of Tregs competes with the effector T cells’ CD28 for B7 molecules (CD80/86) on the surface of antigen-presenting cells (APCs), since CTLA4 has a higher affinity for or B7 molecules (CD80/86), and outcompetes CD28 for B7. This inhibits the ability of APCs to activate effector cells. Tregs could also compete with effector cells for IL-2; IL-2 is necessary for the survival and proliferation of effector T cells (Th1, Th2 cells) as well as Tregs (which could not produce, but heavily rely on IL-2) [24]. Furthermore,

IL-10 – which is predominantly produced by Tregs – suppresses other effector T cells [24]. Overall, Tregs are inhibitors of immune responses; thus, Treg-associated deficiencies lead to autoimmune or allergic diseases such as type 1 diabetes, multiple sclerosis, and rheumatoid arthritis [5]. Recent studies demonstrate that Tregs may also be involved in obesity-associated inflammation and insulin resistance. It is reported that the percentage of VAT Tregs among $CD4^+$ T cells in the adipose tissue is higher than that in the spleen, peripheral lymph nodes, and nonlymphoid tissues like the liver, lungs, and subcutaneous adipose tissue in lean mice [25]. Furthermore, in obese mice, a pro-inflammatory T cell profile is characterized by elevated Th17/Treg and Th1/Treg ratios [29]. VAT is a more potent suppressor of other CD4 and CD8 cells compared to conventional Tregs. All those above indicate that VAT Tregs play a critical role in obesity-associated metabolic disorders.

4.3 The Origin of VAT Tregs

Regulatory T cells can originate thymically or peripherally [23]. Natural Tregs are thymically derived; they bear TCRs specific for self-antigens, which help keep autoimmune and inflammatory responses in check. Alternatively, regulatory T cells can also be generated from conventional naive $CD4^+$ T cells, which are stimulated in the presence of transforming growth factor β, IL-2, or retinoid acid [11]. Since the natural Treg TCR repertoire specifically recognizes self-antigens [22], VAT Tregs appear to be a subpopulation of natural Tregs. Furthermore, activation by self-antigens in the nonlymphoid tissues may aid maintenance of VAT Tregs. Namely, constant activation by self-antigens in situ helps to retain Tregs at peripheral sites. TCR repertoire analysis comparing VAT Tregs and Tregs derived from the lymphoid organs shows that there is very little overlap between them; this suggests that VAT Tregs represent a distinct population. The VAT Treg TCR repertoire also differs from that of conventional $CD4^+$ cells, suggesting that the VAT Treg population does not arise from conventional naive T cells [25]. Finally, there is much repetition in the TCR repertoire within the VAT Tregs, indicating that specific antigens in the adipose tissue may be highly responsible for the accumulation of VAT Tregs. Evidence suggests that VAT Tregs are a subclass of natural Tregs; however, it is still unclear which self-antigens are responsive for activation and maintenance of VAT Tregs in adipose tissues.

4.4 The Role of Tregs in Obesity-Associated Metabolic Disorders

Chronic inflammation in the adipose tissue correlates with metabolic disorders like type 2 diabetes (T2D), insulin resistance, and atherosclerosis [7]. Obesity-induced inflammation in the adipose tissue is predominantly mediated by resident immune cells [16]. Adipose tissue macrophages (ATM) are thought to be the major driver in obesity-associated inflammation [19]. Many fatty acids within the adipose tissue can induce the secretion of TNF-alpha by adipocytes; TNF-alpha then promotes the inflammatory phenotype of M1 macrophages [26]. $CD8^+$ T cells within the adipose tissue can also facilitate the ATM polarization toward the pro-inflammatory M1 phenotype [18]. Recently, it has been reported that NK cells within the adipose tissue are capable of producing TNF-alpha and IFN-γ to promote M1 ATM as well [13]. Apart from ATMs, Th1 and Th17 cells are also reported to promote obesity-associated metabolic disorders [29].

Because regulatory T cells play an essential role in suppressing inflammation, they have an important role in many obesity-associated metabolic diseases [3]. Studies show that induction of VAT Tregs in mice protects the animals from high-fat diet-induced metabolic disorders like insulin resistance, while depletion of VAT Tregs causes susceptibility to adipose tissue inflammation [5, 11]. Th2 cells also appear to be important for metabolic homeostasis; adoptive transfer of GATA3-expressing $CD4^+$ T cells has been reported to rescue insulin resistance. However, since Tregs also express GATA3, they may entirely account for the alleviation of obesity-associated metabolic disorders [27].

Consequently, it is widely accepted that presence of VAT Tregs correlates with lower inflammation, improved glucose metabolism, and reduced insulin resistance. Mice models of type 2 diabetes show a reduction in adipocyte size, number, and weight after adoptive transfer of Tregs [5]. As chronic inflammation is one of the main causes of insulin resistance, VAT Tregs likely prevent this phenomenon by suppressing local inflammation. Inflammatory cytokines like IL-6 and TNF-α mainly secreted by M1 macrophages can induce glucose tolerance and insulin insensitivity by inhibiting the transport of glucose into adipocytes [15]. IL-10 predominantly produced by VAT Tregs can reverse the effects of TNF-α and IL-6 [6]. Furthermore, VAT Tregs can facilitate the conversion of ATM from the M1 to the M2 phenotype, which is associated with immunosuppression [14]. M2 macrophages also prevent infiltration of inflammatory IFNγ-producing $CD8^+$ T and Th1 $CD4^+$ T cells [7]. VAT Tregs may also contribute to the accumulation of Th2 cells, which have been reported to improve insulin sensitivity [12].

4.5 Controversies Related to VAT Tregs

Insulin resistance constitutes two distinct physiologic forms, obesity-associated insulin resistance and age-associated insulin resistance. While VAT Tregs appear to protect individuals from obesity-associated insulin resistance, it has recently been reported that selective depletion of fat Tregs (fTregs) also improves insulin sensitivity in age-associated insulin resistance. As a result, fTregs present a potential target for treatment of age-associated insulin resistance. Surprisingly, depletion of fTregs in mice increases the animals' susceptibility to obesity-associated insulin resistance. The role of VAT Tregs related to T2D or IR in humans is also controversial because some studies show that the percentage of peripheral Tregs greatly decreases in obese individuals compared to their lean counterparts [14]. Likewise, it is also reported that in patients with type 2 diabetes, the percentage of circulating Th17 cells and Th1 cells was significantly higher, but the number of Foxp3$^+$ Tregs was lower [29]. Nevertheless, Yun et al. report elevated Treg population in the visceral adipose tissue of obese humans [28]. One should keep in mind that most studies synonymize Tregs with cells that stain positive for the Foxp3 transcription factor to differentiate Tregs from other effector T cells. However, not all cells that express FoxP3 have equal immunosuppressive potential, and not all immunosuppressive CD4$^+$ T cells express Foxp3. A study that used Helios in addition to FoxP3 to identify VAT Tregs reports that the Foxp3$^+$Helios$^+$ population was smaller in obese individuals compared to lean counterparts [6]. Helios is a transcriptional factor that is controversially used to identify natural from induced Tregs, as even effector T cells can transiently induce Foxp3 expression following TCR stimulation [24]. Thus, if solely Foxp3 is used to identify Treg populations of interest, T cell subsets like Th17 or Th1, which produce inflammatory IL-17 and IFN-γ, respectively, may be falsely accounted for as VAT Tregs.

4.6 The Influence of VAT Microenvironment on Tregs

4.6.1 Chemokines and Chemokine Receptors

Treg trafficking, like that of other immune cells, is mediated by adhesion molecules and chemokines. Th1, Th2, and Th17 cells can be distinguished, in part, from each other by their chemokine receptors. Likewise, it has been shown that Tregs accumulate in different tissues according to signature chemokine receptor profiles. Naïve nTregs that have originated in the thymus express L-selectin and the CC chemokine receptor (CCR) 7 while circulating through the secondary lymphoid organs [28]. Upon activation, these cells lose CCR7 expression and acquire chemokine receptors that promote migration to other tissues. For instance, CCR4 expression is required for the accumulation of Tregs within noninflamed liver, lung, and skin; chemokine ligand (CXC) 12/CXC chemokine receptor (CXCR) 3 and CCR10 are required for the migration of Tregs to the inflamed liver [6]; CXCR4 is required for the accumulation of Tregs in the bone marrow. Regulatory T cells in the adipose tissue also bear a distinct pattern of chemokine receptors in the lean VAT; they overexpress CCR1, CCR2, CCR3, CCR5, and CXCR6 while downregulating CCR6, CCR7, and CXCR3 [10]. In obese mice, upregulation of CCR6 and CXCR3 correlates with a reduction of VAT Tregs, while splenic Treg numbers are increased along with upregulation of CCR7 and CXCR3 [16]. This suggests that the acquisition of different chemokine receptors determines the localization of Tregs. Thus, altered trafficking of Tregs may be responsible for the reduction of Tregs in the adipose tissue of obese individuals.

4.6.2 The Effect of Adiponectin and Leptin on VAT Tregs

In individuals with MS (multiple sclerosis), the number of peripheral Tregs is inversely associated with levels of leptin in the serum [21]. Leptin is predominantly produced by adipocytes and appears to play a pro-inflammatory role in immune response; in leptin-deficient or leptin receptor-deficient mice, the number of peripheral Tregs was significantly reduced [9]. It has been reported that leptin augments the suppressive function of Tregs [20]. Besides adipocytes, Tregs themselves can also secrete leptin and express the leptin receptor in an autocrine manner. One study shows that serum leptin level is elevated, while the number of VAT Tregs is reduced in obese individuals [20]. While leptin inhibits the proliferation of Tregs, it appears to activate Th1 cells. This facilitates their proliferation and secretion of pro-inflammatory cytokines, thus promoting obesity-associated inflammation [9]. It is reported that leptin induces the activation of the mTOR signaling pathway, which downmodulates the activation and proliferation of Tregs [1]. Consistently, increased proliferation of Tregs was observed in leptin receptor-deficient mice. Furthermore, the population of VAT Tregs correlates inversely with the serum level of leptin. Heterozygous ob/+ mice have a much larger VAT Treg population than their lean wild-type counterparts [2]. Another study argues that leptin mediates its inhibitory effect over Tregs through dendritic cells [28]. Yun et al. observed that the expression of maturation markers on dendritic cells is reduced in the absence of leptin. This, in turn, leads to elevation of TGF-β (produced by Tregs) and a reduction of pro-inflammatory cytokines (produced by effector cells).

Adiponectin, a protein hormone predominantly produced by adipocytes, plays an anti-inflammatory role and protects individuals from obesity-associated diseases [21]. Its levels correlate negatively with the body mass index and positively with the number of Tregs in the visceral adipose tissue [20]. The effect of adiponectin on VAT Tregs is unclear, but adiponectin can induce secretion of the anti-inflammatory cytokine IL-10 in M2 macrophages [21]. IL-10, in turn, acts in a paracrine manner to facilitate Foxp3 expression and the suppressive function of Tregs. Furthermore, adiponectin-primed DCs may also promote Treg expansion and function through the enhancement of the programmed death-1/programmed death-1 ligand (PD-1/PD-L1) pathway [20]. The number of VAT Tregs and the level of adiponectin is high in the visceral adipose tissue but not in the subcutaneous and perirenal adipose tissue. This suggests that adipokines account for the accumulation of Tregs in the VAT but not in other fat depots.

4.7 Peroxisome Proliferator-Activated Receptor-γ

Peroxisome proliferator-activated receptor-γ (PPAR-γ) has been demonstrated to contribute to the adipogenesis by promoting fat storage, adipocyte differentiation, and transcription of key lipogenic proteins [4]. PPAR-γ expression in VAT

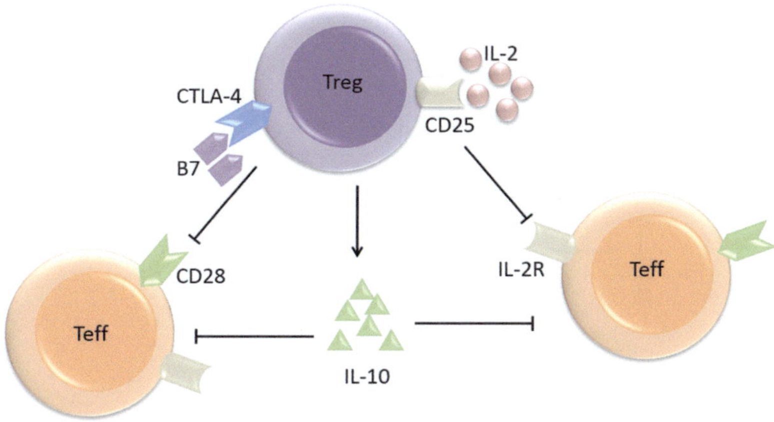

Fig. 4.1 The underlying mechanism of suppressive function of Tregs. CTLA-4 on the membrane surface of Tregs competes with other effector T cells for CD28. IL-2 which is necessary for the survival and proliferation of effector T cells (*Th1*, *Th2* cells) can be seized by Tregs which cannot produce IL-2 but greatly rely on IL-2 to sustain their survival [24]. What is more, IL-10 which is predominantly produced by Tregs exert suppressive functions over other effector T cells

macrophages is required for the phenotypic switch from M1 to M2 macrophages by inhibiting the expression of many inflammatory genes in M1 macrophages [19]. Recently, PPAR-γ (which is highly expressed in VAT Tregs compared to other T cell subsets) was identified as the key orchestrator of recruitment and function Tregs in VAT [4]. In fact, Treg-specific deletion of PPAR-γ decreased the number of Tregs in the VAT but not in the lymphoid and other nonlymphoid organs [4]. A transcriptional analysis of VAT Tregs shows that expression of IL-10 along with other genes correlates highly with that of PPAR-γ [4] (unclear what genes this sentence referred to…). PPAR-γ has been shown to interact with Foxp3 in a co-immunoprecipitation experiment. Furthermore, ectopic expression of PPAR-γ and Foxp3 in conventional $CD4^+$ T cells induced a VAT Treg gene expression profile [4]. VAT Tregs can uptake lipids through CD36, which is induced by PPAR-γ. However, since PPAR-γ signaling occurs in response to an abundance of fatty acids and their metabolites, it is sensible to speculate that VAT Tregs engage in this pathway to keep homeostasis within the adipose tissue in check. It is reported that PPAR-γ phosphorylation at the serine 273 residue by cyclin-dependent kinase 5 in adipocytes is a growth and differentiation checkpoint. The phosphorylation of PPAR-γ can lead to a reduction of adiponectin and high-fat diet-induced obesity. A similar mechanism may play a role in VAT Tregs, but posttranslational modifications of PPAR-γ in VAT Tregs are yet to be identified (Figs. 4.1, 4.2, and 4.3).

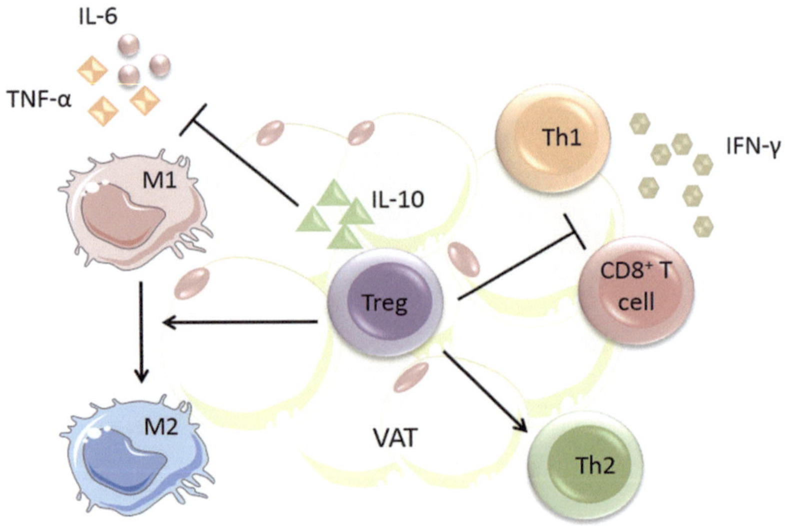

Fig. 4.2 How Tregs contribute their role within the adipose tissue. Tregs can help facilitate the M1 phenotype of adipose resident macrophages which play an anti-inflammatory role; Tregs also secrete much IL-10 to suppress TNF-α-mediated immune responses, what is more, Tregs contain the effector function Th1, Th17 cells while promoting the effector function of Th2 cells

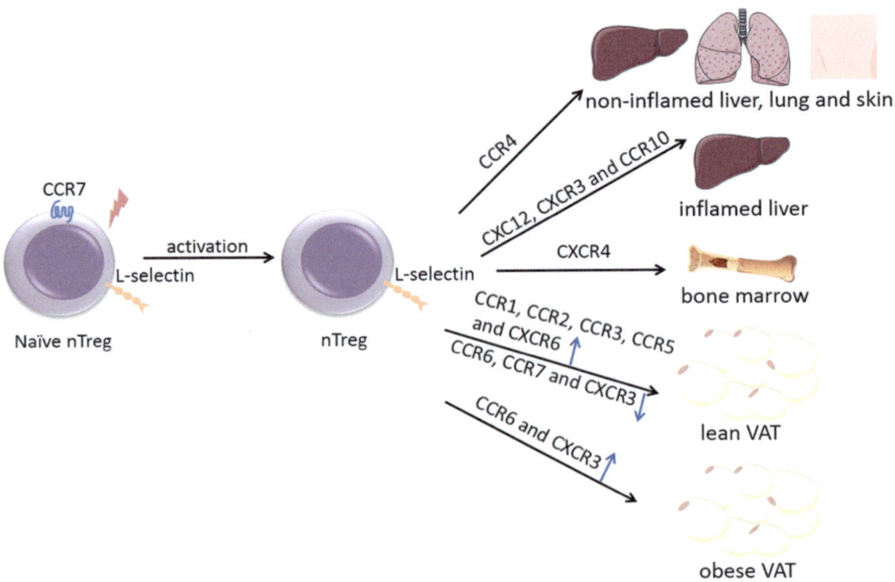

Fig. 4.3 Tregs are attracted to local tissues and organs through chemokine and chemokine receptors. CCR4 expression is required for the accumulation of Tregs within noninflamed liver, lung, and skin; CXC chemokine ligand 12 and CXC chemokine receptor (CXCR) 3 and CCR 10 are required for the migration of Treg cells to the inflamed liver [6]; CXCR4 is required for the accumulation of Treg cells to the bone marrow. Regulatory T cells in the adipose tissue also display a distinct pattern of chemokine receptors in the lean VAT by overexpressing CCR1, CCR2, CCR3, CCR5, and CXCR6 while downregulating CCR6, CCR7, and CXCR3 [10]. However, in obese mice, VAT Treg numbers are decreased by upregulating CCR6 and CXCR3, while splenic Treg numbers are increased with upregulated CCR7 and CXCR3 expression

4.8 Conclusion

Large number of Tregs within the visceral adipose tissue correlates highly with obesity-associated chronic inflammation. The chemokine, adipokine, fatty acid, and metabolite milieu within the adipose tissue is critical for the accumulation and function of VAT Tregs. However, the identification of antigens within the adipose tissue and key signaling molecules besides PPAR-γ that aid trafficking and interactions between adipocytes and VAT Tregs requires further investigation. We expect that new studies on gene engineering mice with the deficiency of key enzymatic subunit of FOXP3 complex in Tregs, which show particular phenotypes of dysfunctional adipocyte metabolism, may provide helpful insights in identifying new therapeutic targets for treating obesity and type 2 diabetes.

References

1. Battaglia M, Stabilini A, Roncarolo MG (2005) Rapamycin selectively expands CD4+CD25+FoxP3+ regulatory T cells. Blood 105:4743–4748
2. Chen X, Wu Y, Wang L (2013) Fat-resident Tregs: an emerging guard protecting from obesity-associated metabolic disorders. Obes Rev: Off J Int Assoc Stud Obes 14:568–578
3. Cipolletta D (2014) Adipose tissue-resident regulatory T cells: phenotypic specialization, functions and therapeutic potential. Immunology 142:517–525
4. Cipolletta D, Feuerer M, Li A, Kamei N, Lee J, Shoelson SE, Benoist C, Mathis D (2012) PPAR-gamma is a major driver of the accumulation and phenotype of adipose tissue Treg cells. Nature 486:549–553
5. Eller K, Kirsch A, Wolf AM, Sopper S, Tagwerker A, Stanzl U, Wolf D, Patsch W, Rosenkranz AR, Eller P (2011) Potential role of regulatory T cells in reversing obesity-linked insulin resistance and diabetic nephropathy. Diabetes 60:2954–2962
6. Feuerer M, Herrero L, Cipolletta D, Naaz A, Wong J, Nayer A, Lee J, Goldfine AB, Benoist C, Shoelson S, Mathis D (2009) Lean, but not obese, fat is enriched for a unique population of regulatory T cells that affect metabolic parameters. Nat Med 15:930–939
7. Glass CK, Olefsky JM (2012) Inflammation and lipid signaling in the etiology of insulin resistance. Cell Metab 15:635–645
8. Hotamisligil GS, Arner P, Caro JF, Atkinson RL, Spiegelman BM (1995) Increased adipose tissue expression of tumor necrosis factor-alpha in human obesity and insulin resistance. J Clin Invest 95:2409–2415
9. Ilan Y, Maron R, Tukpah AM, Maioli TU, Murugaiyan G, Yang K, Wu HY, Weiner HL (2010) Induction of regulatory T cells decreases adipose inflammation and alleviates insulin resistance in ob/ob mice. Proc Natl Acad Sci U S A 107:9765–9770
10. Jagannathan-Bogdan M, McDonnell ME, Shin H, Rehman Q, Hasturk H, Apovian CM, Nikolajczyk BS (2011) Elevated proinflammatory cytokine production by a skewed T cell compartment requires monocytes and promotes inflammation in type 2 diabetes. J Immunol 186:1162–1172
11. Josefowicz SZ, Lu LF, Rudensky AY (2012) Regulatory T cells: mechanisms of differentiation and function. Annu Rev Immunol 30:531–564
12. Kolodin D, van Panhuys N, Li C, Magnuson AM, Cipolletta D, Miller CM, Wagers A, Germain RN, Benoist C, Mathis D (2015) Antigen- and cytokine-driven accumulation of regulatory T cells in visceral adipose tissue of lean mice. Cell Metab 21:543–557

13. Lee BC, Kim MS, Pae M, Yamamoto Y, Eberle D, Shimada T, Kamei N, Park HS, Sasorith S, Woo JR et al (2016) Adipose natural killer cells regulate adipose tissue macrophages to promote insulin resistance in obesity. Cell Metab 23:685
14. Luczynski W, Wawrusiewicz-Kurylonek N, Ilendo E, Bossowski A, Glowinska-Olszewska B, Kretowski A, Stasiak-Barmuta A (2012) Generation of functional T-regulatory cells in children with metabolic syndrome. Arch Immunol Ther Exp 60:487–495
15. Lumeng CN, Bodzin JL, Saltiel AR (2007) Obesity induces a phenotypic switch in adipose tissue macrophage polarization. J Clin Invest 117:175–184
16. Lumeng CN, Saltiel AR (2011) Inflammatory links between obesity and metabolic disease. J Clin Invest 121:2111–2117
17. Miyara M, Yoshioka Y, Kitoh A, Shima T, Wing K, Niwa A, Parizot C, Taflin C, Heike T, Valeyre D et al (2009) Functional delineation and differentiation dynamics of human CD4+ T cells expressing the FoxP3 transcription factor. Immunity 30:899–911
18. Nishimura S, Manabe I, Nagasaki M, Eto K, Yamashita H, Ohsugi M, Otsu M, Hara K, Ueki K, Sugiura S et al (2009) CD8+ effector T cells contribute to macrophage recruitment and adipose tissue inflammation in obesity. Nat Med 15:914–920
19. Odegaard JI, Ricardo-Gonzalez RR, Goforth MH, Morel CR, Subramanian V, Mukundan L, Red Eagle A, Vats D, Brombacher F, Ferrante AW, Chawla A (2007) Macrophage-specific PPARgamma controls alternative activation and improves insulin resistance. Nature 447:1116–1120
20. Ouchi N, Parker JL, Lugus JJ, Walsh K (2011) Adipokines in inflammation and metabolic disease. Nat Rev Immunol 11:85–97
21. Ouchi N, Walsh K (2007) Adiponectin as an anti-inflammatory factor. Clin Chim Acta 380:24–30
22. Rosenblum MD, Gratz IK, Paw JS, Lee K, Marshak-Rothstein A, Abbas AK (2011) Response to self antigen imprints regulatory memory in tissues. Nature 480:538–542
23. Sakaguchi S, Miyara M, Costantino CM, Hafler DA (2010) FOXP3+ regulatory T cells in the human immune system. Nat Rev Immunol 10:490–500
24. Sakaguchi S, Yamaguchi T, Nomura T, Ono M (2008) Regulatory T cells and immune tolerance. Cell 133:775–787
25. Schipper HS, Prakken B, Kalkhoven E, Boes M (2012) Adipose tissue-resident immune cells: key players in immunometabolism. Trends Endocrinol Metab 23:407–415
26. Suganami T, Nishida J, Ogawa Y (2005) A paracrine loop between adipocytes and macrophages aggravates inflammatory changes: role of free fatty acids and tumor necrosis factor alpha. Arterioscler Thromsb Vasc Biol 25:2062–2068
27. Winer S, Chan Y, Paltser G, Truong D, Tsui H, Bahrami J, Dorfman R, Wang Y, Zielenski J, Mastronardi F et al (2009) Normalization of obesity-associated insulin resistance through immunotherapy. Nat Med 15:921–929
28. Yun JM, Jialal I, Devaraj S (2010) Effects of epigallocatechin gallate on regulatory T cell number and function in obese v. lean volunteers. Br J Nutr 103:1771–1777
29. Zeng C, Shi X, Zhang B, Liu H, Zhang L, Ding W, Zhao Y (2012) The imbalance of Th17/Th1/Tregs in patients with type 2 diabetes: relationship with metabolic factors and complications. J Mol Med 90:175–186

Chapter 5
Immune Cell Metabolism in Tumor Microenvironment

Yongsheng Li, Yisong Y. Wan, and Bo Zhu

Abstract Tumor microenvironment (TME) is composed of tumor cells, immune cells, cytokines, extracellular matrix, etc. The immune system and the metabolisms of glucose, lipids, amino acids, and nucleotides are integrated in the tumorigenesis and development. Cancer cells and immune cells show metabolic reprogramming in the TME, which intimately links immune cell functions and edits tumor immunology. Recent findings in immune cell metabolism hold the promising possibilities toward clinical therapeutics for treating cancer. This chapter introduces the updated understandings of metabolic reprogramming of immune cells in the TME and suggests new directions in manipulation of immune responses for cancer diagnosis and therapy.

Keywords Immune cell metabolism • Tumor microenvironment • Cancer stem cells • Clinical diagnosis • Drug repositioning

5.1 Introduction

The tumor microenvironment (TME), first proposed by Lord in 1979, is a complex integrated system for tumor cell growth [1]. This environment is composed of tumor cells, endothelial cells, immune cells, fibroblasts, and extracellular matrix. The immune cells in the TME including macrophages, neutrophils, and lymphocytes play important roles in tumor immune escape, tolerance, and suppression. Traditional theory states that the incidence of cancer is due to failure of immune surveillance [2]. However, this theory was unable to fully explain the interplay between immunity and tumorigenesis. The immunoediting theory, proposed by Schreiber and Dunn,

Y. Li • B. Zhu (✉)
Institute of Cancer, Xinqiao Hospital, Third Military Medical University, Chongqing, China
e-mail: bo.zhu@tmmu.edu.cn

Y.Y. Wan
Lineberger Comprehensive Cancer Center, University of North Carolina at Chapel Hill, Chapel Hill, NC, USA

B. Li, F. Pan (eds.), *Immune Metabolism in Health and Tumor*, Advances in Experimental Medicine and Biology 1011, DOI 10.1007/978-94-024-1170-6_5

elucidated the role of the immune system in cancer development from a new perspective [3]. Depending on the cellular and environmental context, the immune cells may kill tumor cells but also promote tumor development. This theory divides the carcinogenesis into three stages: immune clearance, immune balance, and immune escape. Immune clearance is similar to the immune surveillance. Immune balance is a stalemate stage that the tumor is not completely removed by the immune system, in which the tumor characteristics are reshaped by the immune system. Immune escape refers to the stage that the tumor escapes immune surveillance after tumor cells are edited by the immune system and progress to clinical tumor stage [4].

During the development of cancer, tumor cells and immune cells interact in a dynamic microenvironment that determines the outcome of tumorigenesis [5]. In the immune clearance phase, the immune system removes tumor cells through antigen-specific and nonspecific mechanisms, wherein lymphocytes are the main effector cells. If the immune surveillance stage functions effectively and the tumor cells are cleared, the immunoediting stage will not ensue. However, the immune system is not always effective in removing every tumor cells, especially the ones with low immunogenicity. These tumor cells often escape temporarily from the immune cell-mediated destruction. The immune balance stage then follows [6]. At this stage, the immune system constantly kills high immunogenic tumor cells, whereas it is "blind" to the tumor cells with low immunogenicity, resulting in the gradual emergence of the tumors with low immunogenic and high malignant phenotypes. The immune system continues to impose pressure to select for the tumor cells with accommodating immunological phenotypes. This Darwin's natural selection-like process is referred to as immune remodeling. The surviving tumor cells with low immunogenicity repeatedly stimulate the immune system and eventually induce the immune tolerance [5–7]. Tumor cells remodeled by the immune system can aberrantly produce inhibitory cytokines including interleukin-10 (IL-10), transforming growth factor-β (TGFβ), and vascular endothelial growth factor (VEGF) and induce immunosuppressive cells including regulatory T cells (Tregs) and myeloid-derived suppressor cells (MDSCs), fostering an immunosuppressive TME. In the TME, T cells are tolerant to tumor-associated antigen and functionally suppressive; the professional antigen presentation cells (APCs) are functionally defective, leading to hindered antitumor immune response and the systemic immunosuppression [8]. Consequently, establishing the immune-tolerant TME not only reduces the ability of the immune system to reject tumors but also promotes the immune escape of tumors that should have been rejected.

The metabolomics is an emerging direction of immune and cancer research [9, 10]. The metabolism of carbohydrates, lipids, and amino acids is the material basis of all biological processes. For the development of malignancy, the tumor cells must face two challenges: first, obtaining the nutrients needed for the rapid growth; second, evading the surveillance and attack from the host immune system. Tumor cell's unique metabolic program can be used to meet these challenges. Glycolysis is the major metabolic process used by malignant tumors, even when oxygen supply is adequate, which is termed as "Warburg effect" [11]. Glycolysis decreases the pH value of the TME; therefore, tumor cells can inhibit the activities of antigen-presenting cells (APCs) and cytotoxic T lymphocytes (CTLs) by controlling the acidity of the microenvironment, eventually leading to tumor cell immune escape [12].

Recent studies indicate that tumor-associated immune cells show altered metabolism to affect their differentiation, survival, and function [10]. Since the TME heavily influences carcinogenesis, the metabolic programming and reprogramming of immune cells in the TME and how immune metabolism affects the tumor initiation, development, and metastasis are of great interest and significance [10, 12]. In this chapter, we review the current knowledge on the metabolism of the cell types in the TME, discuss the emerging concept of the metabolic reprogramming in tumor-associated immune cells, and propose the impact of immune metabolism on carcinogenesis and clinical applications.

5.2 Metabolism of Cancer Cells and Cancer Stem Cells in TME

Aberrant proliferation of cancer cells is fueled by altered metabolism (Fig. 5.1). Oncogenic mutations trigger a switch from oxidative phosphorylation (OXPHOS) to glycolysis in tumor cells, and hypoxia further enhances this reprogramming [13, 14]. OXPHOS generates 36 mol ATP from 1 mol glucose, while glycolysis produces only 4 mol ATP [15]. This seeming inefficient metabolic feature relies on abnormal upregulation of the glucose transporters such as GLUT1 [16]. Moreover, mutations in tricarboxylic acid (TCA) cycle enzymes such as succinate dehydrogenase (SDH) or fumarate hydratase (FH) also promote glycolysis and inhibit OXPHOS [14, 17, 18]. Lactic acid is the end product of glycolysis. Lactic acid production can be used as a biomarker of tumor metastasis and overall survival [14, 15]. The secretion of lactic acid by monocarboxylate transporter 4 (MCT4) by cancer cells depends on the intracellular and extracellular concentrations of lactic acid [19, 20]. The extracellular lactate effects include restricting monocyte conversion to dendritic cells (DCs), suppressing cytokine release from DC and CTL, and inhibiting monocyte migration and CTL function. In addition, the release of lactic acid by tumor cells to the extracellular space can block the lactic acid secretion from immune cells to trigger cell death due to excessive intracellular lactic acid [21, 22].

Cancer cells also obtain energy from high levels of glutamine to support the proliferation [23]. Glutaminolysis is the main metabolic pathway regulated in mitochondria, through which glutamate is catabolized into alpha-ketoglutarate (α-KG) and glutamate. α-KG is then converted to pyruvate via TCA cycle and then to lactate [24]. Although combined activation of c-Myc and HIF-1 can induce lactate dehydrogenase A (LDHA) and pyruvate dehydrogenase kinase 1 (PDK1) that contribute to glycolysis [25, 26], mTOR-SIRT4-glutamate dehydrogenase (GDH) axis and c-Myc orchestrate glutaminolysis. Moreover, c-Myc can induce glutamine transporters (SLC5A1) and glutaminase 1 (GLS1) [27, 28]. Glutaminolysis enables

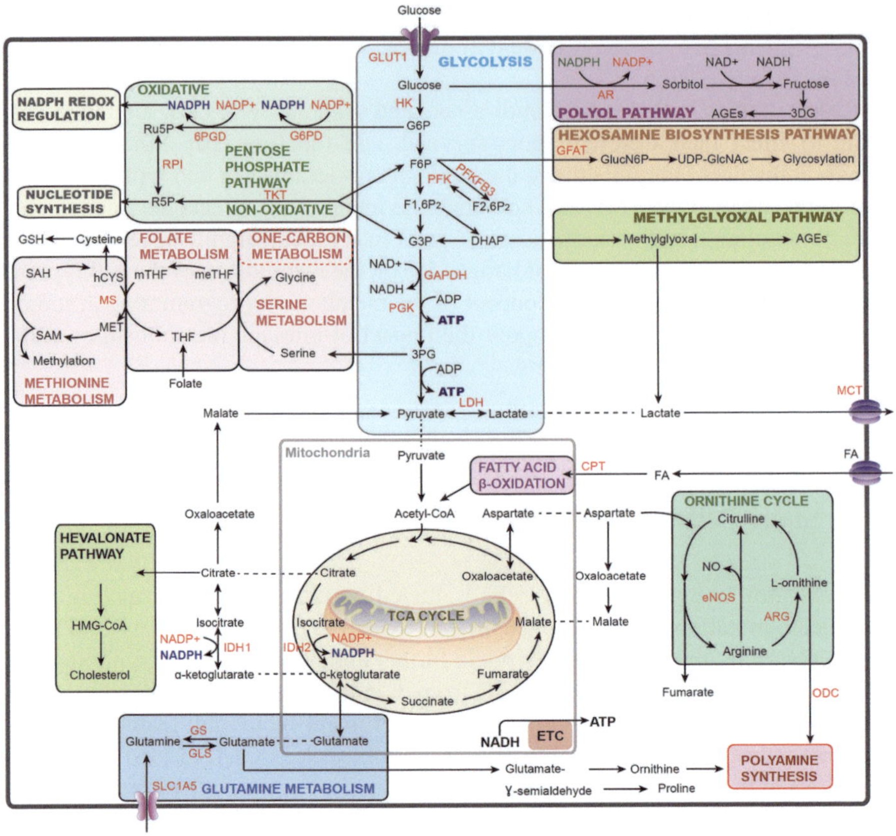

Fig. 5.1 Overview of cancer cell metabolism. This diagram depicts the cellular metabolic pathways in cancer cells. *Abbreviations*: *3DG* 3-deoxyglucosone, *3PG* 3-phosphoglycerate, *6PGD* 6-phosphogluconate dehydrogenase, *AGE* advanced glycation end product, *AR* aldose reductase, *ARG* arginase, *ATP* adenosine triphosphate, *CPT* carnitine palmitoyltransferase, *DHAP* dihydroxyacetone phosphate, *eNOS* endothelial nitric oxide synthase, *ETC* electron transport chain, *F6P* fructose 6-phosphate, *F1,6P2* fructose 1,6-bisphosphate, *F2,6P2* fructose 2,6 bisphosphate, *FA* fatty acid, *G6P* glucose 6-phosphate, *G6PD* glucose 6-phosphate dehydrogenase, *GAPDH* glyceraldehyde 3-phosphate dehydrogenase, *GFAT* glutamine-6-phosphate amidotransferase, *GlucN6P* glucosamine-6-phosphate, *GLS* glutaminase, *GLUT* glucose transporter, *GS* glutamine synthetase, *GSH* glutathione, *hCYS* homocysteine, *HMG-CoA* hydroxymethylglutaryl coenzyme A, *IDH* isocitrate dehydrogenase, *LDH* lactate dehydrogenase, *MCT* monocarboxylate transporter, *ME* malic enzyme, *MET* methionine, *meTHF* 5.10-methylene-tetrahydrofolate, *mTHF* 5-methyltetrahydrofolate, *MS* methionine synthetase, *NAD* nicotinamide adenine dinucleotide, *NADPH* nicotinamide adenine dinucleotide phosphate, *NO* nitric oxide, *ODC* ornithine decarboxylase, *PFK1* phosphofructokinase-1, *PFKFB3* 6-phosphofructo-2-kinase/fructose-2,6-bisphosphatase-3, *PGK* phosphoglycerate kinase, *ROS* reactive oxygen species, *RPI* ribose-5-phosphate isomerase, *SAH* S-adenosylhomocysteine, *SAM* S-adenosylmethionine, *TCA cycle* tricarboxylic acid cycle, *THF* tetrahydrofolate, *TKT* transketolase, *UDP-GlcNAc* uridine diphosphate N-acetylglucosamine

cancer cells to reduce NADP+ to NADPH which is an electron donor for reductive steps in lipid synthesis and nucleotide metabolism and the maintenance of reduced glutathione (GSH); thus, glutaminolysis is essential for cancer cells to regulate redox state [29].

In the presence of extracellular nutrients and enough oxygen, cancer cells synthesize fatty acids. However, under metabolic stress, cancer cells scavenge extracellular lipids to maintain viability and growth. Mechanistically, hypoxia, oncogenic RAS, and mTORC1 stimulate cancer cells to uptake lysophospholipids and desaturated fatty acids [30–32]. Also, fatty acid oxidation is enhanced in cancer cells to enable survival and proliferation [33]. Cancer cells express high levels of monoacylglycerolipase (MAGL), an enzyme that hydrolyze endocannabinoid 2-arachidonoylglycerol (2-AG) and convert monoacylglycerols to free fatty acid and glycerol that are essential for supplying energy for cancer cells [34].

Cancer stem cells (CSCs) are a small group of tumor cells with stem cell characteristics that initiate and maintain tumor growth and correlate with tumor metastasis. CSCs possess unlimited self-renewal, capacity to propagate tumors through asymmetric cell division, and therapeutic resistance [35, 36]. Tumor cells compete with CSCs for the space and energy generating macromolecules in the TME to limit CSC division and to drive tumor dormancy. The death of tumor cells therefore will relieve the CSCs from such competition and lead to the self-renewal and proliferation of CSCs [37]. Accumulating evidence indicates that the metabolic reprogramming is essential for CSCs to maintain stemness [38, 39]. Activated mitochondrial metabolism by genotoxic stress or hypoxia can lead to increased ROS and prostaglandin E2 that awaken dominant G0-phased CSCs to proliferative state [40]. The monocarboxylate transporter 1 (MCT1) is highly expressed in cancer stem-like cells to promote lactate uptake for the self-renewal and invasion [19, 20]. A recent study showed that CSCs express CD36, a scavenger receptor, which uptake oxidative low-density lipoproteins (ox-LDL) to maintain stemness [41]. Redox stress is a hallmark of cancer tissues that mediates robust metabolism in adjacent MCT1-positive proliferating CSCs which utilize lactate derived from glycolytic cancer cells to fuel mitochondrial metabolism. According to the distinct metabolic changes, cancer cells and CSCs dynamically regulate tumor progression.

5.3 Metabolism of Immune Cells in TME

The immune cells infiltrated in TME include cell subsets belong to both innate and adaptive immune systems. The current findings suggest that metabolic reprogramming is a common feature of both cancer cells and immune cells in the TME. The adjustment of the metabolic program regulates the differentiation and functions of tumor-associated immune cells and thus the progression of tumors [9, 10, 12]. Understanding how the immune metabolism changes and how the altered metabolism regulates immune cells in TME is of great interest and vital for developing effective therapies to treat cancer.

5.3.1 Macrophages

Macrophages have both anti- and pro-tumor functions by regulating tumor growth, angiogenesis, invasion, and metastasis [42, 43]. Based on the phenotypes, tumor-associated macrophages were divided arguably into two types, M1 and M2. M1 cells can be induced by interferon (IFN) γ and lipopolysaccharide (LPS), secrete pro-inflammatory cytokines including TNFα and IL-12, and express high levels of MHC molecules and nitric oxide synthase (NOS). M1 cells play crucial roles in pathogen clearance and tumor antigen presentation. M2 cells can be induced by IL-4 and IL-10, express moderate levels of MHC molecules and IL-12, but produce abundant anti-inflammatory cytokine such as IL-10, mannose receptor, and arginases to promote immunosuppression, tumor cell extravasations, and metastasis [44]. Most tumor-promoting tumor-associating macrophages (TAMs) are M2 type, while both M1 and M2 cells do coexist in the TME. The function of TAMs appears plastic as they have been found to display an inflammatory phenotype in the early phase of tumor initiation but exhibit immunosuppressive characteristics during tumor progression and metastasis [42]. These observations are consistent with the immunoediting theory that TME is a dynamically changing system.

Distinct polarization leads to different metabolic modes of macrophages (Fig. 5.2). M1 macrophages undergo glycolysis, while M2 macrophages exhibit increased oxygen consumption rate and show increased OXPHOS and decreased lactate release [45, 46]. TME is heterogeneous and dynamic during carcinogenesis. The glucose metabolism of TAMs varies in distinct phases. Indeed, at the early inflammatory stage of cancer initiation, TAMs favor glycolysis; in contrast, TAMs show OXPHOS at the later stage of tumor progression [47, 48]. This shift is mediated by cytokines and lactic acid. The lactate released from tumor cells promotes hypoxia and induces TAMs by enhancing the expression of arginase 1 (ARG1) to catalyze the metabolism of arginine to ornithine and polymines to promote collagen synthesis and tumor growth [49]. The activation of PI3K-Akt pathway may also contribute to the glucose uptake and glycolysis since it upregulates the expression of glucose transporters (e.g., GLUT1) and key enzymes (e.g., hexokinase and phosphofructokinase-1); promotes acetyl-CoA synthesis to link the metabolism of fatty acids, glucose, cholesterol, and amino acids; and facilitates the diversion of citrate from TCA cycle to acetyl-CoA by phosphorylating and activating ATP citrate lyase [46, 50, 51]. The switch to glycolysis in TAMs is controlled by Akt-mTOR-HIF-1 axis, resulting in the abundant TCA cycle intermediates and succinate accumulation [52, 53]. Factors in TME such as HMGB1 and DAMPs can stimulate Toll-like receptors (TLRs) that activate PI3K-Akt in myeloid cells like TAMs resulting in glycolysis and enhanced inflammation in TME [54]. PI3K-Akt activation in myeloid cells is also involved in resistance to anti-angiogenic therapy [55]. The activation of mTOR, a downstream molecule of PI3K-Akt, counterintuitively promotes M2 cells [52]. Of interest, this activation feedback negatively regulates PI3K [56]. The activation of c-Myc skews macrophage to M2 polarization and promotes tumor-promoting function of TAMs by increasing CCL18, TGFβ, VEGF, and MMPs [57]. However, whether and how c-Myc controls the metabolic reprogramming of TAMs remains unclear.

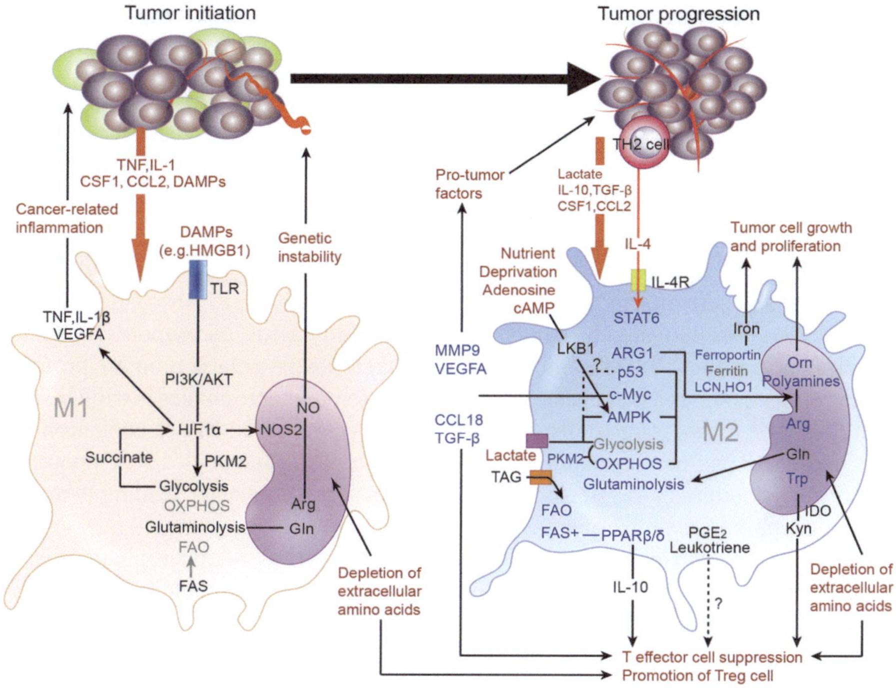

Fig. 5.2 Metabolic reprogramming of TAMs. An overview of the key metabolisms in M1 and M2 macrophages and their roles in tumor initiation and progression. During tumor initiation, macrophages are in M1 phenotype and metabolize through a glycolytic shift, HIF-1 activation, and impaired OXPHOS to mediate the expression of NO, ROI, IL-1β, and TNF, to support genetic instability and cancer-related inflammation that leads to tumorigenesis. HIF-1 also enhances angiogenic molecule VEGF-A. In TAMs (M2 macrophages) in the tumor progression stage, AMPK is activated via nutrient deprivation, Th2-derived IL-4, lactate accumulation, and activated PKM2 suppresses glycolysis while upregulating OXPHOS. This induces immunosuppressive macrophages that promote tumor growth. Amino acid, iron, and fat metabolism that contribute to this process are also shown

Moreover, inflammatory TAMs also express PKM2 which in its inactive dimeric form binds to HIF-1 to promote glycolysis in M1 macrophages, while in its active tetrameric form switches macrophages into M2 phenotype [58]. TAMs enhance cancer-related inflammation via HIF-1 which transcribes several key enzymes in glucose metabolism, such as GLUT1, HK2, and PGK1 [59, 60]. HIF-1 can also induce ROS and RNI production which contributes to the genetic mutation and transformation [61]. Moreover, hypoxic TME-increased glycolysis leads to lactic acid accumulation in the TME and skews TAMs to an M2 phenotype with high expression of ARG1, VEGF-A, Tie-2, and IL-10. These immunosuppressive and pro-angiogenic factors promote TAM-related tumorigenesis [49, 62, 63]. The distinct modes of arginine metabolism in TAMs lead to different functions, with M1 macrophages producing NO showing antitumor effect, while polyamine-producing

M2 macrophages induce cancer cell proliferation, remodeling, and growth [64–66]. TAMs that express ARG1 also contribute to T cell immunosuppression via inducing T cell apoptosis [67].

The lipid metabolism in TME macrophages is altered in response to a variety of stimuli. LPS and IFN-γ, the M1 macrophage inducers, suppress fatty acid intake and oxidation, while M2 macrophages are prone to increase fatty acid oxidation (FAO) [68]. The uptake of lipids, especially triacylglycerol (TAG), is also critical for FAO and M2 activation [69]. The underlying mechanism involves peroxisome proliferator-activated receptors (PPARs), liver X receptors (LXRs), and signal transducer and activator of transcription (STAT) [68, 70–72]. PPARs and LXRs are nuclear receptors activated by lipids, such as free fatty acids, eicosanoids, and cholesterol metabolites. PPARγ mediates M2 macrophage polarization to promote tumor progression and metastasis. PPARδ activation in macrophages is triggered by the clearance of apoptotic cells. The fatty acid synthase and PPAR activation are induced in TAMs to contribute to tumor growth. The phagocytosis of apoptotic tumor cells containing oxysterols activates LXRs in macrophages, leading to an M2 immunosuppressive phenotype.

Arachidonic acid metabolism also mediates the switch of macrophage phenotypes. For example, M1 stimulation leads to increased prostaglandin E2, a cyclooxygenase (COX)-derived eicosanoid, while IL-4 induces the upregulation of 15-lipoxygenase (15-LOX) in macrophages [73, 74]. The alveolar macrophages express high levels of COX-1 and 5-LOX, but TAMs express high levels of COX-2 [72]. Anti-inflammatory factors including IL-10, IL-4, and TGFβ induce AMPK activation which drive TAMs to an immunosuppressive M2 phenotype and induce OXPHOS [75]. These findings indicate distinct metabolic modes in TAMs mediate both anti- and pro-tumor responses.

Glutamine metabolism is another important pathway for the differentiation and functions of TAMs, and macrophages express high levels of glutaminase. Glutamine is required for macrophage phagocytosis and antigen presentation. Also, key enzymes in glutamine metabolism, such as AKG, GPT2, GLUL, and GATM, are enhanced in M2 macrophages [49, 76]. However, the mechanisms by which glutamine metabolism regulates TAMs remain unknown. In addition, M2 macrophages can generate indoleamine-2, 3-dioxygenase (IDO), an enzyme degrading tryptophan [77]. Since IDO upregulates regulatory T cells via tryptophan catabolite, TAMs promotes Treg cell generation to inhibit T cell function and establish an immune-tolerant microenvironment [78].

Polarized macrophages show altered iron metabolism. M1 macrophages express high level of H-ferritin, a protein for iron storage, but M2 macrophages express increased ferroportin, the iron exporter. Thus, M1 macrophages favor iron sequestration and inhibit tumor growth, while M2 macrophages exhibit enhanced iron release which promotes tumor progression [79, 80]. Hemeoxygenase-1 (HO-1), an iron-releasing enzyme metabolizing heme to carbon monoxide (CO), biliverdin, and ferrous iron, is inhibited in M2 but not in M1 macrophages [81]. Importantly, iron is also involved in regulating HIF-1 stability by activating prolyl hydroxylases (PHDs) in TAMs [82, 83]. The evidence suggests the iron metabolism is also crucial for TAM-mediated regulation of carcinogenesis.

5.3.2 Neutrophils

Neutrophils, also known as polymorphonuclear cells (PMNs), account for 50–60 % of the peripheral blood leukocytes, with a potent phagocytic function. Neutrophil to lymphocyte ratio is a risk predictive index for tumor recurrence [84]. In the TME, neutrophils also have two biological phenotypes: antitumor N1 exerts tumor cytotoxicity to reject tumor by enhancing antitumor immune memory; pro-tumor N2 plays the opposing roles, i.e., enhancing tumor growth, invasion, and metastasis, promoting tumor angiogenesis, mediating immunosuppression, and producing enzymes to damage normal tissue cells to facilitate tumor growth, invasion, and metastasis. Such dichotomy of neutrophils is similar to that of macrophages [85, 86]. Tumor-derived TGFβ can switch N1 neutrophils to N2 which blunts $CD8^+$ T cell responses to promote tumor growth [87]. Similar to macrophages, tumor-associated neutrophils (TANs) can produce several factors such as ARG1, ROS, MMPs, IL-6, and IL-1β to promote cancer progression, angiogenesis, and metastasis [85, 86].

The metabolic reprogramming, aerobic glycolysis and pentose phosphate pathway (PPP), controls the functions of neutrophils [88, 89]. Very few mitochondria are in neutrophils; OXPHOS and ATP production are ineffective in these cells. Neutrophils rely on PPP to produce NADPH that is essential for maintaining redox balance and cell survival [90]. The chemotaxis, calcium mobilization, and oxidative burst are driven by glycolysis since G6P deficiency blunts these functions of neutrophils [91]. Glycolysis and PPP are also involved in the formation of neutrophil extracellular traps (NETs), a mixture of DNA, histones, and antimicrobial peptides that traps and kills bacteria [90, 92]. NETs segregate circulating tumor cells and accumulate in the vasculature to promote the inflammatory adherence, contributing cancer-induced organ failure and metastasis [93]. Moreover, neutrophil-derived leukotrienes contribute to the colonization of distant tissues via selectively expanding the sub-pool of cancer cells to retain high tumorigenic potential. Knocking down of 5-lipoxygenase (5-LOX), the key enzyme for leukotriene synthesis, blunts the pro-metastatic activity of neutrophils [94]. Together, these observations suggest that metabolic change regulates the TAN functions and tumor development. Whether and how the lipid and amino acid metabolism are reprogrammed and how these metabolic changes regulate neutrophil functions await further investigation.

5.3.3 Basophils and Eosinophils

Basophils, a cell population derived from myeloid cells, respond to IgE-dependent and IgE-independent stimuli and crosstalk with other immune cells such as lymphocytes, macrophages, and DCs [95]. These cells are involved in Th2 responses through producing cytokines including IL-4, IL-13, and IL-25 and also contribute to immunoglobulin synthesis, tumor angiogenesis, and hematopoiesis by secreting IL-6, GM-CSF, and VEGF and arguably present antigens to T cells. Hence, basophils regulate both innate and adaptive immunity. However, the metabolic reprogramming in tumor-associated basophils remains unknown.

Eosinophils are granulocytic leukocytes derived from hematopoietic progenitors. They interact with both innate and adaptive immune cells. IL-3, IL-5, and GM-CSF are crucial for eosinophil development, while CCL11, CCL24, and CCL26 contribute to eosinophil chemotaxis [96, 97]. Eosinophils express MHC-II and costimulatory molecules such as CD40 and CD80/86 to promote T cell activation and proliferation [98]. Eosinophils also produce IDO and TGFβ to mediate Treg and Th2 polarization [99]. Interestingly, the infiltration of eosinophils in TME associates with improved prognosis in various types of solid tumors [96] but with poor outcome in Hodgkin lymphoma [100]. The antitumor activity of eosinophils is through their degranulation in the tumor. Consistently, eosinophils with CCL11 deficiency exhibit impaired antitumor potential [101]. Also, the necrosis and chemokines in cancer tissues induce the differentiation and migration of eosinophils [102]. Systemic IL-2 and IL-25 therapy promotes eosinophil degranulation [103, 104]. The recruitment of eosinophils in TME is mainly mediated by high-mobility group box 1 (HMGB1), a factor in damage-associated molecular patterns, which elicits eosinophil degranulation by binding to the receptor for advanced glycation end products (RAGE) [105]. Some receptors normally expressed in NK cells such as NKG2D and 2B4 are also expressed in eosinophils to mediate tumor cytotoxicity [106, 107]. These observations suggest that reprogramming the metabolism in eosinophils may trigger the degranulation and improve antitumor immunity.

5.3.4 Mast Cells

Mast cells are derived from bone marrow hematopoietic stem cell-differentiated precursor cells. These precursor cells enter the cavity or mucosa from blood and then mature. As an important class of innate immune cells, mast cells play a key role in allergic diseases such as asthma. After activation, mast cells not only produce and release a variety of cytokines and chemokines, including histamine, serotonin, interleukins, leukotrienes, prostaglandins, and proteases to promote inflammation, but regulate the functional activity of DC, T cells, $CD4^+$ $CD25^+$ regulatory T cells, B cells, and other immune cells in TME [108]. Similar to that in TAMs, PI3K-Akt activation in mast cells correlates with enhanced glycolysis [109]. The metabolic reprogramming of the tumor-associated mast cells is yet unknown.

5.3.5 CAFs

Fibroblasts are a large proportion of cells in the TME. These cells synthesize collagen, laminin, fibronectin, and other matrix components to establish a structural framework in the matrix. Fibroblasts can be activated during wound healing, inflammation, and stress. The TME-activated fibroblasts called cancer-associated fibroblasts (CAF) or myofibroblasts specifically express α-smooth muscle actin (αSMA) [110]. Apart from the local fibroblasts, CAFs can be derived from vascular smooth muscle cells, pericytes, marrow-derived mesenchymal cells, and through epithelial-mesenchymal transformation (EMT) [111]. CAF can reshape the extracellular matrix by secreting matrix-degrading enzymes, particularly metalloproteinases and derivatives to contribute to tumor drug resistance [112]. CAFs promote tumor growth, angiogenesis, and metastasis by secreting several factors, including insulin-like growth factor, hepatocyte growth factor, basic fibroblast growth factor, Wnt ligands and MMPs, as well as cytokines and chemokines such as CCL7, CXCL12, and VEGF-A [110, 111].

It has been observed that CAFs engage in aerobic glycolysis. Mechanistically, IDH3α reduced by TGFβ or platelet-derived growth factor (PDGF) breaks the equilibrium between α-KG and fumarate/succinate that are allosteric regulators of PHDs, thereby in turn increasing HIF-1 transactivation and enhancing glycolysis [113]. CAFs also display increased glutamine metabolism and decreased OXPHOS [114, 115]. They provide lactic acid, amino acids, and ketone bodies to cancer cells, while cancer cells produce ROS to activate HIF-1 in CAFs to maintain the glycolysis [110, 111]. The metabolic change of CAFs impacts the secretion of cytokines and chemokines. For example, α-KG and fumarate/succinate are allosteric regulators of lysine-specific demethylase, PHDs, and methylcytosine demethylase that are epigenetic regulators [116]. Also, glycolysis and glutamine metabolism in CAFs are regulated by p62-mTORC1-c-Myc pathway that promotes ROS and IL-6 production and enhance tumor progression [114, 117].

Distinct from cancer cells, although CAFs favor glycolysis, the proliferation is much slower when compared with normal fibroblasts, suggesting that the biosynthesis of CAFs is not dependent on glycolysis [118]. However, cancer cells uptake the CAF-secreted lactate for tumor anabolic metabolism, growth, and metastasis [117]. GLUT4 is overexpressed on CAFs to release lactic acid, while GLUT1 is upregulated in cancer cells to import glucose and metabolites. The lactate released from CAFs further acidifies the TME to facilitate tumor progression and drug resistance [119]. Hence, CAFs not only secrete growth factors but also fuel cancer cells by providing lactate and other glucose metabolites. The cooperation of metabolites shuttling between CAFs and cancer cells aggregates the TME that facilitate tumor development.

5.3.6 NK and NKT Cells

Natural killer cells (NK cells) are a subtype of lymphocytes in innate immune system. They express Ly49, NCR, and CD16 and play antitumor action by secreting IFN-γ and cytotoxic molecules such as perforin and granzyme or by antibody-dependent cell-mediated cytotoxicity (ADCC) and T cell activation [120, 121]. Under resting state and short-term activation, they favor OXPHOS, while upon prolonged activation by high-dose IL-15, NK cells switch to glycolysis [122]. In the TME, both IL-15 and hypoxia can lead to enhanced glycolysis in NK cells [123]. However, the mechanism by which TME affects the metabolic reprogramming and how the altered metabolism modulates the activity of NK cells remain to be elucidated.

NKT cells are a heterogeneous group of T cells that share properties of both NK cells and T cells. They recognize non-polymorphic CD1d (an APC molecule) that binds self and foreign lipids and glycolipids [124]. Upon activation, these cells produce abundant IL-2, IL-4, IFN-γ, GM-CSF, and IL-21 so that they play antitumor functions in TME [125]. However, the metabolic reprogramming in tumor-associated NKT cells is yet unknown.

5.3.7 Endothelial Cells

In the TME, vasculatures deliver nutrients and oxygen to the tumor, which is the basis of tumor survival and development. Tumor cells release pro-angiogenic signals to drive the metabolic reprogramming of endothelial cells (ECs) [126]. Inhibition of VEGF signaling is a clinically approved strategy, although the benefits are limited since tumors acquire drug resistance within months after treatment [127].

It has been found that the structure and function of vessels in TME and in normal tissues are significantly different. They are dilated, tortuous, and hyperpermeable. The ECs are poorly connected and lack a regular pattern. The basement membrane has nonuniform thickness and composition [128]. These cause deprived oxygen and nutrients. Hypoxia switches cancer cell metabolism away from OXPHOS to glycolysis, from glucose to glutamine as the major substrate for fatty acid synthesis (FAS). Tumor-associated ECs resemble cancer cells to undergo a shifting from quiescence to rapid growth during vessel sprouting. These ECs are highly plastic [126, 129]. They require a baseline glycolysis flux to function as an endothelium and maintain vascular barrier homeostasis. ECs preserve high concentration of oxygen in the blood. Also, ECs protect themselves from oxidative stress using glycolysis. They can also move from normoxic to hypoxic areas [130]. Glycolysis can produce ATP faster than OXPHOS. Hence, similar to cancer cells, glycolysis contributes to vascular sprouting and the survival and proliferation of ECs (Table 5.1) [131, 132]. Lactate dehydrogenase B, GLUT1, and glycolytic enzyme 6-phosphofructo-2-kinase/fructose-2,6-bisphosphatase-3 (PFKFB3) are increased in tumor-associated ECs. PFKFB3 promotes the synthesis of fructose-2,6-bisphosphate(F2,6P2) and the activation of 6-phosphofructo-1-kinase (PFK1) which converts fructose-6-phosphate (F6P) to fructose-1,6-bisphosphate (F1,6P2) [133]. Indeed, interfering with glycolysis by inhibiting PFKFB3 (by 3-(3-pyridinyl)-1-(4-pyridinyl)-2-propen-1-one (3PO)) or other key enzymes blunts angiogenesis-associated tumor growth [134]. However, the systemic complete and permanent inhibition of glycolysis may also induce undesired effects. Fortunately, Partial and transient reduction of glycolysis

Table 5.1 Cancer cell versus endothelial cell metabolism

	Cancer Cell	Endothelial Cell
Glycolysis	Warburg effect Upregulated through oncogenes and tumor hypoxia Switch PKM1 to PKM2	Compartmentalized Activated by growth factors
OXPHOS	Reduced TCA activity	Few mitochondria, low respiration Upregulated in low glucose and stress
Glycogen Metabolism	Glycogen metabolized under hypoxic stress	Glycogen synthesis Stores depleted in low glucose, not in hypoxia
FAO	Source of NADPH Response to oxidative stress	Possibly used for energy Might promote angiogenesis Energy source in low glucose
PPP	Redox homeostasis Regulate cell death via ROS	Redox homeostasis Might promote angiogenesis G6PD regulates ROS signaling High PPP for proliferation
AA Metabolism	Glutamine as nitrogen source for polyamine synthesis and anaplerosis Energy and biosynthesis	Glutamine as alternative energy source under stress GLS blockade reduces proliferation Glutamine inhibits NO production

renders ECs more quiescent without overt detrimental side effects [134]. Glycolytic metabolites such as lactate can be uptaken by ECs through MCT1. Instead of being metabolized, lactate induces HIF-1 activation and competes with α-KG to bind to prolyl hydroxylase 2 (PHD2), thereby enhancing the expression of angiogenesis-related genes [135, 136]. Therefore, the partial and transient reduction of glycolysis may be sufficient to inhibit pathological angiogenesis in the TME. These results indicate that targeting glycolysis in ECs inhibits angiogenesis, but the viability should be concerned.

The decreased supply of glucose in ECs can also be compensated by glycogenolytic production of glucose-1-phosphate (G1P), which can be converted into glucose-6-phosphate (G6P). The glycogenolysis-derived G6P only minimally contributes to energy production in normal or low glucose conditions. G6P catabolism might be important in the oxidative PPP (oxPPP) process to generate energy [137, 138]. PPP also produce ribose-5-phosphate (Rb-5-P) which can feedback into glycolysis through the biogenesis of F6P [139, 140]. Moreover, NADPH produced by PPP protects ECs against ROS, enhances NO synthesis, and contributes to ATP production [141]. Therefore, PPP may promote angiogenesis and regulates redox homeostasis.

ECs have a high level of glutaminolysis to support ATP synthesis and fuel cell proliferation in the conditions of decreased glucose supply (Table 5.1) [142]. ECs can uptake glutamine from extracellular milieu and also produce glutamine from glutamate. Inhibition of glutaminolysis induces EC senescence [143]. Glutamine metabolism also promotes ornithine synthesis, a precursor of mitogenic polyamines [144]. In addition, glutamine metabolite glucosamine can inhibit oxPPP and NO production by reducing NADPH [145]. How glutaminolysis mediate angiogenesis has not been described.

Most ECs contain mitochondria that compose less than 5 % of the cell volume. The mitochondria-derived ROS activates HIFs by inhibiting PHDs, which in turn enhances glycolytic metabolism and angiogenesis [146]. Glycolysis-derived ATP also in part contributes to maintaining EC mitochondrial network. However, the role of mitochondria in tumorigenesis is yet unknown. FAO is induced in ECs upon glucose deprivation, in which process AMPK is activated [147]. Whether and how FAO influences angiogenesis is also unclear.

5.3.8 Dendritic Cells

DCs are divided into immature (imDCs), semi-mature (smDCs), and mature (mDCs). imDCs show low expression of MHC class I molecules, lack of B7 costimulatory molecules, etc. These cells induce immune tolerance since they cannot effectively activate T cells [148]. Recent studies showed that semi-mature and mDCs can also induce immune tolerance. smDCs have a unique feature which can be induced by granulocyte-macrophage colony-stimulating factor (GM-CSF), IL-4, and TNF with bone marrow cells in vitro. smDCs obtain a significantly different molecular phenotype, as compared with mDCs and imDCs, showing high expression of MHC class I molecules and moderate expression of costimulatory molecules, but they do not secrete inflammatory cytokines such as IL-1, IL-6, and IL-12 [149, 150]. These cells produce inflammation-inhibiting factors including IL-10 and can enhance the activation of $CD4^{+}CD25^{+}$ Tregs [151].

As a major population of professional antigen-presenting cells, DCs are activated and mature by sensing pathogen-associated or damage-related stimuli, displaying upregulation of MHC molecules and costimulatory molecules (e.g., CD80/86, CD40), cytokines (e.g., IL-12), and chemokine receptors (e.g., CCR7). Then DCs migrate to lymphoid organs to present antigens and activate T cells. However, tumor-associated DCs are immunosuppressive [152–154]. Moreover, increased imDC numbers and decreased mDC and DC numbers with impaired functions were observed in cancer microenvironment [149, 152].

The differentiation and function of DCs also rely on the metabolic reprogramming. Under resting state, DCs favor OXPHOS, while they switch to glycolysis after activation [155]. Also, glycolysis is essential for upregulated costimulatory molecules (CD80/86, CD40) and cytokines (IL-12) and DC survival [50]. Indeed, the activated DCs show increased NO production, enhanced PI3K-Akt activation, and impaired OXPHOS [50, 155].

In the hypoxic TME, HIF-1 drives the transcription of mTOR and glycolysis-associated genes, thereby promoting DC glycolysis, maturation, and activation [156]. However, hypoxia also inhibits the recruitment of monocyte-derived DCs from peripheral blood [157]. Hypoxia upregulates adenosine receptor (A2b), and adenosine can bind this receptor to blunt DC differentiation and switch them to a

Th2-promoting profile, i.e., express IL-10, TGFβ, and COX-2 that lead to abnormal differentiation of myeloid cells thereby causing Treg activation and DC defection [158]. Mechanistically, adenosine can induce AMP-activated protein kinase (AMPK) that promotes OXPHOS and inhibits glycolysis [50]. The lactic acid also contributes to inhibit DCs by inhibiting glycolysis [159]. These impaired DCs also secrete VEGF-A to promote angiogenesis and tumor growth [160].

PI3K-Akt activation also contributes to enhanced expression of GLUT1, the glucose transporter key for glycolysis, and disabled-2 adaptor (DAB2) on DCs which suppresses T cell response against tumors [161]. The mTOR inhibition by rapamycin in BMDCs promotes their antitumor activity, while mTOR inhibition in human monocyte-derived and plasmacytoid DCs blunts the immunostimulatory actions and T cell response [155].

Similar to macrophages, tumor-associated DCs (TADCs) can express enhanced ARG1, IDO, and iNOS that deplete arginine and tryptophan in the TME so that suppression of $CD8^+$ T cell function and survival is mediated [162]. Moreover, the retinoic acid metabolism in TADCs can promote Treg-mediated immunosuppression [163]. These amino acid metabolism pathways suggest potential targets for cancer immunotherapy. Recently, it was found that vaccines can trigger GCN2, a nutrient sensor, in DCs that in turn activate the antigen-presenting function of DCs and CTL response [164].

TADCs are also regulated by lipid metabolism. During DC activation, the FAS can promote the antigen presentation by increasing ER and Golgi expansion [165, 166]. The scavenging receptors such as CD36 and MSR1 upregulated in TADCs contribute to the uptake of lipids [167]. The lipid accumulation subsequently attenuates DC function and T cell activation. Hence, it has been proposed that the switch of glucose metabolism from glycolysis to OXPHOS in the TADCs may turn on the FAS and lipid uptake, therefore impairing DCs in TME and inducing an immunotolerogenic condition [155, 165, 168]. Furthermore, the limited nutrients in TME can induce ER stress that will induce excessive lipid accumulation via TCA cycle in TADCs and further inhibit CTL priming [169].

Abovementioned finding suggests that the altered metabolic programming in TADCs impairs their activation and antitumor function. Rectifying the abnormal metabolism may restore DCs' ability to reject tumor in the TME.

5.3.9 Myeloid-Derived Suppressor Cells (MDSCs)

MDSCs, originally discovered in the 1970s, are a group of myeloid immature cells in tumor-bearing mice [170, 171]. The morphology of these cells is similar to the granulocyte-monocyte progenitor cells that are yet to differentiate into macrophages, DCs, or granulocytes. These cells are $CD11b^+Gr1^+$ and can be subcategorized into monocytic ($Ly6C^{high}Ly6G^-$) and granulocytic ($Ly6C^{low}Ly6G^+$) based on the expression of Ly6C and Ly6G. In humans, MDSCs also express hematopoietic stem cell surface markers CD34 and immature marker CD31, hardly expressing

MHC class I molecules and maturation markers [172, 173]. MDSCs can secrete MMP9 and undergo endothelialization to induce angiogenesis and promote tumor survival and invasion [174]. With the development and progression of tumor, large amounts of cytokines, such as VEGF, IL-10, and TGFβ, can be released in the TME. These cytokines not only recruit MDSCs to tumor but also promote their differentiation into immunosuppressive cells [175]. On the other hand, MDSCs are immunosuppressive through multiple means: MDSCs can suppress the maturation and antigen presentation function of DCs directly or indirectly [176]. By direct interaction, MDSCs suppress IFN-γ production by CD8 T cells [177]. MDSCs secrete TGFβ, arginine enzymes, and reactive oxygen and nitrogen species to inhibit T cell activation and proliferation [177] and to promote Treg cell generation [178].

Glycolysis, glutaminolysis, and TCA cycle metabolism and arginine metabolism are upregulated during MDSC maturation. Fatty acid uptake and FAO mediate the immunosuppressive function of MDSCs in TME [179]. Amino acid metabolism and oxidative stress also mediate MDSC immunosuppressive functions. By expressing IDO, MDSCs metabolize tryptophan to kynurenine, induce Treg expansion, and inhibit T cell functions [180]. L-arginine and L-cysteine can be depleted by MDSCs, leading to the downregulation of CD3ζ and the inhibition of T cell activation [181, 182]. MDSCs express high levels of ARG and iNOS. ROS such as peroxynitrites are produced under the conditions of limited L-arginine availability. ROS induces T cell apoptosis by nitrotyrosylating and preventing tyrosine phosphorylation of key signaling proteins for T cell activation [183]. Peroxynitrites also nitrate TCR, IL-2R, and CD8 molecules, leading to T cell signaling disruption [184, 185]. Recent studies indicated that a lipid mediator derived from arachidonic acid via cyclooxygenase 2 (COX-2), i.e., prostaglandin E_2 (PGE_2), can be secreted by tumor cells and MDSCs [186, 187]. PGE_2 enhances MDSC development by inducing ARG, iNOS, and IDO and promotes MDSC recruitment to the TME by inducing CXCL12 [188]. Taken together, the metabolic reprogramming of MDSCs provided a potential target for regulating the immunosuppressive network in the TME.

5.3.10 T Cells

The infiltration and activation of T cells in TME control tumor progression. $CD8^+$ T cells are the major effector cells in tumor immunity. Upon activation by APCs, $CD8^+$ T cells migrate to the tumor tissue to kill target cells through perforin (to damage cell membranes), granzymes (to enter target cells and degrade DNA), and FasL. In addition, CD8 T cells secrete cytokines such as IFN-γ and TNFα to promote antitumor immune response [189]. $CD4^+$ T cells differentiate into distinct subtypes to promote or repress tumorigenesis. $CD4^+$ Th1 T cells produce IFN-γ to promote tumor immune rejection. Th1 cell infiltration in TME is associated with good clinical prognosis. In contrast, Th2 cells and Tregs temper tumor rejection and facilitate tumor immune escape. Th17 cells may both promote and inhibit tumor progression in a tumor-type and stage-dependent manner [190].

Table 5.2 T cell metabolism

	Naïve CD8	EffectorCD8	Memory CD8	Exhausted CD8	Th1	Th2	Treg	Th17
Glycolysis		+++			+++	+++		++
OXPHOS	+++		+	++			+++	
PPP		++			++			++
FAS		++						++
FAO	++		++	++			++	
Glutaminolysis	+	++		+	++	++		++
Tryptophan metabolism		++			++	++		++

For activated T cells to proliferate and release cytotoxic factors and cytokines, they switch the metabolism to aerobic glycolysis to increase the uptake of glucose and glutamine [191, 192]. However, the activation of PD-1 can inhibit the uptake and utilization of these nutrients and promote FAO, thereby inducing T cell anergy, exhaustion, and autophagy [193–195]. Since H^+ secretion increases when tumor cells release lactic acid, TME becomes acidic, leading to reduced T cell function [196].

In the TME, tumor cells compete for nutrients to hinder antitumor functions of T cells and lead to the metabolic reprogramming of T cells (Table 5.2). The efficiency in tumor cell uptake of glucose is ten times of that in activated T cells. Because glucose is the sole source of the energy required for effector T cells [192]. The lack of glucose severely affects IFN-γ production and the cytolytic activity of CTLs. Moreover, the accumulation of metabolic wastes in TME, such as lactate and kynurenine, can also inhibit T cell function [197]. Acidification and hypoxia of TME also impairs the proliferation and function of CTL [192, 196]. De novo FAS is critical for the development of effector T cells, while the generation and survival of memory T cells need FAO and OXPHOS [198].

Naive T cells get energy from OXPHOS, fatty acid oxidation (FAO), and low levels of glutaminolysis, whereas these cells need much more nutrient for activation [191, 192, 198]. Therefore, activated T cells show enhanced glycolysis, PPP, and glutaminolysis and decreased FAO (Table 5.2). Glycolysis is required for the functions of effector T cells, while PPP and glutamine metabolism are involved in biosynthesis. Upon activation, signaling pathways involving PI3K-Akt, mTOR, HIF-1, and c-Myc are triggered in $CD8^+$, Th1, Th2, and Th17 cells to promote the expression of key factors in nutrient metabolism, such as GLUT1, PDK1, and HK2 [191–193, 198]. This in turn leads to further enhanced glycolysis and glutaminolysis. In addition, the activated T cells exhibit a noncanonical Myc-dependent transcriptome coupling glycolysis and glutaminolysis to polyamine biosynthesis to maintain T cell proliferation [198, 199]. The mTOR activation regulates the balance between effector and memory T cells by modulating T-bet, a key transcription factor for Th1 cell differentiation [200]. The energy sensor AMPK, activated by increased ratio of AMP and ATP, nutrient deprivation, and anti-inflammatory cytokines (e.g., IL-4,

IL-10, and TGFβ), suppresses IFN-γ and granzyme B production and induces OXPHOS and glutamine-dependent mitochondrial metabolism in T cells to suppress T cell-mediated antitumor response [201, 202]. Tumor-derived oxysterol induces LXR activation to inhibit neutrophil recruitment and DC migration, switches on M2 polarization, and suppresses T cell response to aid tumor immune tolerance [70].

The metabolism of amino acids, such as arginine, tryptophan, glutamine, and cysteine, is important for TIL functions (Table 5.2). For example, deficiency of arginine impairs protein synthesis in TILs, leading to reduced TIL activation [203, 204]. L-arginine metabolism is dependent on the activities of NOS and ARG. NOS converts arginine to NO and citrulline, and ARG hydrolyzes arginine into urea and ornithine. Administration of ARG and NOS-specific inhibitors can activate TILs [205]. Mechanistically, NO can react with ROS to produce reactive nitrogen species (RNS) such as peroxynitrite which induce lymphocyte anergy and apoptosis by nitration of tyrosine residues or the mitochondrial permeability transition pore (voltage-dependent anion channel) [184, 206]. Moreover, RNS modifies chemokines such as CCL2, by nitration or nitrosylation, to inhibit T cell infiltration into the tumor. Indeed, drugs targeting nitration in TME induce T cell infiltration [207]. Effector $CD8^+$ T cells can also impair the CAF-mediated chemoresistance by inhibiting cysteine and glutathione metabolism in fibroblasts [208], suggesting a novel intersection for combined chemotherapy and immunotherapy in cancer treatment.

As mentioned earlier, tryptophan deprivation in TME contributes to tumor progression. IDO, the rate-limiting enzyme in tryptophan metabolism, inhibits the proliferation of effector T cells by depleting tryptophan in the TME [209, 210]. IDO is mainly expressed in mesenchymal cells, such as ECs, macrophages, and DCs. Tumor cells also express IDO upon IFN stimulation [211, 212]. By metabolizing tryptophan, IDO leads to the release of kynurenine. The reduction of tryptophan and the increase of kynurenine synergistically inhibit the activation and proliferation of antitumor T cells [209]. Therefore, IDO upregulation in cancer patients correlates with impaired T cell accumulation, proliferation, and function and poor prognosis. Agents inhibiting IDO, such as INCB024360 and 1MT, can promote the antitumor T cell function [209, 213].

A recent study revealed a new mechanism that cholesterol metabolism regulates the antitumor responses of $CD8^+$ T cells [214]. Cholesterol is abundant in the plasma membrane, which is key for the TCR clustering and immunological synapse formation. The deficiency of ACAT1, a key cholesterol esterification enzyme, led to potentiated effector function and proliferation of $CD8^+$ T cells. However, ACAT1 knockout in $CD8^+$ T cells could not affect the glycolysis, OXPHOS, and FAO levels. Inhibition of PD-1 did not alter the expressions of ACAT1 and other cholesterol esterification genes. Combination of anti-PD-1 antibody and ACAT1 inhibition synergistically blunted the tumor development.

Tregs are generally divided into natural regulatory T cells (nTregs, $CD4^+CD25^+$) and induced regulatory T cells (iTregs, $CD4^+CD25^-$), both of them express Foxp3. They express IL-10, TGFβ, and IL-2 receptor α-chain (CD25), but do not produce IL-2. Tregs suppress immune response and T cell activation through cell-cell interaction and cytokines [215, 216]. Tregs are present in various tumors and play an immunosuppressive role. AhR-enhanced IDO1 and kynurenine mediate Treg generation [217], while mTOR activation inhibits Treg development [218]. Thus the elimination of Tregs in the TME, to some extent and excluding in colorectal cancer, inhibits tumor growth. The function of regulatory T cells (Tregs) is not affected by lactic acid and acidic environment. Of interest, Tregs favor OXPHOS, FAO, and activated AMPK for the nutrient metabolism (Table 5.2) [218–220]. Hence they survive well in the nutrient-deprived TME. Moreover, the HIF-1 upregulation in the hypoxic TME promotes Treg expansion [221]. Interestingly, HIF-1 shows an important metabolic checkpoint for the differentiation of Tregs or Th17 cells [222].

Th17 cells are a newly discovered class of T helper cell subsets. Naive $CD4^+$ T cells preferentially differentiate to Th17 under the stimulation of TGFβ and IL-6. In addition, IL-23 is a key factor for the maintenance and expansion of Th17 cells [223]. Recent studies found that the presence of Th17 in the TME antagonizes the IFN-γ-producing Th1 cells to favor tumor growth [224]. Th17 cells also rely on glycolysis and FAS for differentiation and activation [225] (Table 5.2).

5.3.11 B Cells

By producing antibodies and immune complexes, B cells can regulate the functions of myeloid cells to promote tumor growth. It has been reported that c-Myc, but not HIF-1, mediates LPS and antigen-stimulated activation of B cells and triggers the glycolysis and mitochondrial metabolic activity. Tumor cells express BAFF which can also induce the glycolysis and antibody production of B cells by a GLUT1-dependent manner [226]. Therefore, metabolic reprogramming of B cells will be also of interest to be investigated to better understand the immunoediting in tumor progression.

5.4 Clinical Diagnostic and Therapeutic Applications

5.4.1 PET/CT and PET/MRI

Cancer cells favor glycolysis to metabolize glucose, regardless of oxygen tension, which is termed as Warburg effect. Anaerobic glycolysis (fermentation) is more rapid but less efficient than OXPHOS to generate ATP. This process produces lactate and contributes to immunosuppression. At present, Warburg effect has offered an opportunity to diagnose and monitor therapy response in many clinical cancers. Position emission tomography-computed tomography/magnetic resonance imaging (PET/CT or PET/MRI) is a clinical imaging technique combining PET and CT/MRI [227–230]. PET imaging shows the spatial distribution of metabolic activity (especially the Warburg effect), while CT/MRI precisely aligns the anatomic imaging. PET with distinct radiotracers can evaluate altered metabolisms of glucose, fatty acids, amino acids, and other cancer markers. Hence, molecular imaging with PET is very precise for detection and directing therapy and has been applied widely in several cancer types.

5.4.2 PD-L1, PD-1, CTLA-4, and IDO

Recent breakthrough in cancer immunotherapy based on the clinical application of monoclonal antibodies targeting T cell immune checkpoints, including PD-1 and CTLA-4, clearly demonstrates the significance of effector T cell activation in antitumor response [231]. PD-L1:PD-1 and CTLA-4 signaling dampens antitumor responses. The expression of PD-L1, a key immune checkpoint, can be induced by hypoxia in tumor-associated DCs. PD-L1 is a target gene of HIF-1 and NF-κB, two central transcriptional factors in hypoxic responses. PD-L1/PD-1 pathway inhibits glycolysis and promotes FAO and lipolysis to mediate T cell metabolic reprogramming [232–234]. Of interest, PD-1 is also expressed in cancer cells such as melanoma cells. Activation of melanoma-PD-1 promotes tumor progression by mTOR pathway [235]. Preclinical data suggested that inhibition of PD-1 and prostaglandin E synthases synergistically promotes tumor eradication [236]. Therefore, it has been proposed to use PD-1 as a radiolabeled PET imaging tracer to efficiently distinguish PD-L1-positive and PD-L1-negative tumors.

CTLA-4 is a target gene of Foxp3 and has a major role in enhancing Treg activity and suppressing T helper cells [237]. IDO activity can be induced by CTLA-4 in plasmacytoid DCs via reversing CD80 signaling [238]. The increase of IDO, PD-L1, and CTLA-4 in the peripheral blood of cancer patients correlates with advanced disease and poor outcome, independent of the stages of cancer [239, 240]. Therefore, combination treatments targeting several of these markers to modulate metabolisms in immune cells may have a synergistic effect.

5.4.3 CAR-T

Cancer immunotherapy based on the adoptive transfer of autologous T cells has shown promising efficacies. Chimeric antigen receptor (CAR)-T cells have been used to exert potent antitumor effect [241, 242]. CARs consist of cytoplasmic domain of the Fc receptor γ chain or CD3ζ modules and that of costimulatory cytoplasmic domains such as CD28, 4-1BB, and ICOS [243]. Little is known about the metabolic reprogramming of CAR-T cells. As described in the above section, naïve and memory T cells rely on fatty acid oxidation, while activated effector T cells shift to glycolysis and enhanced OXPHOS. A recent report showed that CD28 or 4-1BB CD3ζ CAR-T cells exhibited increased survival and proliferation, promoting central memory T cells. CD28-CD3ζ CAR-T cells favored aerobic glycolysis, while 4-1BB-CD3ζ CAR-T cells preferred FAO [243]. The choice of CAR impacted the T cell metabolic reprogramming and differentiation, suggesting that modulating the metabolism in T cells may also be important to enhance the antitumor effects of CAR-T cells.

5.4.4 Drug Repositioning

Drug repositioning, also known as drug repurposing, re-profiling, re-tasking, or therapeutic switching, is the application of known drugs and compounds to new indications. It is an emerging and important application in drug development for cancer therapy [244]. The computational approaches can enhance the efficiency and success rates, particularly in terms of high-throughput shotgun repurposing. For example, proton pump inhibitor (PPI) is an acid-activated drug that inhibits H/K--ATPase to treat gastric cancer. PPI can synergistically modulate the acidic TME with chemotherapy and improve chemoresistance [245, 246]. Terfenadine is a histamine receptor H1 antagonist, which can prevent VEGF secretion from mast cells in hypoxic microenvironment and induce ROS-mediated apoptosis and autophagy of melanoma cells [247]. Simvastatin specifically inhibits HMG-CoA reductase so that it restrains p53 mutation from activating mevalonate pathway for cholesterol synthesis in breast and ovarian cancer cells [248]. System xc^- cystine/glutamate antiporter, a heterodimer composed of the 4F2 heavy chain (SLC3A2) and the light chain xCT, is a membrane amino acid transporter that mediates the exchange of extracellular cystine and intracellular glutamate [249]. Sulfasalazine is a specific inhibitor of xCT cystine transporter. It blocks the reduced GSH synthesis, leading to oxidative stress in cancer cells, resulting in the suppression of NSCLC and gastric tumor progression and breast cancer metastasis [249–251]. Metformin is an oral drug used to treat type 2 diabetes mellitus by suppressing glucose production by the liver [252]. Recently, metformin was reported to inhibit tumor progression and ameliorate the prognosis [253–255]. By inhibiting ATP-binding cassette subfamily G member 1 (ABCG2) and ectonucleotide pyrophosphatase/phosphodiesterase family member 1 (ENPP1), metformin suppresses cancer cell chemoresistance to drugs [256]. These drugs are being applied in clinical trials and show promising results.

5.4.5 Metabolism-Based Antiangiogenic Therapy

The current paradigm for anti-angiogenic therapy is to block VEGF and VEGFRs [257]. However, the tumor cells can rely on other signaling pathways for pro-angiogenesis. The hypoxic conditions caused by treatment often lead to the outgrowth of resistant tumor clones [258]. The EC metabolism requirement potentially provides novel anti-angiogenesis therapeutic opportunities. Silencing PFKFB3 by

3PO is capable to reduce EC glycolysis and vessel sprouting, without switching to aerobic respiration [134]. More importantly, the effect of 3PO on glycolysis is reversible, as normal sprouting was recovered after administration for 6 h in vivo The moderate reduction in glycolysis by 3PO is sufficient for increasing the fraction of quiescent ECs and reducing EC proliferation and migration [132, 133]. Combination of 3PO and VEGFR tyrosine kinase inhibitor SU5416 significantly impaired angiogenesis, as compared with the optimal doses of any one of these inhibitors alone [134]. In addition, 3PO is a chemotherapeutic agent to block tumor proliferation [259]. Together these findings indicate that targeting tumor-associated EC metabolism is a potential therapeutic strategy.

5.5 Concluding Remarks

The metabolism of cancer cells and immune cells in TME is instrumental for tumor initiation, progression, and metastasis. The deprivation of nutrients from the environment suppresses antitumor immune cells, such as $CD8^+$ T cells, M1 macrophages, and N1 neutrophils, promoting the differentiation and activation of pro-tumor immune cells, including MDSCs, M2 macrophages, and Tregs [260, 261]. The extent to which metabolism pathways represent true vulnerabilities for tumor development remains unclear. Targeting glycolysis, glutaminolysis, and FAO has provided clinical benefits; the strategies integrating redox homeostasis and PPP may also generate new opportunities [9, 10, 119].

The most serious challenge in reshaping the immune profiles in TME is to understand the metabolic heterogeneity which is extremely complex depending not only on tumor and immune cell types but also on tumor stages and etiology (Fig. 5.3). Activation of signaling pathways including PI3K-Akt, mTOR, HIF-1, c-Myc, etc. in tumor-associated immune cells regulates their metabolism for survival, differentiation, and pro- or antitumor functions [10]. For example, HIF-1 activation in cancer cells and immune cells may upregulate glycolytic metabolism and enhance cancer-related inflammation during the initiation. In the tumor progression stage, HIF-1 elevation in TAMs, TADCs, MDSCs, and Tregs contributes to immunosuppression and angiogenesis by PD-L1 expression, lactate release, and adenosine-adenosine receptor interaction that facilitate tumor growth [32, 46, 58, 60, 136, 221, 232, 233]. Other factors and molecules, such as noncoding RNAs, complements, and coagulation-related factors, also regulate immune cell survival, differentiation, and functions in TME [262–264]. However, whether and how these molecules regulate metabolic reprogramming in tumor-associated immune cells will be of interest to be investigated.

The metabolite exchange adds an essential dimension of heterogeneity in the TME to contribute to tumor growth, metastasis and clinical resistance [12, 260, 261, 265]. Nonetheless, we posit that targeting the immune cell metabolism in TME in addition to the traditional cancer therapies will lead to more precise and efficient diagnosis and treatment of these fatal diseases.

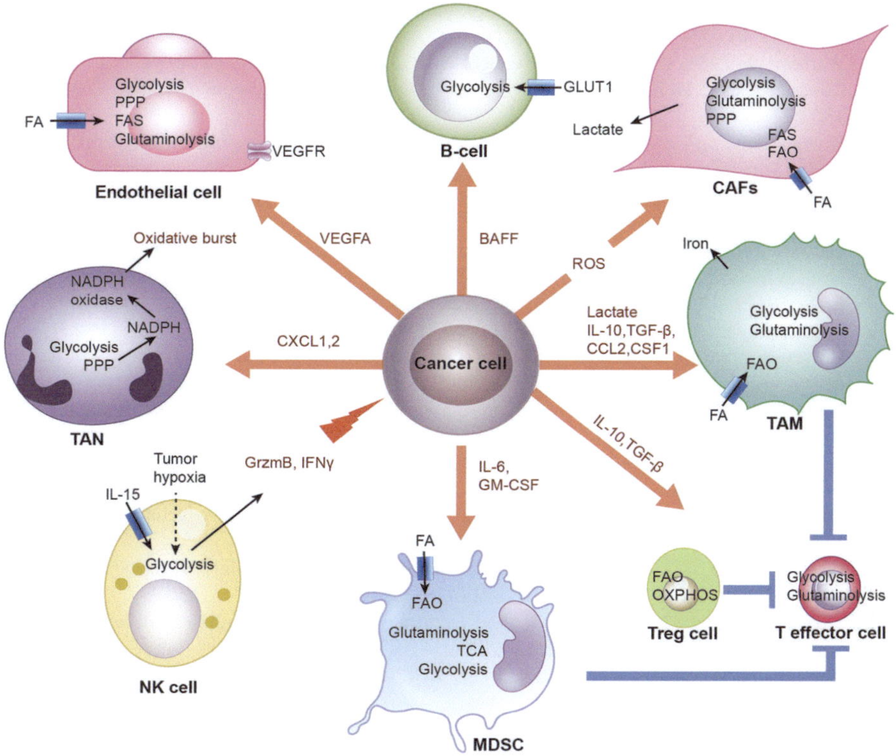

Fig. 5.3 Metabolic crosstalk between cancer cells and TME-immune cells

References

1. Lord EM, Penney DP, Sutherland RM, Cooper RA Jr (1979) Morphological and functional characteristics of cells infiltrating and destroying tumor multicellular spheroids in vivo. Virchows Arch B Cell Pathol Incl Mol Pathol 31:103–116
2. Shearer WT, Fink MP (1977) Immune surveillance system: its failure and activation. Prog Hematol 10:247–310
3. Dunn GP, Old LJ, Schreiber RD (2004) The three Es of cancer immunoediting. Annu Rev Immunol 22:329–360
4. Teng MW, Galon J, Fridman WH, Smyth MJ (2015) From mice to humans: developments in cancer immunoediting. J Clin Invest 125:3338–3346
5. Schreiber RD, Old LJ, Smyth MJ (2011) Cancer immunoediting: integrating immunity's roles in cancer suppression and promotion. Science 331:1565–1570
6. Galon J, Angell HK, Bedognetti D, Marincola FM (2013) The continuum of cancer immunosurveillance: prognostic, predictive, and mechanistic signatures. Immunity 39:11–26
7. Desrichard A, Snyder A, Chan TA (2016) Cancer neoantigens and applications for immunotherapy. Clin Cancer Res 22:807–812
8. Inman BA, Frigola X, Dong H, Kwon ED (2007) Costimulation, coinhibition and cancer. Curr Cancer Drug Targets 7:15–30
9. Boroughs LK, DeBerardinis RJ (2015) Metabolic pathways promoting cancer cell survival and growth. Nat Cell Biol 17:351–359

10. Biswas S, Metabolic K (2015) Reprogramming of immune cells in cancer progression. Immunity 43:435–449
11. Warburg O (1956) On the origin of cancer cells. Science 123:309–314
12. Chiarugi P, Cirri P (2015) Metabolic exchanges within tumor microenvironment. Cancer Lett 380:272–280
13. Krebs H, The A (1972) Pasteur effect and the relations between respiration and fermentation. Essays Biochem 8:1–34
14. Cairns RA, Harris IS, Mak TW (2011) Regulation of cancer cell metabolism. Nat Rev Cancer 11:85–95
15. Garcia-Heredia JM, Carnero A (2015) Decoding Warburg's hypothesis: tumor-related mutations in the mitochondrial respiratory chain. Oncotarget 6:41582–41599
16. Chan DA et al (2011) Targeting GLUT1 and the Warburg effect in renal cell carcinoma by chemical synthetic lethality. Sci Transl Med 3:94ra70
17. Lussey-Lepoutre C et al (2015) Loss of succinate dehydrogenase activity results in dependency on pyruvate carboxylation for cellular anabolism. Nat Commun 6:8784
18. Clark GR et al (2014) Germline FH mutations presenting with pheochromocytoma. J Clin Endocrinol Metab 99:E2046–E2050
19. Sonveaux P et al (2008) Targeting lactate-fueled respiration selectively kills hypoxic tumor cells in mice. J Clin Invest 118:3930–3942
20. Semenza GL (2008) Tumor metabolism: cancer cells give and take lactate. J Clin Invest 118:3835–3837
21. Fischer K et al (2007) Inhibitory effect of tumor cell-derived lactic acid on human T cells. Blood 109:3812–3819
22. Gottfried E et al (2006) Tumor-derived lactic acid modulates dendritic cell activation and antigen expression. Blood 107:2013–2021
23. De Vitto H, Perez-Valencia J, Radosevich JA (2016) Glutamine at focus: versatile roles in cancer. Tumour Biol 37:1541–1558
24. McKeehan WL (1982) Glycolysis, glutaminolysis and cell proliferation. Cell Biol Int Rep 6:635–650
25. Mushtaq M, Darekar S, Klein G, Kashuba E (2015) Different mechanisms of regulation of the Warburg effect in lymphoblastoid and Burkitt lymphoma cells. PLoS One 10:e0136142
26. Dang CV, Kim JW, Gao P, Yustein J (2008) The interplay between MYC and HIF in cancer. Nat Rev Cancer 8:51–56
27. Gao P et al (2009) c-Myc suppression of miR-23a/b enhances mitochondrial glutaminase expression and glutamine metabolism. Nature 458:762–765
28. Wise DR et al (2008) Myc regulates a transcriptional program that stimulates mitochondrial glutaminolysis and leads to glutamine addiction. Proc Natl Acad Sci U S A 105:18782–18787
29. DeBerardinis RJ et al (2007) Beyond aerobic glycolysis: transformed cells can engage in glutamine metabolism that exceeds the requirement for protein and nucleotide synthesis. Proc Natl Acad Sci U S A 104:19345–19350
30. Haissaguerre M, Saucisse N, Cota D (2014) Influence of mTOR in energy and metabolic homeostasis. Mol Cell Endocrinol 397:67–77
31. Salloum D, Mukhopadhyay S, Tung K, Polonetskaya A, Foster DA (2014) Mutant ras elevates dependence on serum lipids and creates a synthetic lethality for rapamycin. Mol Cancer Ther 13:733–741
32. Bensaad K et al (2014) Fatty acid uptake and lipid storage induced by HIF-1alpha contribute to cell growth and survival after hypoxia-reoxygenation. Cell Rep 9:349–365
33. Carracedo A, Cantley LC, Pandolfi PP (2013) Cancer metabolism: fatty acid oxidation in the limelight. Nat Rev Cancer 13:227–232
34. Nomura DK et al (2010) Monoacylglycerol lipase regulates a fatty acid network that promotes cancer pathogenesis. Cell 140:49–61
35. Sugiarto S et al (2011) Asymmetry-defective oligodendrocyte progenitors are glioma precursors. Cancer Cell 20:328–340

36. Borah A, Raveendran S, Rochani A, Maekawa T, Kumar DS (2015) Targeting self-renewal pathways in cancer stem cells: clinical implications for cancer therapy. Oncogenesis 4:e177
37. Takeishi S, Nakayama KI (2016) To wake up cancer stem cells, or to let them sleep, that is the question. Cancer Sci 107:875–881
38. Viale A et al (2014) Oncogene ablation-resistant pancreatic cancer cells depend on mitochondrial function. Nature 514:628–632
39. Kurtova AV et al (2015) Blocking PGE2-induced tumour repopulation abrogates bladder cancer chemoresistance. Nature 517:209–213
40. Zhuang ZH et al (2008) Early upregulation of cyclooxygenase-2 in human papillomavirus type 16 and telomerase-induced immortalization of human esophageal epithelial cells. J Gastroenterol Hepatol 23:1613–1620
41. Hale JS et al (2014) Cancer stem cell-specific scavenger receptor CD36 drives glioblastoma progression. Stem Cells 32:1746–1758
42. Franklin RA et al (2014) The cellular and molecular origin of tumor-associated macrophages. Science 344:921–925
43. Condeelis J, Pollard JW (2006) Macrophages: obligate partners for tumor cell migration, invasion, and metastasis. Cell 124:263–266
44. Mantovani A et al (2004) The chemokine system in diverse forms of macrophage activation and polarization. Trends Immunol 25:677–686
45. Yang K et al (2013) T cell exit from quiescence and differentiation into Th2 cells depend on Raptor-mTORC1-mediated metabolic reprogramming. Immunity 39:1043–1056
46. Cheng SC et al (2014) mTOR- and HIF-1alpha-mediated aerobic glycolysis as metabolic basis for trained immunity. Science 345:1250684
47. Baardman J, Licht I, de Winther MP, Van den Bossche J (2015) Metabolic-epigenetic crosstalk in macrophage activation. Epigenomics 7:1155–1164
48. Bosca L et al (2015) Metabolic signatures linked to macrophage polarization: from glucose metabolism to oxidative phosphorylation. Biochem Soc Trans 43:740–744
49. Colegio OR et al (2014) Functional polarization of tumour-associated macrophages by tumour-derived lactic acid. Nature 513:559–563
50. Krawczyk CM et al (2010) Toll-like receptor-induced changes in glycolytic metabolism regulate dendritic cell activation. Blood 115:4742–4749
51. Chang M, Hamilton JA, Scholz GM, Elsegood CL (2009) Glycolytic control of adjuvant-induced macrophage survival: role of PI3K, MEK1/2, and Bcl-2. J Leukoc Biol 85:947–956
52. Byles V et al (2013) The TSC-mTOR pathway regulates macrophage polarization. Nat Commun 4:2834
53. Howell JJ, Manning BD (2011) mTOR couples cellular nutrient sensing to organismal metabolic homeostasis. Trends Endocrinol Metab 22:94–102
54. Schmid MC et al (2011) Receptor tyrosine kinases and TLR/IL1Rs unexpectedly activate myeloid cell PI3kgamma, a single convergent point promoting tumor inflammation and progression. Cancer Cell 19:715–727
55. Rivera LB et al (2015) Intratumoral myeloid cells regulate responsiveness and resistance to antiangiogenic therapy. Cell Rep 11:577–591
56. Rozengurt E, Soares HP, Sinnet-Smith J (2014) Suppression of feedback loops mediated by PI3K/mTOR induces multiple overactivation of compensatory pathways: an unintended consequence leading to drug resistance. Mol Cancer Ther 13:2477–2488
57. Pello OM et al (2012) Role of c-MYC in alternative activation of human macrophages and tumor-associated macrophage biology. Blood 119:411–421
58. Palsson-McDermott EM et al (2015) Pyruvate kinase M2 regulates Hif-1alpha activity and IL-1beta induction and is a critical determinant of the warburg effect in LPS-activated macrophages. Cell Metab 21:65–80
59. Semenza GL, Roth PH, Fang HM, Wang GL (1994) Transcriptional regulation of genes encoding glycolytic enzymes by hypoxia-inducible factor 1. J Biol Chem 269:23757–23763
60. Semenza G, Targeting L (2003) HIF-1 for cancer therapy. Nat Rev Cancer 3:721–732

61. Biswas SK, Mantovani A (2010) Macrophage plasticity and interaction with lymphocyte subsets: cancer as a paradigm. Nat Immunol 11:889–896
62. Lewis CE, De Palma M, Naldini L (2007) Tie2-expressing monocytes and tumor angiogenesis: regulation by hypoxia and angiopoietin-2. Cancer Res 67:8429–8432
63. Lemke G, Lu Q (2003) Macrophage regulation by Tyro 3 family receptors. Curr Opin Immunol 15:31–36
64. Chang CI, Liao JC, Kuo L (2001) Macrophage arginase promotes tumor cell growth and suppresses nitric oxide-mediated tumor cytotoxicity. Cancer Res 61:1100–1106
65. Ho VW, Sly LM (2009) Derivation and characterization of murine alternatively activated (M2) macrophages. Methods Mol Biol 531:173–185
66. Stuehr DJ, Nathan CF (1989) Nitric oxide. A macrophage product responsible for cytostasis and respiratory inhibition in tumor target cells. J Exp Med 169:1543–1555
67. Ellyard JI, Quah BJ, Simson L, Parish CR (2010) Alternatively activated macrophage possess antitumor cytotoxicity that is induced by IL-4 and mediated by arginase-1. J Immunother 33:443–452
68. Odegaard JI, Chawla A (2011) Alternative macrophage activation and metabolism. Annu Rev Pathol 6:275–297
69. Huang SC et al (2014) Cell-intrinsic lysosomal lipolysis is essential for alternative activation of macrophages. Nat Immunol 15:846–855
70. Traversari C, Sozzani S, Steffensen KR, Russo V (2014) LXR-dependent and -independent effects of oxysterols on immunity and tumor growth. Eur J Immunol 44:1896–1903
71. Li Y et al (2011) Pleiotropic regulation of macrophage polarization and tumorigenesis by formyl peptide receptor-2. Oncogene 30:3887–3899
72. Park H, Lee J, Park T, Lee S, Yi W (2015) Enhancement of photo-current conversion efficiency in a CdS/CdSe quantum-dot-sensitized solar cell incorporated with single-walled carbon nanotubes. J Nanosci Nanotechnol 15:1614–1617
73. Martinez FO, Gordon S, Locati M, Mantovani A (2006) Transcriptional profiling of the human monocyte-to-macrophage differentiation and polarization: new molecules and patterns of gene expression. J Immunol 177:7303–7311
74. Biswas SK, Mantovani A (2012) Orchestration of metabolism by macrophages. Cell Metab 15:432–437
75. Zhou D et al (2014) Macrophage polarization and function with emphasis on the evolving roles of coordinated regulation of cellular signaling pathways. Cell Signal 26:192–197
76. Jha AK et al (2015) Network integration of parallel metabolic and transcriptional data reveals metabolic modules that regulate macrophage polarization. Immunity 42:419–430
77. Zhao Q et al (2012) Activated CD69+ T cells foster immune privilege by regulating IDO expression in tumor-associated macrophages. J Immunol 188:1117–1124
78. Frumento G, Piazza T, Di Carlo E, Ferrini S (2006) Targeting tumor-related immunosuppression for cancer immunotherapy. Endocr Metab Immune Disord Drug Targets 6:233–237
79. Cairo G, Recalcati S, Mantovani A, Locati M (2011) Iron trafficking and metabolism in macrophages: contribution to the polarized phenotype. Trends Immunol 32:241–247
80. Recalcati S et al (2010) Differential regulation of iron homeostasis during human macrophage polarized activation. Eur J Immunol 40:824–835
81. Deng R et al (2013) Inhibition of tumor growth and alteration of associated macrophage cell type by an HO-1 inhibitor in breast carcinoma-bearing mice. Oncol Res 20:473–482
82. Wilkinson N, Pantopoulos K (2013) IRP1 regulates erythropoiesis and systemic iron homeostasis by controlling HIF2alpha mRNA translation. Blood 122:1658–1668
83. Koskenkorva-Frank TS, Weiss G, Koppenol WH, Burckhardt S (2013) The complex interplay of iron metabolism, reactive oxygen species, and reactive nitrogen species: insights into the potential of various iron therapies to induce oxidative and nitrosative stress. Free Radic Biol Med 65:1174–1194
84. Kayadibi H, Sertoglu E, Uyanik M, Tapan S (2014) Neutrophil-lymphocyte ratio is useful for the prognosis of patients with hepatocellular carcinoma. World J Gastroenterol 20:9631–9632

85. Galdiero MR et al (2013) Tumor associated macrophages and neutrophils in cancer. Immunobiology 218:1402–1410
86. Galdiero MR, Garlanda C, Jaillon S, Marone G, Mantovani A (2013) Tumor associated macrophages and neutrophils in tumor progression. J Cell Physiol 228:1404–1412
87. Fridlender ZG et al (2009) Polarization of tumor-associated neutrophil phenotype by TGF-beta: "N1" versus "N2" TAN. Cancer Cell 16:183–194
88. Federzoni EA et al (2012) PU.1 is linking the glycolytic enzyme HK3 in neutrophil differentiation and survival of APL cells. Blood 119:4963–4970
89. Ahmed N, Weidemann MJ (1994) Purine metabolism in promyelocytic HL60 and dimethylsulphoxide-differentiated HL60 cells. Leuk Res 18:441–451
90. Azevedo EP et al (2015) A metabolic shift toward pentose phosphate pathway is necessary for amyloid fibril- and phorbol 12-myristate 13-acetate-induced Neutrophil Extracellular Trap (NET) formation. J Biol Chem 290:22174–22183
91. Suzuki H, Kakinuma K (1983) Evidence that NADPH is the actual substrate of the oxidase responsible for the "respiratory burst" of phagocytosing polymorphonuclear leukocytes. J Biochem 93:709–715
92. Rodriguez-Espinosa O, Rojas-Espinosa O, Moreno-Altamirano MM, Lopez-Villegas EO, Sanchez-Garcia FJ (2015) Metabolic requirements for neutrophil extracellular traps formation. Immunology 145:213–224
93. Cools-Lartigue J et al (2013) Neutrophil extracellular traps sequester circulating tumor cells and promote metastasis. J Clin Invest. 123:3446–3458
94. Wculek SK, Malanchi I (2015) Neutrophils support lung colonization of metastasis-initiating breast cancer cells. Nature 528:413–417
95. Karasuyama H, Yamanishi Y (2014) Basophils have emerged as a key player in immunity. Curr Opin Immunol 31:1–7
96. Davis BP, Rothenberg ME (2014) Eosinophils and cancer. Cancer Immunol Res 2:1–8
97. Berek C (2016) Eosinophils: important players in humoral immunity. Clin Exp Immunol 183:57–64
98. Shi HZ (2004) Eosinophils function as antigen-presenting cells. J Leukoc Biol 76:520–527
99. Cho KS et al (2014) Adipose-derived stem cells ameliorate allergic airway inflammation by inducing regulatory T cells in a mouse model of asthma. Mediat Inflamm 2014:436476
100. Pinto A et al (1997) The role of eosinophils in the pathobiology of Hodgkin's disease. Ann Oncol 8(Suppl 2):89–96
101. Nolen BM, Lokshin AE (2010) Targeting CCL11 in the treatment of ovarian cancer. Expert Opin Ther Targets 14:157–167
102. Vendramini-Costa DB, Carvalho JE (2012) Molecular link mechanisms between inflammation and cancer. Curr Pharm Des 18:3831–3852
103. Simon HU et al (2003) Interleukin-2 primes eosinophil degranulation in hypereosinophilia and Wells' syndrome. Eur J Immunol 33:834–839
104. Nadif R, Zerimech F, Bouzigon E, Matran R (2013) The role of eosinophils and basophils in allergic diseases considering genetic findings. Curr Opin Allergy Clin Immunol 13:507–513
105. Lotfi R et al (2009) Eosinophils oxidize damage-associated molecular pattern molecules derived from stressed cells. J Immunol 183:5023–5031
106. Elishmereni M, Bachelet I, Nissim Ben-Efraim AH, Mankuta D, Levi-Schaffer F (2013) Interacting mast cells and eosinophils acquire an enhanced activation state in vitro. Allergy 68:171–179
107. Kataoka S, Konishi Y, Nishio Y, Fujikawa-Adachi K, Tominaga A (2004) Antitumor activity of eosinophils activated by IL-5 and eotaxin against hepatocellular carcinoma. DNA Cell Biol 23:549–560
108. Krystel-Whittemore M, Dileepan KN, Wood JG (2015) Mast cell: a multi-functional master cell. Front Immunol 6:620

109. Sekar Y, Moon TC, Slupsky CM, Befus AD (2010) Protein tyrosine nitration of aldolase in mast cells: a plausible pathway in nitric oxide-mediated regulation of mast cell function. J Immunol 185:578–587
110. Kalluri R, Zeisberg M (2006) Fibroblasts in cancer. Nat Rev Cancer 6:392–401
111. Shiga K et al (2015) Cancer-associated fibroblasts: their characteristics and their roles in tumor growth. Cancers (Basel) 7:2443–2458
112. Johansson AC et al (2012) Cancer-associated fibroblasts induce matrix metalloproteinase-mediated cetuximab resistance in head and neck squamous cell carcinoma cells. Mol Cancer Res 10:1158–1168
113. Zhang D et al (2015) Metabolic reprogramming of cancer-associated fibroblasts by IDH3alpha downregulation. Cell Rep 10:1335–1348
114. Guido C et al (2012) Mitochondrial fission induces glycolytic reprogramming in cancer-associated myofibroblasts, driving stromal lactate production, and early tumor growth. Oncotarget 3:798–810
115. Lisanti MP, Martinez-Outschoorn UE, Sotgia F (2013) Oncogenes induce the cancer-associated fibroblast phenotype: metabolic symbiosis and "fibroblast addiction" are new therapeutic targets for drug discovery. Cell Cycle 12:2723–2732
116. Janzer A et al (2012) The H3K4me3 histone demethylase Fbxl10 is a regulator of chemokine expression, cellular morphology, and the metabolome of fibroblasts. J Biol Chem 287:30984–30992
117. Martinez-Outschoorn UE, Lisanti MP, Sotgia F (2014) Catabolic cancer-associated fibroblasts transfer energy and biomass to anabolic cancer cells, fueling tumor growth. Semin Cancer Biol 25:47–60
118. Shi H et al (2015) Overexpression of monocarboxylate anion transporter 1 and 4 in T24-induced cancer-associated fibroblasts regulates the progression of bladder cancer cells in a 3D microfluidic device. Cell Cycle 14:3058–3065
119. Xing Y, Zhao S, Zhou BP, Mi J (2015) Metabolic reprogramming of the tumour microenvironment. FEBS J 282:3892–3898
120. Orr MT, Lanier LL (2010) Natural killer cell education and tolerance. Cell 142:847–856
121. Vivier E et al (2011) Innate or adaptive immunity? The example of natural killer cells. Science 331:44–49
122. Keppel MP, Saucier N, Mah AY, Vogel TP, Cooper MA (2015) Activation-specific metabolic requirements for NK Cell IFN-gamma production. J Immunol 194:1954–1962
123. Velasquez SY et al (2016) Short-term hypoxia synergizes with interleukin 15 priming in driving glycolytic gene transcription and supports human natural killer cell activities. J Biol Chem 291:12960–12977
124. Le Nours J et al (2016) Atypical natural killer T-cell receptor recognition of CD1d-lipid antigens. Nat Commun 7:10570
125. Dahlberg CI, Sarhan D, Chrobok M, Duru AD, Alici E (2015) Natural killer cell-based therapies targeting cancer: possible strategies to gain and sustain anti-tumor activity. Front Immunol 6:605
126. Hida K, Maishi N, Torii C, Hida Y (2016) Tumor angiogenesis-characteristics of tumor endothelial cells. Int J Clin Oncol 21:206–212
127. Ribatti D (2016) Tumor refractoriness to anti-VEGF therapy. Oncotarget 7:46668–46677
128. Goel S et al (2011) Normalization of the vasculature for treatment of cancer and other diseases. Physiol Rev 91:1071–1121
129. Dudley AC (2012) Tumor endothelial cells. Cold Spring Harb Perspect Med 2:a006536
130. Lee SL, Fanburg BL (1987) Glycolytic activity and enhancement of serotonin uptake by endothelial cells exposed to hypoxia/anoxia. Circ Res 60:653–658
131. Jakobsson L et al (2010) Endothelial cells dynamically compete for the tip cell position during angiogenic sprouting. Nat Cell Biol 12:943–953
132. De Bock K, Georgiadou M, Carmeliet P (2013) Role of endothelial cell metabolism in vessel sprouting. Cell Metab 18:634–647

133. De Bock K et al (2013) Role of PFKFB3-driven glycolysis in vessel sprouting. Cell 154:651–663
134. Schoors S et al (2014) Partial and transient reduction of glycolysis by PFKFB3 blockade reduces pathological angiogenesis. Cell Metab 19:37–48
135. Vegran F, Boidot R, Michiels C, Sonveaux P, Feron O (2011) Lactate influx through the endothelial cell monocarboxylate transporter MCT1 supports an NF-kappaB/IL-8 pathway that drives tumor angiogenesis. Cancer Res 71:2550–2560
136. Sonveaux P et al (2012) Targeting the lactate transporter MCT1 in endothelial cells inhibits lactate-induced HIF-1 activation and tumor angiogenesis. PLoS One 7:e33418
137. Leopold JA, Zhang YY, Scribner AW, Stanton RC, Loscalzo J (2003) Glucose-6-phosphate dehydrogenase overexpression decreases endothelial cell oxidant stress and increases bioavailable nitric oxide. Arterioscler Thromb Vasc Biol 23:411–417
138. Tian WN et al (1998) Importance of glucose-6-phosphate dehydrogenase activity for cell growth. J Biol Chem 273:10609–10617
139. Locasale JW et al (2011) Phosphoglycerate dehydrogenase diverts glycolytic flux and contributes to oncogenesis. Nat Genet 43:869–874
140. Jiang P et al (2011) p53 regulates biosynthesis through direct inactivation of glucose-6-phosphate dehydrogenase. Nat Cell Biol 13:310–316
141. Anastasiou D et al (2011) Inhibition of pyruvate kinase M2 by reactive oxygen species contributes to cellular antioxidant responses. Science 334:1278–1283
142. Spolarics Z, Lang CH, Bagby GJ, Spitzer JJ (1991) Glutamine and fatty acid oxidation are the main sources of energy for Kupffer and endothelial cells. Am J Phys 261:G185–G190
143. Unterluggauer H et al (2008) Premature senescence of human endothelial cells induced by inhibition of glutaminase. Biogerontology 9:247–259
144. Li H et al (2001) Regulatory role of arginase I and II in nitric oxide, polyamine, and proline syntheses in endothelial cells. Am J Physiol Endocrinol Metab 280:E75–E82
145. Wu G, Haynes TE, Li H, Yan W, Meininger CJ (2001) Glutamine metabolism to glucosamine is necessary for glutamine inhibition of endothelial nitric oxide synthesis. Biochem J 353:245–252
146. Kim YW, Byzova TV (2014) Oxidative stress in angiogenesis and vascular disease. Blood 123:625–631
147. Dagher Z, Ruderman N, Tornheim K, Ido Y (2001) Acute regulation of fatty acid oxidation and amp-activated protein kinase in human umbilical vein endothelial cells. Circ Res 88:1276–1282
148. Wang Q et al (2006) Induction of allospecific tolerance by immature dendritic cells genetically modified to express soluble TNF receptor. J Immunol 177:2175–2185
149. Dudek AM, Martin S, Garg AD, Agostinis P (2013) Immature, semi-mature, and fully mature dendritic cells: toward a DC-cancer cells interface that augments anticancer immunity. Front Immunol 4:438
150. Perrot I et al (2007) Dendritic cells infiltrating human non-small cell lung cancer are blocked at immature stage. J Immunol 178:2763–2769
151. Cahill EF, Tobin LM, Carty F, Mahon BP, English K (2015) Jagged-1 is required for the expansion of CD4+ CD25+ FoxP3+ regulatory T cells and tolerogenic dendritic cells by murine mesenchymal stromal cells. Stem Cell Res Ther 6:19
152. Melief CJ (2008) Cancer immunotherapy by dendritic cells. Immunity 29:372–383
153. Steinman RM, Banchereau J (2007) Taking dendritic cells into medicine. Nature 449:419–426
154. Apetoh L, Locher C, Ghiringhelli F, Kroemer G, Zitvogel L (2011) Harnessing dendritic cells in cancer. Semin Immunol 23:42–49
155. Dong H, Bullock TN (2014) Metabolic influences that regulate dendritic cell function in tumors. Front Immunol 5:24
156. Kumar V, Gabrilovich DI (2014) Hypoxia-inducible factors in regulation of immune responses in tumour microenvironment. Immunology 143:512–519

157. Mancino A et al (2008) Divergent effects of hypoxia on dendritic cell functions. Blood 112:3723–3734
158. Yang M et al (2010) HIF-dependent induction of adenosine receptor A2b skews human dendritic cells to a Th2-stimulating phenotype under hypoxia. Immunol Cell Biol 88:165–171
159. Nasi A, Rethi B (2013) Disarmed by density: a glycolytic break for immunostimulatory dendritic cells? Oncoimmunology 2:e26744
160. Novitskiy SV et al (2008) Adenosine receptors in regulation of dendritic cell differentiation and function. Blood 112:1822–1831
161. Ahmed MS et al (2015) Dab2, a negative regulator of DC immunogenicity, is an attractive molecular target for DC-based immunotherapy. Oncoimmunology 4:e984550
162. Tran Janco JM, Lamichhane P, Karyampudi L, Knutson KL (2015) Tumor-infiltrating dendritic cells in cancer pathogenesis. J Immunol 194:2985–2991
163. Galvin KC et al (2013) Blocking retinoic acid receptor-alpha enhances the efficacy of a dendritic cell vaccine against tumours by suppressing the induction of regulatory T cells. Cancer Immunol Immunother 62:1273–1282
164. Ravindran R et al (2014) Vaccine activation of the nutrient sensor GCN2 in dendritic cells enhances antigen presentation. Science 343:313–317
165. Everts B et al (2014) TLR-driven early glycolytic reprogramming via the kinases TBK1-IKKvarepsilon supports the anabolic demands of dendritic cell activation. Nat Immunol 15:323–332
166. Rehman A et al (2013) Role of fatty-acid synthesis in dendritic cell generation and function. J Immunol 190:4640–4649
167. Harshyne LA, Zimmer MI, Watkins SC, Barratt-Boyes SM (2003) A role for class A scavenger receptor in dendritic cell nibbling from live cells. J Immunol 170:2302–2309
168. Malinarich F et al (2015) High mitochondrial respiration and glycolytic capacity represent a metabolic phenotype of human tolerogenic dendritic cells. J Immunol 194:5174–5186
169. Cubillos-Ruiz JR et al (2015) ER stress sensor XBP1 controls anti-tumor immunity by disrupting dendritic cell homeostasis. Cell 161:1527–1538
170. Li S, Champlin R, Fitchen JH, Gale RP (1985) Abnormalities of myeloid progenitor cells after "successful" bone marrow transplantation. J Clin Invest 75:234–241
171. Strober S (1984) Natural suppressor (NS) cells, neonatal tolerance, and total lymphoid irradiation: exploring obscure relationships. Annu Rev Immunol 2:219–237
172. Young MR, Kolesiak K, Wright MA, Gabrilovich DI (1999) Chemoattraction of femoral CD34+ progenitor cells by tumor-derived vascular endothelial cell growth factor. Clin Exp Metastasis 17:881–888
173. Bronte V et al (2000) Identification of a CD11b(+)/Gr-1(+)/CD31(+) myeloid progenitor capable of activating or suppressing CD8(+) T cells. Blood 96:3838–3846
174. Tartour E et al (2011) Angiogenesis and immunity: a bidirectional link potentially relevant for the monitoring of antiangiogenic therapy and the development of novel therapeutic combination with immunotherapy. Cancer Metastasis Rev 30:83–95
175. Morse MA, Hall JR, Plate JM (2009) Countering tumor-induced immunosuppression during immunotherapy for pancreatic cancer. Expert Opin Biol Ther 9:331–339
176. Maeda A, Kawamura T, Ueno T, Usui N, Miyagawa S (2014) Monocytic suppressor cells derived from human peripheral blood suppress xenogenic immune reactions. Xenotransplantation 21:46–56
177. Obermajer N, Muthuswamy R, Lesnock J, Edwards RP, Kalinski P (2011) Positive feedback between PGE2 and COX2 redirects the differentiation of human dendritic cells toward stable myeloid-derived suppressor cells. Blood 118:5498–5505
178. Draghiciu O, Lubbers J, Nijman HW, Daemen T (2015) Myeloid derived suppressor cells-an overview of combat strategies to increase immunotherapy efficacy. Oncoimmunology 4:e954829

179. Hossain F et al (2015) Inhibition of fatty acid oxidation modulates immunosuppressive functions of myeloid-derived suppressor cells and enhances cancer therapies. Cancer Immunol Res 3:1236–1247
180. Zoso A et al (2014) Human fibrocytic myeloid-derived suppressor cells express IDO and promote tolerance via Treg-cell expansion. Eur J Immunol 44:3307–3319
181. Srivastava MK, Sinha P, Clements VK, Rodriguez P, Ostrand-Rosenberg S (2010) Myeloid-derived suppressor cells inhibit T-cell activation by depleting cystine and cysteine. Cancer Res 70:68–77
182. Nagaraj S, Schrum AG, Cho HI, Celis E, Gabrilovich DI (2010) Mechanism of T cell tolerance induced by myeloid-derived suppressor cells. J Immunol 184:3106–3116
183. Lu T et al (2011) Tumor-infiltrating myeloid cells induce tumor cell resistance to cytotoxic T cells in mice. J Clin Invest 121:4015–4029
184. Brito C et al (1999) Peroxynitrite inhibits T lymphocyte activation and proliferation by promoting impairment of tyrosine phosphorylation and peroxynitrite-driven apoptotic death. J Immunol 162:3356–3366
185. Kasic T et al (2011) Modulation of human T-cell functions by reactive nitrogen species. Eur J Immunol 41:1843–1849
186. Wu AA, Drake V, Huang HS, Chiu S, Zheng L (2015) Reprogramming the tumor microenvironment: tumor-induced immunosuppressive factors paralyze T cells. Oncoimmunology 4:e1016700
187. Crook KR et al (2015) Myeloid-derived suppressor cells regulate T cell and B cell responses during autoimmune disease. J Leukoc Biol 97:573–582
188. Obermajer N, Muthuswamy R, Odunsi K, Edwards RP, Kalinski P (2011) PGE(2)-induced CXCL12 production and CXCR4 expression controls the accumulation of human MDSCs in ovarian cancer environment. Cancer Res 71:7463–7470
189. Zhang N, Bevan MJ (2011) CD8(+) T cells: foot soldiers of the immune system. Immunity 35:161–168
190. Liao W, Lin JX, Leonard WJ (2013) Interleukin-2 at the crossroads of effector responses, tolerance, and immunotherapy. Immunity 38:13–25
191. Siska PJ, Rathmell JC (2015) T cell metabolic fitness in antitumor immunity. Trends Immunol 36:257–264
192. MacIver NJ, Michalek RD, Rathmell JC (2013) Metabolic regulation of T lymphocytes. Annu Rev Immunol 31:259–283
193. Zheng Y, Delgoffe GM, Meyer CF, Chan W, Powell JD (2009) Anergic T cells are metabolically anergic. J Immunol 183:6095–6101
194. Patsoukis N et al (2015) PD-1 alters T-cell metabolic reprogramming by inhibiting glycolysis and promoting lipolysis and fatty acid oxidation. Nat Commun 6:6692
195. Rabinowitz JD, White E (2010) Autophagy and metabolism. Science 330:1344–1348
196. Calcinotto A et al (2012) Modulation of microenvironment acidity reverses anergy in human and murine tumor-infiltrating T lymphocytes. Cancer Res 72:2746–2756
197. Chang CH et al (2013) Posttranscriptional control of T cell effector function by aerobic glycolysis. Cell 153:1239–1251
198. Maciolek JA, Pasternak JA, Wilson HL (2014) Metabolism of activated T lymphocytes. Curr Opin Immunol 27:60–74
199. Wang Q, Liang B, Shirwany NA, Zou MH (2011) 2-Deoxy-D-glucose treatment of endothelial cells induces autophagy by reactive oxygen species-mediated activation of the AMP-activated protein kinase. PLoS One 6:e17234
200. Zeng H, Chi H (2014) mTOR signaling and transcriptional regulation in T lymphocytes. Transcription 5:e28263
201. Fernandez-Ramos AA, Poindessous V, Marchetti-Laurent C, Pallet N, Loriot MA (2016) The effect of immunosuppressive molecules on T-cell metabolic reprogramming. Biochimie 127:23–36

202. Blagih J et al (2015) The energy sensor AMPK regulates T cell metabolic adaptation and effector responses in vivo. Immunity 42:41–54
203. Bronte V, Zanovello P (2005) Regulation of immune responses by L-arginine metabolism. Nat Rev Immunol 5:641–654
204. Rodriguez PC et al (2004) Arginase I production in the tumor microenvironment by mature myeloid cells inhibits T-cell receptor expression and antigen-specific T-cell responses. Cancer Res 64:5839–5849
205. Bronte V et al (2005) Boosting antitumor responses of T lymphocytes infiltrating human prostate cancers. J Exp Med 201:1257–1268
206. Aulak KS et al (2001) Proteomic method identifies proteins nitrated in vivo during inflammatory challenge. Proc Natl Acad Sci U S A 98:12056–12061
207. Molon B et al (2011) Chemokine nitration prevents intratumoral infiltration of antigen-specific T cells. J Exp Med 208:1949–1962
208. Wang W et al (2016) Effector T cells abrogate stroma-mediated chemoresistance in ovarian cancer. Cell 165:1092–1105
209. Liu X et al (2010) Selective inhibition of IDO1 effectively regulates mediators of antitumor immunity. Blood 115:3520–3530
210. Uyttenhove C et al (2003) Evidence for a tumoral immune resistance mechanism based on tryptophan degradation by indoleamine 2,3-dioxygenase. Nat Med 9:1269–1274
211. Carbotti G et al (2015) IL-27 induces the expression of IDO and PD-L1 in human cancer cells. Oncotarget 6:43267–43280
212. Leung BS, Stout LE, Shaskan EG, Thompson RM (1992) Differential induction of indoleamine-2,3-dioxygenase (IDO) by interferon-gamma in human gynecologic cancer cells. Cancer Lett 66:77–81
213. Sorensen RB et al (2011) Spontaneous cytotoxic T-cell reactivity against indoleamine 2,3-dioxygenase-2. Cancer Res 71:2038–2044
214. Yang W et al (2016) Potentiating the antitumour response of CD8(+) T cells by modulating cholesterol metabolism. Nature 531:651–655
215. Mucida D et al (2007) Reciprocal TH17 and regulatory T cell differentiation mediated by retinoic acid. Science 317:256–260
216. Liu Z et al (2015) Immune homeostasis enforced by co-localized effector and regulatory T cells. Nature 528:225–230
217. Vacchelli E et al (2014) Trial watch: IDO inhibitors in cancer therapy. Oncoimmunology 3:e957994
218. Michalek RD et al (2011) Cutting edge: distinct glycolytic and lipid oxidative metabolic programs are essential for effector and regulatory CD4+ T cell subsets. J Immunol 186:3299–3303
219. Lochner M, Berod L, Sparwasser T (2015) Fatty acid metabolism in the regulation of T cell function. Trends Immunol 36:81–91
220. Beier UH et al (2015) Essential role of mitochondrial energy metabolism in Foxp3(+) T-regulatory cell function and allograft survival. FASEB J 29:2315–2326
221. Bollinger T et al (2014) HIF-1alpha- and hypoxia-dependent immune responses in human CD4+CD25high T cells and T helper 17 cells. J Leukoc Biol 96:305–312
222. Shi LZ et al (2011) HIF1alpha-dependent glycolytic pathway orchestrates a metabolic checkpoint for the differentiation of TH17 and Treg cells. J Exp Med 208:1367–1376
223. McGeachy MJ, Cua DJ (2008) Th17 cell differentiation: the long and winding road. Immunity 28:445–453
224. Leung S et al (2010) The cytokine milieu in the interplay of pathogenic Th1/Th17 cells and regulatory T cells in autoimmune disease. Cell Mol Immunol 7:182–189
225. Berod L et al (2014) De novo fatty acid synthesis controls the fate between regulatory T and T helper 17 cells. Nat Med 20:1327–1333

226. Caro-Maldonado A et al (2014) Metabolic reprogramming is required for antibody production that is suppressed in anergic but exaggerated in chronically BAFF-exposed B cells. J Immunol 192:3626–3636
227. Bashir U et al (2015) PET/MRI in oncological imaging: state of the art. Diagnostics (Basel) 5:333–357
228. Flechsig P, Mehndiratta A, Haberkorn U, Kratochwil C, Giesel FL (2015) PET/MRI and PET/CT in lung lesions and thoracic malignancies. Semin Nucl Med 45:268–281
229. Mehanna H et al (2016) PET-CT Surveillance versus neck dissection in advanced head and neck cancer. N Engl J Med 374:1444–1454
230. Radford J et al (2015) Results of a trial of PET-directed therapy for early-stage Hodgkin's lymphoma. N Engl J Med 372:1598–1607
231. McGranahan N et al (2016) Clonal neoantigens elicit T cell immunoreactivity and sensitivity to immune checkpoint blockade. Science 351:1463–1469
232. Palazon A et al (2012) The HIF-1alpha hypoxia response in tumor-infiltrating T lymphocytes induces functional CD137 (4-1BB) for immunotherapy. Cancer Discov 2:608–623
233. Noman MZ et al (2014) PD-L1 is a novel direct target of HIF-1alpha, and its blockade under hypoxia enhanced MDSC-mediated T cell activation. J Exp Med 211:781–790
234. Gowrishankar K et al (2015) Inducible but not constitutive expression of PD-L1 in human melanoma cells is dependent on activation of NF-kappaB. PLoS One 10:e0123410
235. Kleffel S et al (2015) Melanoma cell-intrinsic PD-1 receptor functions promote tumor growth. Cell 162:1242–1256
236. Zelenay S et al (2015) Cyclooxygenase-dependent tumor growth through evasion of immunity. Cell 162:1257–1270
237. Chen C, Rowell EA, Thomas RM, Hancock WW, Wells AD (2006) Transcriptional regulation by Foxp3 is associated with direct promoter occupancy and modulation of histone acetylation. J Biol Chem 281:36828–36834
238. Onodera T et al (2009) Constitutive expression of IDO by dendritic cells of mesenteric lymph nodes: functional involvement of the CTLA-4/B7 and CCL22/CCR4 interactions. J Immunol 183:5608–5614
239. Chevolet I et al (2015) Characterization of the immune network of IDO, tryptophan metabolism, PD-L1, and in circulating immune cells in melanoma. Oncoimmunology 4:e982382
240. Hellmann MD, Friedman CF, Wolchok JD (2016) Combinatorial cancer immunotherapies. Adv Immunol 130:251–277
241. Dai H, Wang Y, Lu X, Han W (2016) Chimeric antigen receptors modified T-cells for cancer therapy. J Natl Cancer Inst 108:djv439
242. Jackson HJ, Rafiq S, Brentjens RJ (2016) Driving CAR T-cells forward. Nat Rev Clin Oncol 13:370–383
243. Kawalekar OU et al (2016) Distinct signaling of coreceptors regulates specific metabolism pathways and impacts memory development in CAR T cells. Immunity 44:380–390
244. Wurth R et al (2016) Drug-repositioning opportunities for cancer therapy: novel molecular targets for known compounds. Drug Discov Today 21:190–199
245. Wang BY et al (2015) Intermittent high dose proton pump inhibitor enhances the antitumor effects of chemotherapy in metastatic breast cancer. J Exp Clin Cancer Res 34:85
246. Luciani F et al (2004) Effect of proton pump inhibitor pretreatment on resistance of solid tumors to cytotoxic drugs. J Natl Cancer Inst 96:1702–1713
247. Nicolau-Galmes F et al (2011) Terfenadine induces apoptosis and autophagy in melanoma cells through ROS-dependent and -independent mechanisms. Apoptosis 16:1253–1267
248. Strandberg TE et al (2004) Mortality and incidence of cancer during 10-year follow-up of the Scandinavian Simvastatin Survival Study (4S). Lancet 364:771–777
249. Ishimoto T et al (2011) CD44 variant regulates redox status in cancer cells by stabilizing the xCT subunit of system xc(-) and thereby promotes tumor growth. Cancer Cell 19:387–400
250. Yae T et al (2012) Alternative splicing of CD44 mRNA by ESRP1 enhances lung colonization of metastatic cancer cell. Nat Commun 3:883

251. Yoshikawa M et al (2013) xCT inhibition depletes CD44v-expressing tumor cells that are resistant to EGFR-targeted therapy in head and neck squamous cell carcinoma. Cancer Res 73:1855–1866
252. Ferrannini E (2014) The target of metformin in type 2 diabetes. N Engl J Med 371:1547–1548
253. Psutka SP et al (2015) The association between metformin use and oncologic outcomes among surgically treated diabetic patients with localized renal cell carcinoma. Urol Oncol 33(67):e15–e23
254. Zhang Y et al (2014) Involvement of metformin and AMPK in the radioresponse and prognosis of luminal versus basal-like breast cancer treated with radiotherapy. Oncotarget 5:12936–12949
255. Chen H et al (2015) Synergistic effects of metformin in combination with EGFR-TKI in the treatment of patients with advanced non-small cell lung cancer and type 2 diabetes. Cancer Lett 369:97–102
256. Takahashi RU et al (2015) Loss of microRNA-27b contributes to breast cancer stem cell generation by activating ENPP1. Nat Commun 6:7318
257. Singh M, Ferrara N (2012) Modeling and predicting clinical efficacy for drugs targeting the tumor milieu. Nat Biotechnol 30:648–657
258. Bergers G, Hanahan D (2008) Modes of resistance to anti-angiogenic therapy. Nat Rev Cancer 8:592–603
259. Clem B et al (2008) Small-molecule inhibition of 6-phosphofructo-2-kinase activity suppresses glycolytic flux and tumor growth. Mol Cancer Ther 7:110–120
260. Whiteside TL (2006) The role of immune cells in the tumor microenvironment. Cancer Treat Res 130:103–124
261. Gajewski TF, Schreiber H, Fu YX (2013) Innate and adaptive immune cells in the tumor microenvironment. Nat Immunol 14:1014–1022
262. Mauge L, Terme M, Tartour E, Helley D (2014) Control of the adaptive immune response by tumor vasculature. Front Oncol 4:61
263. Kohlhapp FJ, Mitra AK, Lengyel E, Peter ME (2015) MicroRNAs as mediators and communicators between cancer cells and the tumor microenvironment. Oncogene 34:5857–5868
264. Pio R, Ajona D, Lambris JD (2013) Complement inhibition in cancer therapy. Semin Immunol 25:54–64
265. Yuneva MO et al (2012) The metabolic profile of tumors depends on both the responsible genetic lesion and tissue type. Cell Metab 15:157–170

Chapter 6
Regulation of Metabolism Across Different Subsets of T Cells in Cancer

Xuerui Luo, Fuxiang Zhu, Zuojia Chen, Dan Li, and Bin Li

Abstract T cells play a critical role to defend against tumor and maintain immune homeostasis. The diverse functions of T cells require precise regulation of metabolic pathways. Recent studies reveal that metabolic changes are tightly linked to the activation and function of T cells. Given the importance of these cells in tumor progression, it is important to understand how the tumor microenvironment regulates metabolism of T cells and how the metabolic reprogramming of T cells affects tumor growth. Here, we review new findings and discuss how metabolic reprogramming of different types of T cells affects the immune response in tumors.

Keyword Metabolic regulation • Tumors • T Cells

X. Luo • F. Zhu
Key Laboratory of Molecular Virology & Immunology, Institut Pasteur of Shanghai, Shanghai Institutes for Biological Sciences, University of Chinese Academy of Sciences, Shanghai 200031, China

Z. Chen
National Cancer Institute, Bethesda, MD 20892, USA

D. Li
Shanghai Institute of Immunology, Shanghai JiaoTong University School of Medicine, Shanghai 200025, China

Department of Immunology and Microbiology, Shanghai Jiao Tong University School of Medicine, Shanghai 200025, China

B. Li (✉)
Shanghai Institute of Immunology, Department of Immunology and Microbiology, Shanghai Jiao Tong University, Shanghai 200025, China

School of Medicine, Shanghai Jiao Tong University, Shanghai 200025, China

CAS Center for Excellence in Molecular Cell Science, CAS Key Laboratory of Molecular Virology and Immunology, Unit of Molecular Immunology, Institut Pasteur of Shanghai, Shanghai Institutes for Biological Sciences, University of the Chinese Academy of Sciences, Chinese Academy of Sciences, Shanghai 200025, China
e-mail: binli@shsmu.edu.cn

B. Li, F. Pan (eds.), *Immune Metabolism in Health and Tumor*, Advances in Experimental Medicine and Biology 1011, DOI 10.1007/978-94-024-1170-6_6

Inflammation and infiltration of leukocytes are a vital event during tumor progression. The immune cells that can be found in the tumor include those from the innate immune system as well as those from the adaptive immune system. While many types of immune cells (B cells, NK cells, and dendritic cells) regulate cancer progression, T cells are a major class of tumor-infiltrating lymphocytes. In fact, T cells have been shown to affect survival, growth, and proliferation of cancer cells [1, 2].

Increasing evidence reveals significant changes in glucose, lipid, and amino acid in T cells upon activation, highlighting the importance of metabolic reprogramming in these cells. Metabolic reprogramming to accommodate proliferation and adaptation to the immune microenvironment has also been well recognized in cancer cells [3, 4]. However, it remains to be elucidated whether tumor-associated T cells also undergo metabolic changes during tumor progression and how these alterations regulate malignancy. Here, we review current understanding of T cell metabolism in the tumor microenvironment to improve future design of immunotherapies.

6.1 Metabolism in Cancer Cells

6.1.1 Glycolysis

Otto Heinrich Warburg, the recipient of the 1931 Nobel Prize in Physiology, pioneered cancer metabolism research. Thus, the observation that tumor cells have a high rate of glycolysis despite the absence of oxygen is today known as the Warburg effect. The conversion from mitochondrial oxidative phosphorylation to glycolysis provides both energy (due to high consumption of glucose) and metabolic intermediates that are used by the tumor cells for growth and proliferation [4, 5]. This switch is also useful to survival in the hypoxic tumor microenvironment. It has been reported that a key mechanism in this process involves the activation of the transcription factor hypoxia-inducible factor 1 (HIF1) through the phosphoinositol 3-kinase (PI3K)-AKT signaling pathway. HIF1 induces the expression of glycolytic genes such as glucose transporter-1 (GLUT1), monocarboxylate transporter 4 (MCT4), and lactate dehydrogenase A (LDHA) through cooperation with other transcription factors or oncogenes (c-Myc, Oct1, p53) [6].

6.1.2 Lactate Metabolism

Lactate, which is a product of glycolysis, accumulates in the tumor microenvironment. High levels of lactate have been observed to increase cancer cell stemness, migration, and metastasis. Lactate also induces the expression of transforming growth factor (TGFβ), hyaluronic acid, and CD44, which are crucial for tumor progression.

Lactate also affects immune cells through inhibition of p38 and JNK/c-Jun signaling-dependent proliferation, cytokine production, and activity of cytotoxic T lymphocytes (CTL) [7]. Recently, Claudio Mauro's lab showed that extracellular sodium lactate and lactic acid inhibit the motility of $CD4^+$ and $CD8^+$ T cells via subtype-specific transporters (Slc5a12 and Slc16a1, respectively). Sodium lactate-mediated inhibition of $CD4^+$ T cell motility is caused by disruption of glycolysis, while lactic acid effect on $CD8^+$ T cell motility is independent of glycolysis. More importantly, sodium lactate induces Th17 differentiation through the induction of IL-17, whereas lactic acid causes the loss of the cytolysis function in CD8+ T cells [8].

6.1.3 *Other Metabolites*

Glutamine, the most abundant amino acid, is also important for tumor cell proliferation. Glutaminolysis, which describes the conversion of glutamine to glutamate or lactate, is high in tumor cells. In fact, the oncogene c-Myc has been shown to upregulate glutaminolysis in tumor cells [4, 5].

Lipid metabolic network contains import or export of lipids, fatty acid β-oxidation (FAO) pathway, and de novo synthesis pathways (e.g., lipogenesis and cholesterol synthesis). Fatty acids supply more than twice as much ATP per mole to cells compared with glucose or amino acids. Many cancer cells show strong uptake and synthesis of lipids, such as prostate adenocarcinoma and diffuse large B cell lymphoma. Furthermore, fatty acid oxidation seems to help cancer cells survive during tumor regression. In KRAS-driven pancreatic cancer model, mitochondrial respiration highly affects tumor cell regression [3, 4, 9]. Besides, increased monoacylglycerol lipase (MAGL) in tumor cells is associated with pro-tumorigenic growth. [10] Collectively, metabolic reprogramming of tumor cells is a critical characteristic of tumor progression [4, 5].

6.2 Metabolism in T Cells

As key players in the adaptive immune response, T cells can be divided into $CD4^+$ and $CD8^+$ subsets. Activated $CD4^+$ T cells differentiate into different subtypes of effector cells and induced regulatory T cells (iTreg) to regulate tumor growth. There are several subtypes of effector T cells: Th1 cells activate CTL, macrophages, and NK cells to induce an anti-tumor response through secretion of cytokine interferon-gamma (IFNγ), tumor necrosis factor (TNF), and IL-2; Th2 cells limit CTL proliferation through IL-10, IL-4, and IL-5; Th17 cells can promote or inhibit tumor progression in different conditions dependent on IL-17; and regulatory T cells (Treg) promote tumor-induced immunosuppression through IL-10, TGFβ, or inhibitory receptors [11]. $CD8^+$ T cells, on the other hand, promote tumor cell lysis through perforin, granzymes, and Fas-Fas ligand-mediated initiation of apoptosis [5]. Changes in T cell metabolism during activation and differentiation have been studied extensively, and we summarize established findings as below (Fig. 6.1).

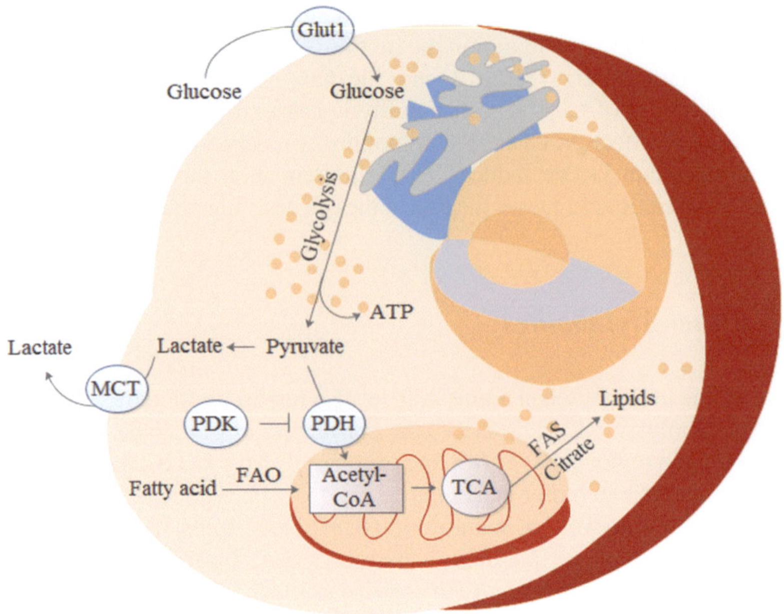

Fig. 6.1 The regulatory network of metabolisms in T cells

6.2.1 *Naïve T Cells*

Quiescent naïve T cells utilize oxidative phosphorylation, mitochondrial fatty acid β-oxidation (FAO), and little glutamine metabolism to maintain viability. Although naïve T cells have low metabolic requirements, the migration and maintenance of naïve T cells in secondary lymphocyte organs still require ATP consumption.

Survival of naïve $CD4^+$ T cells is maintained, in part, by interleukin-7 (IL-7) - interleukin-7 receptor (IL7R) signaling. Glucose uptake through Glut1 is essential for survival of naïve cells, and IL-7/IL7R signaling through PI3K/Akt/mTOR increases the expression and translocation of Glut1 to the cell surface [12–14]. IL-7/IL7R signaling through JAK3 and PI3K/Akt pathways also promotes phosphorylation and activation of STAT5. Activated STAT5 translocates to the nucleus, where it upregulates expression of genes necessary for survival and proliferation of T cells [14]. Forkhead box P1 (Foxp1) and forkhead box protein O1 (Foxo1) are also important in this signaling pathway. By antagonizing the enhancer of IL-7Rα signaling - Foxo1 - Foxp1 keeps naïve T cells in quiescent state [15]. Direct signaling through the TCR also increases cell-surface expression of Glut1 to promote glucose uptake [16].

6.2.2 Effector T Cells

Upon activation, metabolic reprogramming of naïve T cells is required to support rapid proliferation [4]. Increased OXPHOS, rate of glycolysis in comparison to rate of OXPHOS, glutaminolysis, and decreased FAO are all observed during activation. Although OXPHOS provides ATP to support T cells' proliferation, glycolysis generates metabolic intermediates that can be shuttled to produce macromolecules necessary for effector function and proliferation. For instance, metabolic switch from primarily OXPHOS to OXPHOS and aerobic glycolysis is required for secretion of cytokines like IFN-γ and IL-2 (which are important for anti-tumor responses) in effector T cells [17]. Although glycolysis requires more glucose to yield the same amount of ATP, it produces ATP much faster than OXPHOS. The importance of glucose metabolism has been shown in the acute graft-versus-host disease (GvHD) and inflammatory bowel disease (IBD) mouse model, where depletion of Glut1 inhibits growth, proliferation, survival, and differentiation of Teff cells, but does not affect the suppressive function of Treg cells [18]. Therefore, glycolysis is necessary to meet the metabolic demands of effector T cells [16].

TCR and CD28 signaling activate HIF1 and c-Myc through the PI3K/Akt/mTOR pathway. In turn, increased c-Myc directly regulates key genes including Glut1, LDHA, and PKM2 and amino acid transporters (SLC3A2, SLC5A1, SLC7A1) involved in metabolism [19, 20]. Metabolism is also implicated in T cell differentiation. It was recently reported that an increase in amino acid transporter ASCT2 promotes development of Th1, Th17, and overall inflammatory T cell response after activation [21]. Similar to c-Myc, mTOR signaling is also critical for metabolic shift from primarily OXPHOS to OXPHOS and glycolysis. mTOR signaling complexes are composed of mTOR complex 1 (mTORC1) and mTOR complex 2 (mTORC2). mTORC1 Rheb-deficient T cells are unable to differentiate into Th1 or Th17 cells, but Th2 cells persist. However, mTORC1 is also important in Th2 cell differentiation. In mTORC1 raptor-deficient T cells, Th2 differentiation was significantly impaired; these T cells failed to generate IL-2 [22]. In turn, mTORC2 rictor-deficient T cells enhanced Th1 and Th17 but not Th2 differentiation after activation [23]. Apart from mTOR, Th17 cell differentiation is also dependent on HIF1 to induce glycolysis; HIF-1 regulates expression of glycolytic genes to facilitate metabolic changes in Th17 cells [24].

Recently, single-cell RNA sequencing is used to investigate the molecular mechanisms governing heterogeneity of T cells. Computational analysis of Th17 cells isolated from the central nervous system (CNS) and lymph nodes (LN) at the peak of autoimmune encephalomyelitis (EAE) or differentiated in vitro under either pathogenic or nonpathogenic polarization conditions revealed elevation of genes governing pathogenicity and disease susceptibility. These included Gpr65, Plzp, Toso, and Cd5l/AIM. CD5L/AIM is expressed specifically in nonpathogenic Th17 cells. Loss of CD5L converted nonpathogenic Th17 cells into pathogenic cells that induced autoimmunity. CD5L modulated the intracellular lipidome, altered fatty acid composition, and restricted cholesterol biosynthesis. This suggests that lipid

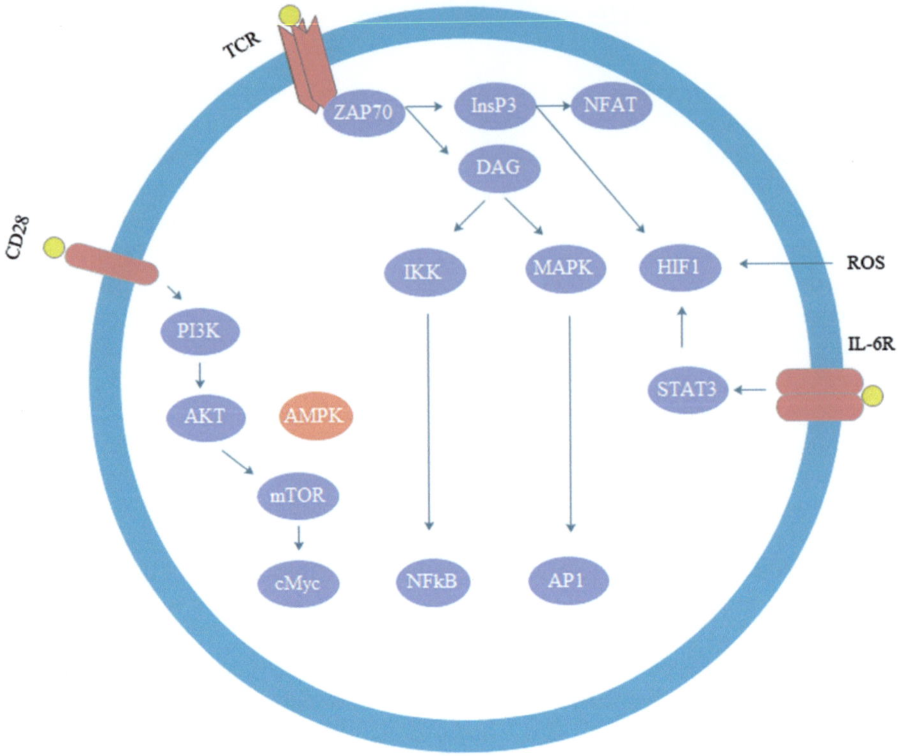

Fig. 6.2 Metabolic signaling programs in T cells

metabolism is important for balance of immune protection and autoimmunity in Th17 cells. Altogether, metabolic-associated signaling (mTOR, TCR) and transcription factors (c-Myc, HIF1) regulate the differentiation of effector T cells [25, 26] (Fig. 6.2).

6.2.3 Regulatory T Cells

Unlike other effector T cells, Treg cells preferentially utilize adenosine 5-monophosphate (AMP)-activated protein kinase (AMPK)-driven FAO to fuel their growth and suppressive function. Rapid proliferation of tumor cells and poor vascularization create a nutrient-deprived tumor environment that impairs TCR signaling, OXPHOS signaling, and overall anti-tumor immune responses. However, Treg cells are less susceptible to such metabolic hurdles because they primarily utilize FAO rather than glycolysis to survive [7].

Studies also show that glycolysis is a key metabolic checkpoint during differentiation of Th17 and Treg cells. Deficiency of hypoxia-inducible factor 1 (HIF1) in T cells controls T cell differentiation through decreased HIF1-dependent glycolytic pathway and altered mTOR signaling [27]. HIF-1 is a transcription factor that consists of an oxygen-sensitive HIF-1a subunit and a constitutively expressed HIF-1b subunit [28]. HIF-1a is hydroxylated at prolines (Pro) 402 and 564 sites by prolyl-hydroxylases PHD2 under normal condition. Prolyl-hydroxylated HIF-1a is bound by the von Hippel-Lindau (VHL) tumor-suppressor protein, which recruits the elongin C-elongin B-cullin 2-E3-ubiquitin-ligase complex, leading to proteasomal degradation of HIF-1a [29]. Hypoxia signal-induced HIF1 expression in Treg cells promotes Treg cell plasticity through binding to FOXP3 and proteasome-mediated degradation of FOXP3. HIF-1a-deficient mice exhibit impaired Th17 cell differentiation and IL-17 production, however increased Treg cell percentage [30]. Besides, the treatment with glycolysis inhibitor 2-deoxyglucose (2-DG) dampens development of Th17 cells but promotes differentiation of Treg cells. It has also been observed that inhibition of mTOR signaling with rapamycin promotes Treg development but does not affect their function [31]. Raptor/mTORC, but not mTORC2, promotes cholesterol and lipid metabolism to regulate expression of CTLA4 and ICOS [32].

On the other hand, loss of tuberous sclerosis 1 (TSC1) in T cells (a negative regulator of mTOR) enhanced Th1 and Th17 and impaired Treg cell differentiation. TSC1-null Foxp3$^+$ cells (transcription factor commonly used to identify Treg cells) exhibited elevated IL-17 production, which was reversed by in-vivo knockdown of the mTOR target S6K1. This suggests that TSC1 is important for the metabolic changes that occur during development of naïve CD4$^+$ T cells into Treg and Th17 cells [33].

AMPK acts as a sensor of nutrient stress and is important for the expansion of Treg cells. AMPK mediates the metabolic reprogramming from glycolysis to FAO. High phosphorylation of AMPK is seen in both nTreg and iTreg cells. Furthermore, AMPK activator metformin (Met) promotes Treg cell expansion in mice through Glut1 and mTOR signaling pathway, suggesting Treg cells utilize different ratio of OXPHOS, glycolysis, and AMPK-driven FAO compared with other effecter T cells [34].

Recently, glycolysis has also been identified to control the expression of the Foxp3-E2 variant through the glycolytic enzyme enolase-1, which is important for Treg function. The Foxp3-E2-related suppressive activity of iTreg cells is altered in human autoimmune diseases, including multiple sclerosis (MS) and type I diabetes (T1D). This is associated with impaired glycolysis and signaling via IL-2-STAT5, revealing that glycolysis is necessary for Treg cells to be able to perform their suppressive functions [35]. Besides, FOXP3 also has been shown to be sufficient to suppress glycolysis and promote oxidative metabolism. FOXP3-overexpressed pro-B cells exhibit decreased ECAR, glucose uptake, and glycolytic rate, while increased OCR. On the other hand, Toll-like receptor (TLR) signals promote glycolysis, proliferation of Treg cells through PI(3)K-AKT-mTOR signal, and enhanced expression level of Glut1. However, TLR signals impair Treg cells' suppressive

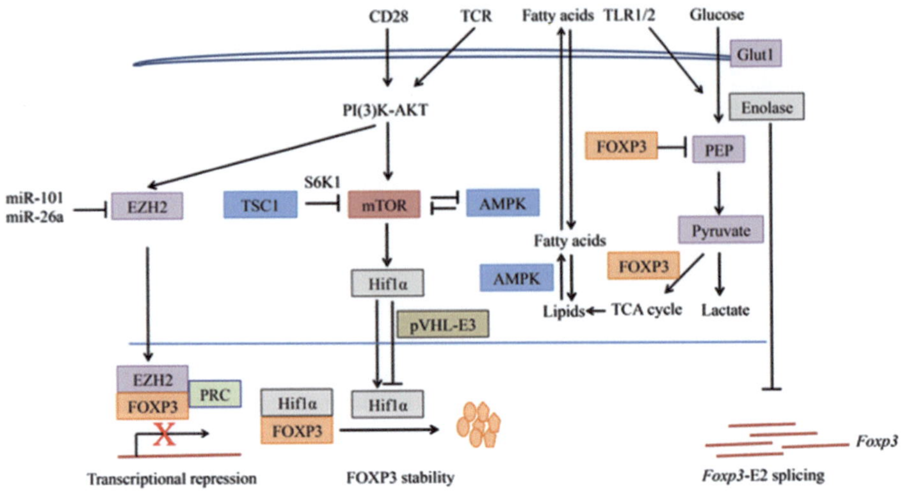

Fig. 6.3 Effects of metabolism in Treg cell generation and function

function. Glut1 transgenic mice show increased Treg cells but reduced FOXP3 expression level and impaired suppressive function, suggesting inflammatory signals and FOXP3 balance Treg cell proliferation and function [36] (Fig. 6.3).

6.2.4 *Type I Regulatory T Cells*

Type I regulatory T cells (Tr1 cells), a kind of FOXP3-negative regulatory CD4$^+$ T cell, control inflammation through IL-10 production. The transcription factor aryl hydrocarbon receptor (AHR) regulates expression of IL-10 and IL-21 in Tr1 cells. Unlike Treg cells but similar to Teff cells, Tr1 cells favor glycolysis [37]. Aerobic glycolysis controls Tr1 cell differentiation through HIF1-α and AHR. At early metabolic reprograming of Tr1 cells, HIF1-α controls glycolysis. However, AHR promotes HIF1-α degradation to regulate Tr1 cell metabolism at later time points. Moreover, extracellular ATP (eATP) and hypoxia suppress Tr1 cell differentiation through the induction of HIF1-α and inactivation of AHR. On the other hand, CD39 promotes Tr1 cell differentiation by limiting eATP level [37].

6.2.5 *CD8$^+$ T Cells*

CD8$^+$ T cells play an important role in the adaptive immune response to cancer. Adoptive transfer of purified CD8$^+$ T cell populations has revealed that less differentiated Tscms and Tcms (Tscm: stem memory T cells; Tcm: central memory T cells) and enhanced anti-tumor and anti-viral responses, due to increased proliferative and survival capacities [38, 39]. Thus, metabolic reprogramming of $CD8^+$ T cells may be an important regulator of the anti-tumor response.

Upon activation and differentiation into effector T cells, naïve CD8 T cells divide. c-Myc has been reported to asymmetrically localize to the proximal daughter cell during division of activated CD8 T cells. This may be a mechanism to control proliferation, metabolism, and differentiation of CD8 T cells. Asymmetric distribution of amino acids and mTORC1 activity are required to sustain asymmetric c-Myc levels in daughter T cells [40].

It has recently been elucidated that hypoxia regulates CD8 cytotoxic T lymphocytes (CTL) through hypoxia-inducible factors (HIFs). Loss of the von Hippel-Lindau tumor-suppressor VHL in CD8 T cells, the negative regulator of HIF, induces HIF-dependent cell death after persistent infection. Increased HIF activity in VHL-deficient mice upregulates glycolysis and effector and co-stimulatory molecule expression, all of which enhance clearance of viruses and tumors [41].

6.2.6 Memory T Cell Metabolism

Generation of long-lived memory T cells after a primary immune response is now beginning to be defined. After stimulation, $CD8^+$-naïve T cells differentiate into effector T cells (Teffs) and distinct memory T cell subsets, including stem cell memory T cells (Tscms), central memory T cells (Tcms), and effector memory T cells (Tems) [42]. Mitochondrial fatty acid oxidation (FAO) metabolism greatly contributes to the differentiation of memory T cells. In 2009, Erika L. Pearce et al. found that tumor necrosis factor (TNF) receptor-associated factor 6 (TRAF6) regulated CD8 memory T cell development via fatty acid metabolism. TRAF6-deficient has fewer CD8 memory cells and impaired fatty acid metabolism; this effect, however, can be rescued by metformin. In vivo, metformin treatment protected TRAF6-deficient mice after infection with *L. monocytogenes* and tumor challenge through increased memory T cell generation [43].

While naïve $CD8^+$ T cells primarily use fatty acid as a primary source of energy, aerobic glycolysis is important for survival of long-term memory $CD8^+$ T cells. For instance, overexpression of the glycolytic enzyme phosphoglycerate mutase-1 severely impaired the ability of $CD8^+$ T cells to form long-term memory. Conversely, activation of $CD8^+$ T cells in the presence of an inhibitor of glycolysis, 2-deoxyglucose (2-DG), enhanced generation of memory cells and anti-tumor response [44]. TRAF6 is specifically required for the formation of antigen-specific $CD8^+$ memory T cells. The absence of TRAF6 in $CD8^+$ memory T cells is associated with decreased AMPK activity, which reduces FAO [43]. In addition to TRAF6, a novel orphan protein named lymphocyte expansion molecule (LEM) has been identified; it appears to promote CD8 T cells' response to chronic lymphocytic choriomeningitis virus (LCMV) infection. LEM regulates OXPHOS through interaction with CR6-interacting factor (CRIF1) to promote the production of mitochondrial reactive oxygen species (mROS). Thus, it appears that LEM controls CTL expansion and memory cell generation by increasing mROS production after activation [(45).]

Effector memory (EM) CD8 T cells experience increased glycolytic flux. To begin, effector memory CD8 T cells exhibit distinct oxidative and glycolytic signatures compared to naïve T cells, such as increased glyceraldehyde-3-phosphate dehydrogenase (GAPDH) activity. Stimulation of T cell antigen receptor and CD28 induces a rapid glycolytic switch and IFNγ production in EM CD8 T cells; these events are dependent on Akt and mTORC2 signaling [46].

6.3 Metabolism of T Cell in Diseases

The prospect of fine-tuning T cell metabolism to control immune responses in disease is attracting much attention to the field of immunometabolism. For instance, glucose-deprived tumor microenvironment suppresses aerobic glycolysis of infiltrating T cells. Glycolytic metabolite phosphoenolpyruvate (PEP) suppresses sarco/ER Ca^{2+}-ATPase (SERCA)-mediated ER calcium reuptake and modulates the Ca^{2+}-NFAT1 signaling in CD4+ T cells, suggesting the critical role of immune responses in disease microenvironment [47]. The PGK1 enzyme transfers phosphate from 1,3-BPG to ADP, generating 3-PG and ATP during glycolysis. Hypoxia induces ERK activation and increases mitochondrial localization of PGK1 in U87 cells through the EGFR signaling pathway. Phosphorylated PGK1 (S203) (which was activated by ERK) binds to PIN1 and the TOM complex to enter into mitochondria. In the mitochondria, the phosphorylated PGK1 (S203) phosphorylates PDHK1 at T338 site and activates PDHK1. Activated PDHK1 inhibits PDH, a key enzyme in the TCA cycle. This shunts the TCA cycle, but increases glycolysis and tumor cell proliferation [48].

It appears that cholesterol metabolism is also important for modulation of antitumor responses by CD8 T cells. A key cholesterol esterification enzyme - ACAT1 - appears to potentiate anti-tumor activity of CD8 T cells. ACAT1 inhibitor avasimibe alone or in combination with anti-PD-1 blocking antibody exhibits a good antitumor effect in the murine melanoma model [49].

Monocytes and macrophages of patients with atherosclerotic coronary artery disease (CAD) produce excessive IL-6 and IL-1β, which are ROS dependent. In patient-derived monocytes and macrophages, increased glucose uptake potentiates ROS production, which - in turn - promotes dimerization of the PKM2 and enables its nuclear translocation. Nuclear PKM2 phosphorylates STAT3 to induce IL-6 and IL-1β production. On the other hand, reducing glycolysis or PKM2 tetramerization dampens the pro-inflammatory phenotype of CAD macrophages [50].

In T cell acute lymphoblastic leukemia (T-ALL), the abnormal T cells exhibit an increased rate of glycolysis compared to activated T cells; this is believed to be important for the survival and proliferation of the T-ALL cells. Notch signaling regulates glycolytic and mitochondrial metabolism by driving metabolic stress and AMPK activation in T-ALL. AMPK signaling negatively controls mTORC1 to inhibit glycolysis in T-ALL and promote mitochondrial oxidative metabolism. Such regulation of complex I maintains T-ALL cell viability and allows disease progression [51].

Immunometabolism is a growing field with many discoveries remaining to be had. Due to the critical role of metabolism in tumor-infiltrating lymphocytes, better understanding of the pathways that regulate the proliferation, function, and differentiation of these cells holds much promise for clinical translation. It is becoming clear that metabolic reprogramming of immune cells can be exploited for therapeutic purposes, but it is also important to consider effects that such treatments may have on tumor cells themselves.

References

1. Coussens LM, Werb Z (2002) Inflammation and cancer. Nature 420:860–867
2. Disis ML (2010) Immune regulation of cancer. J Clin Oncol 28:4531–4538
3. Hanahan D, Weinberg RA (2011) Hallmarks of cancer: the next generation. Cell 144:646–674
4. Ward PS, Thompson CB (2012) Metabolic reprogramming: a cancer hallmark even warburg did not anticipate. Cancer Cell 21:297–308
5. Cairns RA, Harris IS, Mak TW (2011) Regulation of cancer cell metabolism. Nat Rev Cancer 11:85–95
6. Semenza GL (2003) Targeting HIF-1 for cancer therapy. Nat Rev Cancer 3:721–732
7. Koppenol WH, Bounds PL, Dang CV (2011) Otto Warburg's contributions to current concepts of cancer metabolism. Nat Rev Cancer 11:325–337
8. Haas R, Smith J, Rocher-Ros V, Nadkarni S, Montero-Melendez T, D'Acquisto F, Bland EJ, Bombardieri M, Pitzalis C, Perretti M, Marelli-Berg FM, Mauro C (2015) Lactate regulates metabolic and pro-inflammatory circuits in control of T cell migration and effector functions. PLoS Biol 13:e1002202
9. Viale A, Pettazzoni P, Lyssiotis CA, Ying H, Sanchez N, Marchesini M, Carugo A, Green T, Seth S, Giuliani V, Kost-Alimova M, Muller F, Colla S, Nezi L, Genovese G, Deem AK, Kapoor A, Yao W, Brunetto E, Kang Y, Yuan M, Asara JM, Wang YA, Heffernan TP, Kimmelman AC, Wang H, Fleming JB, Cantley LC, DePinho RA, Draetta GF (2014) Oncogene ablation-resistant pancreatic cancer cells depend on mitochondrial function. Nature 514:628–632
10. Qin H, Ruan ZH (2014) The role of monoacylglycerol lipase (MAGL) in the cancer progress. Cell Biochem Biophys 70:33–36
11. Zhu J, Yamane H, Paul WE (2010) Differentiation of effector CD4 T cell populations (*). Annu Rev Immunol 28:445–489
12. Jacobs SR, Michalek RD, Rathmell JC (2010) IL-7 is essential for homeostatic control of T cell metabolism in vivo. J Immunol 184:3461–3469
13. Rathmell JC, Farkash EA, Gao W, Thompson CB (2001) IL-7 enhances the survival and maintains the size of naive T cells. J Immunol 167:6869–6876
14. Wofford JA, Wieman HL, Jacobs SR, Zhao Y, Rathmell JC (2008) IL-7 promotes Glut1 trafficking and glucose uptake via STAT5-mediated activation of Akt to support T-cell survival. Blood 111:2101–2111
15. Feng X, Wang H, Takata H, Day TJ, Willen J, Hu H (2011) Transcription factor Foxp1 exerts essential cell-intrinsic regulation of the quiescence of naive T cells. Nat Immunol 12:544–550
16. McCarthy SA, Mufson RA, Pearce EJ, Rathmell JC, Howcroft TK (2013) Metabolic reprogramming of the immune response in the tumor microenvironment. Cancer Biol Ther 14:315–318
17. Wieman HL, Wofford JA, Rathmell JC (2007) Cytokine stimulation promotes glucose uptake via phosphatidylinositol-3 kinase/Akt regulation of Glut1 activity and trafficking. Mol Biol Cell 18:1437–1446

18. Macintyre AN, Gerriets VA, Nichols AG, Michalek RD, Rudolph MC, Deoliveira D, Anderson SM, Abel ED, Chen BJ, Hale LP, Rathmell JC (2014) The glucose transporter Glut1 is selectively essential for CD4 T cell activation and effector function. Cell Metab 20:61–72
19. Wang R, Dillon CP, Shi LZ, Milasta S, Carter R, Finkelstein D, McCormick LL, Fitzgerald P, Chi H, Munger J, Green DR (2011) The transcription factor Myc controls metabolic reprogramming upon T lymphocyte activation. Immunity 35:871–882
20. Frauwirth KA, Riley JL, Harris MH, Parry RV, Rathmell JC, Plas DR, Elstrom RL, June CH, Thompson CB (2002) The CD28 signaling pathway regulates glucose metabolism. Immunity 16:769–777
21. Nakaya M, Xiao Y, Zhou X, Chang JH, Chang M, Cheng X, Blonska M, Lin X, Sun SC (2014) Inflammatory T cell responses rely on amino acid transporter ASCT2 facilitation of glutamine uptake and mTORC1 kinase activation. Immunity 40:692–705
22. Yang K, Shrestha S, Zeng H, Karmaus PW, Neale G, Vogel P, Guertin DA, Lamb RF, Chi H (2013) T cell exit from quiescence and differentiation into Th2 cells depend on Raptor-mTORC1-mediated metabolic reprogramming. Immunity 39:1043–1056
23. Delgoffe GM, Pollizzi KN, Waickman AT, Heikamp E, Meyers DJ, Horton MR, Xiao B, Worley PF, Powell JD (2011) The kinase mTOR regulates the differentiation of helper T cells through the selective activation of signaling by mTORC1 and mTORC2. Nat Immunol 12:295–303
24. Pan F, Barbi J, Pardoll DM (2012) Hypoxia-inducible factor 1: a link between metabolism and T cell differentiation and a potential therapeutic target. Oncoimmunology 1:510–515
25. Wang C, Yosef N, Gaublomme J, Wu C, Lee Y, Clish CB, Kaminski J, Xiao S, Meyer Zu Horste G, Pawlak M, Kishi Y, Joller N, Karwacz K, Zhu C, Ordovas-Montanes M, Madi A, Wortman I, Miyazaki T, Sobel RA, Park H, Regev A, Kuchroo VK (2015) CD5L/AIM regulates lipid biosynthesis and restrains Th17 cell pathogenicity. Cell 163:1413–1427
26. Gaublomme JT, Yosef N, Lee Y, Gertner RS, Yang LV, Wu C, Pandolfi PP, Mak T, Satija R, Shalek AK, Kuchroo VK, Park H, Regev A (2015) Single-cell genomics unveils critical regulators of Th17 cell pathogenicity. Cell 163:1400–1412
27. Shi LZ, Wang R, Huang G, Vogel P, Neale G, Green DR, Chi H (2011) HIF1alpha-dependent glycolytic pathway orchestrates a metabolic checkpoint for the differentiation of TH17 and Treg cells. J Exp Med 208:1367–1376
28. Wang GL, Jiang BH, Rue EA, Semenza GL (1995) Hypoxia-inducible factor 1 is a basic-helix-loop-helix-PAS heterodimer regulated by cellular O2 tension. Proc Natl Acad Sci U S A 92:5510–5514
29. Kaelin WG Jr, Ratcliffe PJ (2008) Oxygen sensing by metazoans: the central role of the HIF hydroxylase pathway. Mol Cell 30:393–402
30. Dang EV, Barbi J, Yang HY, Jinasena D, Yu H, Zheng Y, Bordman Z, Fu J, Kim Y, Yen HR, Luo W, Zeller K, Shimoda L, Topalian SL, Semenza GL, Dang CV, Pardoll DM, Pan F (2011) Control of T(H)17/T(reg) balance by hypoxia-inducible factor 1. Cell 146:772–784
31. Kopf H, de la Rosa GM, Howard OM, Chen X (2007) Rapamycin inhibits differentiation of Th17 cells and promotes generation of FoxP3+ T regulatory cells. Int Immunopharmacol 7:1819–1824
32. Zeng H, Yang K, Cloer C, Neale G, Vogel P, Chi H (2013) mTORC1 couples immune signals and metabolic programming to establish T(reg)-cell function. Nature 499:485–490
33. Park Y, Jin HS, Lopez J, Elly C, Kim G, Murai M, Kronenberg M, Liu YC (2013) TSC1 regulates the balance between effector and regulatory T cells. J Clin Invest 123:5165–5178
34. Michalek RD, Gerriets VA, Jacobs SR, Macintyre AN, MacIver NJ, Mason EF, Sullivan SA, Nichols AG, Rathmell JC (2011) Cutting edge: distinct glycolytic and lipid oxidative metabolic programs are essential for effector and regulatory CD4+ T cell subsets. J Immunol 186:3299–3303
35. De Rosa V, Galgani M, Porcellini A, Colamatteo A, Santopaolo M, Zuchegna C, Romano A, De Simone S, Procaccini C, La Rocca C, Carrieri PB, Maniscalco GT, Salvetti M, Buscarinu MC, Franzese A, Mozzillo E, La Cava A, Matarese G (2015) Glycolysis controls the induction of human regulatory T cells by modulating the expression of FOXP3 exon 2 splicing variants. Nat Immunol 16:1174–1184

36. Gerriets VA, Kishton RJ, Johnson MO, Cohen S, Siska PJ, Nichols AG, Warmoes MO, de Cubas AA, MacIver NJ, Locasale JW, Turka LA, Wells AD, Rathmell JC (2016) Foxp3 and Toll-like receptor signaling balance Treg cell anabolic metabolism for suppression. Nat Immunol 17:1459–1466
37. Mascanfroni ID, Takenaka MC, Yeste A, Patel B, Wu Y, Kenison JE, Siddiqui S, Basso AS, Otterbein LE, Pardoll DM, Pan F, Priel A, Clish CB, Robson SC, Quintana FJ (2015) Metabolic control of type 1 regulatory T cell differentiation by AHR and HIF1-alpha. Nat Med 21:638–646
38. Gattinoni L, Zhong XS, Palmer DC, Ji Y, Hinrichs CS, Yu Z, Wrzesinski C, Boni A, Cassard L, Garvin LM, Paulos CM, Muranski P, Restifo NP (2009) Wnt signaling arrests effector T cell differentiation and generates CD8+ memory stem cells. Nat Med 15:808–813
39. Gattinoni L, Lugli E, Ji Y, Pos Z, Paulos CM, Quigley MF, Almeida JR, Gostick E, Yu Z, Carpenito C, Wang E, Douek DC, Price DA, June CH, Marincola FM, Roederer M, Restifo NP (2011) A human memory T cell subset with stem cell-like properties. Nat Med 17:1290–1297
40. Verbist KC, Guy CS, Milasta S, Liedmann S, Kaminski MM, Wang R, Green DR (2016) Metabolic maintenance of cell asymmetry following division in activated T lymphocytes. Nature 532:389–393
41. Doedens AL, Phan AT, Stradner MH, Fujimoto JK, Nguyen JV, Yang E, Johnson RS, Goldrath AW (2013) Hypoxia-inducible factors enhance the effector responses of CD8(+) T cells to persistent antigen. Nat Immunol 14:1173–1182
42. Gattinoni L, Klebanoff CA, Restifo NP (2012) Paths to stemness: building the ultimate antitumour T cell. Nat Rev Cancer 12:671–684
43. Pearce EL, Walsh MC, Cejas PJ, Harms GM, Shen H, Wang LS, Jones RG, Choi Y (2009) Enhancing CD8 T-cell memory by modulating fatty acid metabolism. Nature 460:103–107
44. Sukumar M, Liu J, Ji Y, Subramanian M, Crompton JG, Yu Z, Roychoudhuri R, Palmer DC, Muranski P, Karoly ED, Mohney RP, Klebanoff CA, Lal A, Finkel T, Restifo NP, Gattinoni L (2013) Inhibiting glycolytic metabolism enhances CD8+ T cell memory and antitumor function. J Clin Invest 123:4479–4488
45. Okoye I, Wang L, Pallmer K, Richter K, Ichimura T, Haas R, Crouse J, Choi O, Heathcote D, Lovo E, Mauro C, Abdi R, Oxenius A, Rutschmann S, Ashton-Rickardt PG (2015) T cell metabolism. The protein LEM promotes CD8(+) T cell immunity through effects on mitochondrial respiration. Science 348:995–1001
46. Gubser PM, Bantug GR, Razik L, Fischer M, Dimeloe S, Hoenger G, Durovic B, Jauch A, Hess C (2013) Rapid effector function of memory CD8+ T cells requires an immediate-early glycolytic switch. Nat Immunol 14:1064–1072
47. Ho PC, Bihuniak JD, Macintyre AN, Staron M, Liu X, Amezquita R, Tsui YC, Cui G, Micevic G, Perales JC, Kleinstein SH, Abel ED, Insogna KL, Feske S, Locasale JW, Bosenberg MW, Rathmell JC, Kaech SM (2015) Phosphoenolpyruvate is a metabolic checkpoint of anti-tumor T cell responses. Cell 162:1217–1228
48. Li X, Jiang Y, Meisenhelder J, Yang W, Hawke DH, Zheng Y, Xia Y, Aldape K, He J, Hunter T, Wang L, Lu Z (2016) Mitochondria-translocated PGK1 functions as a protein kinase to coordinate glycolysis and the TCA cycle in tumorigenesis. Mol Cell 61:705–719
49. Yang W, Bai Y, Xiong Y, Zhang J, Chen S, Zheng X, Meng X, Li L, Wang J, Xu C, Yan C, Wang L, Chang CC, Chang TY, Zhang T, Zhou P, Song BL, Liu W, Sun SC, Liu X, Li BL, Xu C (2016) Potentiating the antitumour response of CD8(+) T cells by modulating cholesterol metabolism. Nature 531:651–655
50. Shirai T, Nazarewicz RR, Wallis BB, Yanes RE, Watanabe R, Hilhorst M, Tian L, Harrison DG, Giacomini JC, Assimes TL, Goronzy JJ, Weyand CM (2016) The glycolytic enzyme PKM2 bridges metabolic and inflammatory dysfunction in coronary artery disease. J Exp Med 213:337–354
51. Kishton RJ, Barnes CE, Nichols AG, Cohen S, Gerriets VA, Siska PJ, Macintyre AN, Goraksha-Hicks P, de Cubas AA, Liu T, Warmoes MO, Abel ED, Yeoh AE, Gershon TR, Rathmell WK, Richards KL, Locasale JW, Rathmell JC (2016) AMPK is essential to balance glycolysis and mitochondrial metabolism to control T-ALL cell stress and survival. Cell Metab 23:649–662

Chapter 7
Innate and Adaptive Immune Cell Metabolism in Tumor Microenvironment

Duojiao Wu

Abstract During an immune response, leukocytes undergo major changes in growth and function that are tightly coupled to dynamic shifts in metabolic processes. Immunometabolism is an emerging field that investigates the interplay between immunological and metabolic processes. The immune system has a key role to play in controlling cancer initiation and progression. Increasing evidence indicates the immunosuppressive nature of the local environment in tumor. In tumor microenvironment, immune cells collectively adapt in a dynamic manner to the metabolic needs of cancer cells, thus prompting tumorigenesis and resistance to treatments. Here, we summarize the latest insights into the metabolic reprogramming of immune cells in tumor microenvironment and their potential roles in tumor progression and metastasis. Manipulating metabolic remodeling and immune responses may provide an exciting new option for cancer immunotherapy.

Keywords Immunometabolism • Tumor microenvironment • Dendritic cell • T cell • Macrophage

7.1 Introduction

As cancers edit and escape this initial immune detection, they also generate an immunosuppressive microenvironment which restricts T-cell infiltration, activation, and effector function both through direct repression (via cytokines, nutrients restriction, etc.) and the recruitment of immunosuppressive populations [1].

Tumors display altered metabolism relative to benign tissues [2, 3]. The common feature of this altered metabolism is the increased glucose uptake and fermentation of glucose to lactate. This phenomenon is observed even in the presence of completely functioning mitochondria and, together, is known as the "Warburg effect"

D. Wu (✉)
Zhongshan Hospital Institute of Clinical Science, Fudan University, Shanghai Institute of Clinical Bioinformatics, Shanghai, China
e-mail: wu.duojiao@zs-hospital.sh.cn

B. Li, F. Pan (eds.), *Immune Metabolism in Health and Tumor*, Advances in Experimental Medicine and Biology 1011, DOI 10.1007/978-94-024-1170-6_7

[4]. These changes support abnormal survival and growth of malignant cells by providing energy, macromolecular precursors, and reducing equivalents [5]. Recently, increasing evidence indicates that cancer cells subvert the metabolic characteristics of the tumor microenvironment to shape immune responses within tumors [6]. Papers by Chang et al. [7] and Ho et al. [8] show that Warburg metabolism enables tumor cells to restrict glucose availability to T cells, suppressing antitumor immunity.

Immunometabolism is an emerging field of investigation dissecting the contribution of key metabolic pathways to immune cell development, fate, and behavior [9, 10]. The present article aims at understanding immune cells' metabolism in tumor microenvironment and its potential role in tumor progression and metastasis. Characterizing the reciprocal metabolic interplay between immune and cancer cells will provide a better understanding of treatment efficacy and resistance and also help develop a new strategy for cancer immunotherapy.

7.2 Innate Cells in the Tumor Microenvironment

7.2.1 Macrophage

Tumor-associated macrophages (TAM) are among the most abundant inflammatory cells in tumors. Within the same tumor, the coexistence of two distinct TAM subpopulations has been shown, both derived from tumor-infiltrating inflammatory monocytes: M2-like MHC-II^{low} TAM that reside in the hypoxic regions of the tumor and perform angiogenic, immunosuppressive, and protumoral activities and M1-like MHC-II^{high} TAM that are present in the normoxic tumor regions and possess proinflammatory and antitumoral characteristics [11, 12]. Recent studies on intracellular metabolism in macrophages provide new insights on the functioning of these critical controllers of innate and adaptive immunity [13, 14]. The metabolic reprogramming of M1 and M2 has been summarized in Fig. 7.1.

7.2.1.1 Classically Activated Macrophages (M1)

M1 macrophage metabolism is characterized by aerobic glycolysis, fatty acid synthesis, and a truncated tricarboxylic acid (TCA) cycle. Although the importance of glycolysis in inflammatory activation of macrophages was first noted almost a century ago, its biochemical and bioenergetic importance had not been appreciated until recently. The activated M1macrophages have high rates of glucose and glutamine uptake and lactic acid production. However, neither glycolysis nor glutaminolysis are necessary for ATP generation. The decreased respiration and a broken TCA cycle of M1 macrophages lead to accumulation of succinate, citrate, and nitric oxide (NO), which support the production of key M1 cellular products or act as signals to alter immune function [15, 16]. The endogenous metabolites can adopt

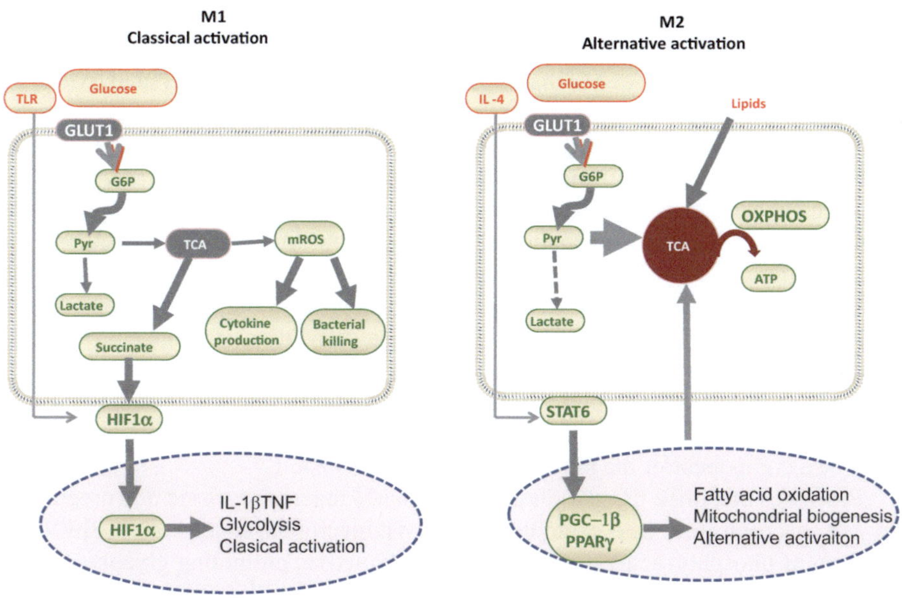

Fig. 7.1 Metabolic reprogramming of macrophages in tumor microenvironment. Macrophages in a tumor microenvironment have been characterized as M1- and M2-polarized subtypes. In M1 macrophages, there is decreased respiration and a broken Krebs cycle, leading to accumulation of succinate and citrate, which act as signals to alter immune function. Downstream of TLR signaling, mitochondrial ROS (mROS) can also support the function of M1. In M2 macrophages, the Krebs cycle and oxidative phosphorylation are intact and FAO is utilized. Type 2 cytokine, such as IL-4, activates STAT6 transcription factor. Then STAT6 promotes the metabolic transition to oxidative metabolism by inducing genes of FAO and mitochondrial biogenesis. In addition, STAT6 transcriptionally induces PGC-1β, PPARγ, which synergize with STAT6 to enhance oxidative metabolism.*GLUT1* glucose transporter 1, *G6P* glucose 6-phosphate, *HIF-1α* hypoxia-induced factor 1α, *IL-1β* interleukin-1β, *IL-4* interleukin-4, *OXPHOS* oxidative phosphorylation, *Pyr* pyruvate, *PGC-1β* PPARγ coactivator-1β, *PPAR* peroxisome proliferation-activated receptor, *ROS* reactive oxygen species, *STAT* signal transducer and activator of transcription, *TCA* tricarboxylic acid cycle, *TLR* Toll-like receptor, *TNF* tumor necrosis factor

regulatory roles that govern specific aspects of inflammatory response, as recently shown for succinate, which regulates the proinflammatory interleukin-1β(IL-1β) hypoxia-inducible factor-1α(HIF-1α) axis. Furthermore, Lampropoulou V et al. [17] report that itaconate modulates macrophage metabolism and effector functions by inhibiting SDH-mediated oxidation of succinate. Collectively, these studies demonstrate that endogenous metabolite-derived signals might be important integrators and effectors of host immunity in tumor microenvironment. Both the increased mitochondrial oxidation of succinate via succinate dehydrogenase (SDH) and an elevation of mitochondrial membrane potential combine drive mitochondrial reactive oxygen species (ROS) production [13]. Blocking ROS production by uncoupling mitochondria inhibits this inflammatory phenotype. Therefore, remodeling of the TCA cycle is a metabolic adaptation accompanying inflammatory macrophage activation.

7.2.1.2 Alternatively Activated Macrophages (M2)

The metabolic signature of M2 macrophages is characterized by fatty acid oxidation (FAO) and an oxidative TCA cycle. Tumor-promoting M2 macrophages are induced under the influence of interleukin-4(IL-4), IL-13, IL-10, and macrophage colony-stimulating factor (M-CSF) and lack the cytotoxicity of M1 macrophages. M2 macrophages appear to contribute to immune suppression through the production of IL-10 and transforming growth factor-β (TGF-β) [18]. Myeloid-derived suppressor cells are immature myeloid-lineage cells that also can be immunosuppressive in the tumor microenvironment. Major mechanisms of suppression include the expression and functional activity of arginase [19] and the nitrosylation of surface proteins on infiltrating T cells, including the T-cell receptor (TCR) [20]. Gene expression profiling studies of human melanoma have revealed arginase transcripts expressed in a subset of non-T-cell-infiltrated tumors [21], so myeloid-derived suppressor cells may be a component of the phenotype of T-cell exclusion.

Processes that drive the glycolytic switch in M1 macrophages are downregulated in M2 macrophages. One example is that M1 macrophages express u-PFK2, an isoform of phosphofructokinase-2 that is highly active, promoting glycolysis [15]. In contrast, M2 macrophages express a different isoform, PFK FB1, which is much less active [22, 23]. Another example is that the activation of the key metabolic regulator pyruvate kinase M2 (PKM2) attenuated an LPS-induced proinflammatory M1 macrophage phenotype while promoting traits typical of an M2 macrophage [24]. The activation of PKM2 using two well-characterized small molecules, DASA-58 and TEPP-46, inhibited LPS-induced HIF-1α and IL-1β, the important genes involved in glycolysis. The mechanistic target of rapamycin complex 2(mTORC2) operated in parallel with the IL-4Rα-signal transducer and activator of transcription 6(Stat6) pathway to facilitate increased glycolysis during M2 activation via the induction of the transcription factor IRF4. IRF4 expression required both mTORC2 and Stat6 pathways, providing an underlying mechanism to explain how glucose utilization is increased to support M2 activation [25].

Since macrophage metabolism is certainly connected to its functionality, metabolic reprogramming of M2-like TAM might be a new strategy to repolarize TAM toward an antitumoral phenotype and thus dampen tumor growth and metastasis. Although a system-level understanding of TAM metabolism is currently absent and rather limited, there is an emerging evidence that unraveling the TAM phenotype might lead to the identification of alternative, novel targets for TAM-directed intervention.

7.2.2 Dendritic Cell

Dendritic cells (DCs) display different phenotypes and activity in tumors and exhibit distinct pro-tumorigenic and anti-tumorigenic functions. DCs are supposed to play a key role in inducing and maintaining the antitumor immunity. However, their antigen-presenting function is jeopardized in tumor microenvironment. Under circumstance, DCs are polarized into tolerogenic phenotype with immunosuppressive function, which limit antitumor activity of effector T cells (TE). The metabolic switch of DCs from the anti-tumorigenic phenotype to the tolerogenic phenotype has been summarized in Fig. 7.2.

7.2.2.1 Immunogenic DCs

Catabolic metabolism centered around mitochondrial oxidative phosphorylation (OXPHOS) is associated with cellular longevity and quiescence of DCs, whereas cellular activation and proliferation are accompanied by a switch to glycolytic metabolism to support anabolic pathways needed for biosynthesis [26]. The phosphatidylinositol 3′-kinase/Akt(PI3K/Akt) pathway, which could be antagonized by the adenosine monophosphate (AMP)-activated protein kinase (AMPK), is required for DC maturation [27]. The switch from OXPHOS to glycolysis is a direct consequence of TLR-induced inducible nitric oxide synthase (iNOS) expression that through the production of NO poisons the mitochondrial respiratory chain in an autocrine fashion [28]. Bart Everts et al. [29] found that DC glycolytic flux increased within minutes of exposure to TLR agonists and that this served an essential role in supporting the de novo synthesis of fatty acids for the expansion of the endoplasmic reticulum and Golgi required for the production and secretion of proteins that are integral to DC activation. Signaling via the kinases TBK1, IKKε, and Akt was essential for the TLR-induced increase in glycolysis by promoting the association of the glycolytic enzyme HK-II with mitochondria [29] (Fig. 7.2a).

7.2.2.2 Tolerogenic DCs

In contrast to immunogenic DCs, tolerogenic DCs favor OXPHOS and FAO. Tolerogenic DCs, as opposed to immunogenic DCs, are maturation resistant and express increased levels of immunoregulatory factors, important for controlling regulatory T-cell (Treg) responses [30]. Proteomic analysis of human DCs treated with immunosuppressive drugs dexamethasone and vitamin D3 revealed increased expression of genes associated with mitochondrial metabolism and OXPHOS along with enhanced tolerogenic phenotypes [31–33](Fig. 7.2b). The direct inhibition of glycolysis in TLR-activated DCs favors the induction of forkhead box P3(Foxp3)-expressing helper T cells [29]. Consistent with a role for peroxisome proliferator-activated receptor gamma (PPARγ) coactivator 1α(PGC1α) in regulating DC activation, resveratrol, a drug favoring catabolic metabolism through activation of the histone deacetylase (HDAC) sirtuin 1, inhibits the expression of hypoxia-inducible factor 1α(HIF1α) and enhances PGC1α expression in DCs, which render these cells more tolerogenic phenotypes [34–37].

Plasmacytoid dendritic cells (pDCs) are a distinct lineage of DCs that are more specialized for cytokine production, particularly type I interferon (IFN) production.

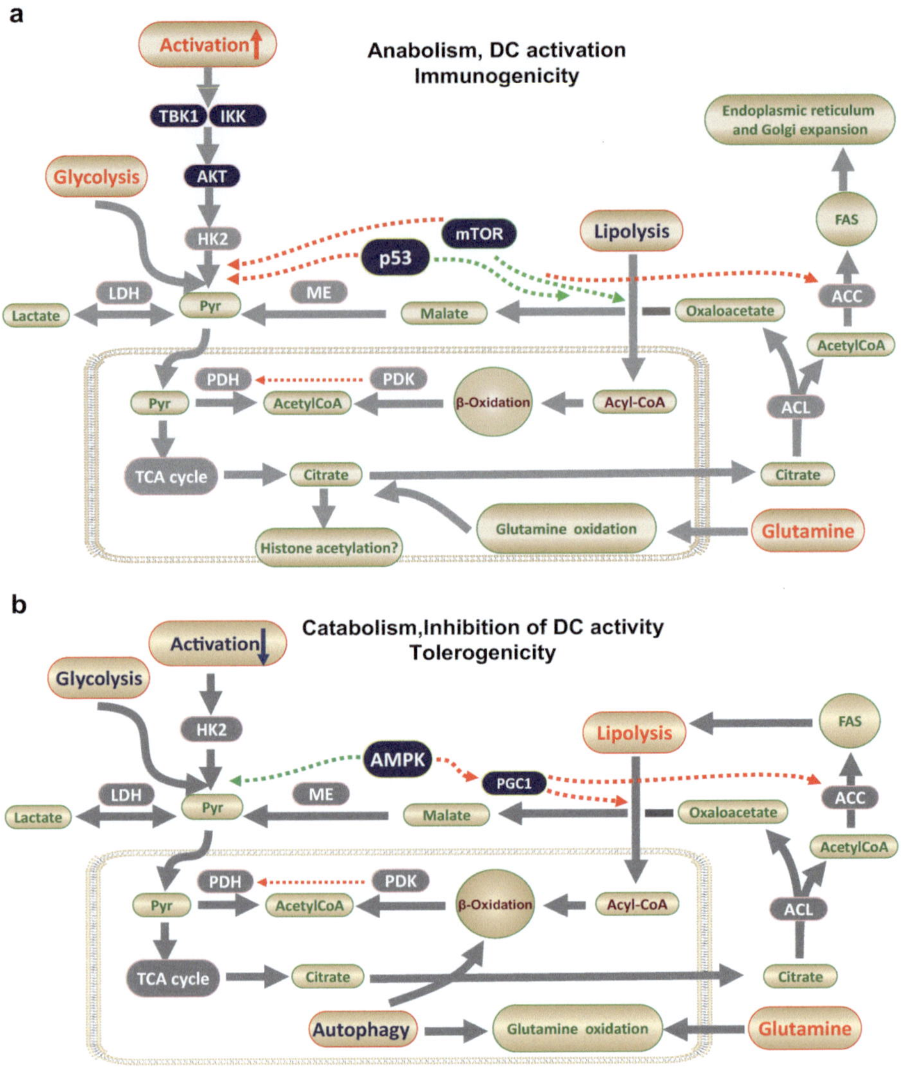

Fig. 7.2 Metabolic reprogramming of dendritic cells in tumor microenvironment. Metabolic switch from anabolic metabolism to catabolic metabolisms consistent with DC function transferring from immunogenicity to tolerogenicity. (**a**) Demonstrates activated DCs that need metabolic reprogramming to provide the bioenergetic and biosynthetic support. The PI3K/Akt/mTOR pathway, which could be antagonized by the AMPK, is required for DCs maturation. Signaling via the kinases TBK1, IKKε, and Akt was essential for the TLR-induced increase in glycolysis by promoting the association of the glycolytic enzyme HK-II with mitochondria. Acetyl-CoA converted from citrate is a major acetyl donor for the acetylation pathway. (**b**) Illustrates that catabolic metabolism centered around OXPHOS is associated with cellular longevity and quiescence of DCs. Enhances catabolic metabolism through activation of the histone deacetylase (HDAC) sirtuin 1,inhibits the expression of hypoxia-inducible factor 1α(HIF1α), and upregulates PGC1α expression in DCs. In contrast to immunogenic DC, autophagy and the oxidation of fatty acids and glutamine render these cells more tolerogenic phenotypes. *AMPK* AMP-activated protein kinase, *ATP* adenosine triphosphate, *ACC* acetyl-CoA carboxylase, *ACL* ATP citrate lyase, *AKT* protein kinase B, *FAS* fatty acid synthesis, *GLUT1* glucose transporter 1, *G6P* glucose 6-phosphate, *HK* hexokinase, *LDH* lactate dehydrogenase, *mTOR* mammalian target of rapamycin, *PDH* pyruvate dehydrogenase, *PDK* pyruvate dehydrogenase kinase, *Pyr* pyruvate, *PGC-1* PPARγ coactivator-1, *TBK1* TANK-binging kinase 1, *IKKε* inhibitor of nuclear factor-κB kinase subunit-ε, *TCA* tricarboxylic acid cycle

In vivo pDCs depletion delayed tumor growth, showing that tumor-associated pDC provides an immune-subversive environment, most likely through Treg activation, thus favoring tumor progression [38]. Wu et al. [39] recently report that through an autocrine type 1 IFN receptor-dependent pathway, induced changes in pDCs of cellular metabolism are characterized by increased FAO and OXPHOS. Direct inhibition of FAO and of pathways prevented full pDC activation [39].

How the effects of fatty acid synthesis differ so markedly in DCs isolated from tumors compared with those TLR-activated DCs is an important unanswered question. However, scientists speculate that accumulated fatty acids are supporting FAO and therefore tolerogenicity in the cancer setting [40]. These adaptations allow tolerogenic DCs a metabolic advantage in low-glucose, lactate-rich environments; they resist suppression of DCs function and proliferation in competition of nutrients.

7.3 Adaptive Cells in Tumor Microenvironment

Among the tumor-infiltrated lymphocytes (TILs),T cells are the most abundant cells [41]. $CD8^+$ T cells have a central role in antitumor immunity, but their activity is suppressed in the tumor microenvironment. Reactivating the cytotoxicity of $CD8^+$ T cells is of great clinical interest in cancer immunotherapy.

7.3.1 Effector T Cells(TE)

Most tumor cells express antigens that can mediate recognition by host $CD8^+$ T cells [41]. Resting $CD8^+$ T cells undergo dynamic shifts in cell metabolism and switch from an oxidative metabolism to aerobic glycolysis upon activation. This transition is essential to support growth and differentiation into cytotoxic T cells capable of dividing every 6–8 h and of producing inflammatory cytokines and the cytolytic granules perforin and granzyme-B [42]. Chang et al. [7] report that glucose consumption by tumors metabolically restricts T cells, leading to their dampened mTOR activity, glycolytic capacity, and IFN-γ production. Targeting aerobic glycolysis in the tumor has increased the supply of glucose to TILs, thus boosting their effector function. The checkpoint blockade antibodies against CTLA-4, PD-1, and PD-L1 restore glucose in tumor microenvironment, permitting T-cell glycolysis and IFN-γ production [7](Fig. 7.3). Recently, Ho et al. [8] uncovered the glycolytic metabolite phosphoenolpyruvate (PEP) as a new metabolic checkpoints for T-cell activity. PEP plays an important role in sustaining T-cell receptor-mediated $Ca2^+$-nuclear factor of activated T cell (NFAT) signaling and effector functions by repressing sarco/ER $Ca2^+$-ATPase (SERCA) activity. Together these data suggest that both nutrients and substrates concentration in a local microenvironment can have a marked impact on immune cell function.

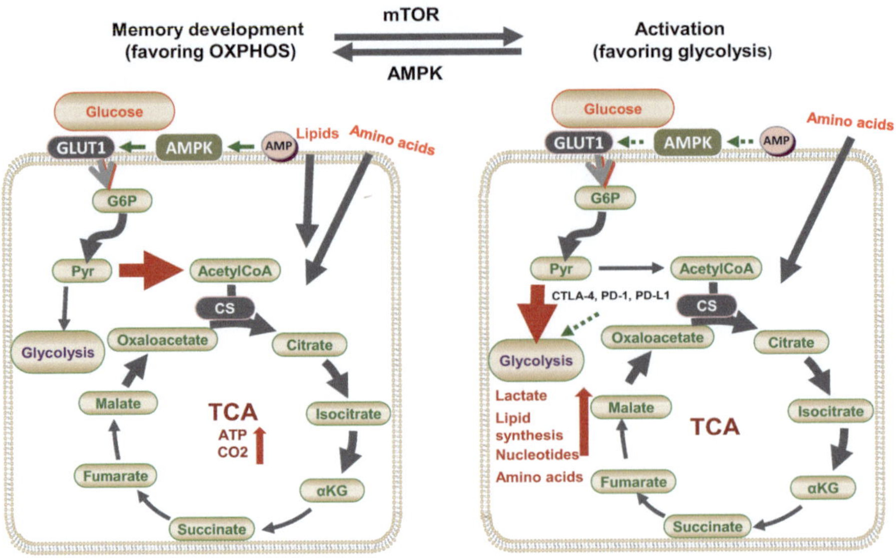

Fig. 7.3 Metabolic reprogramming of T lymphocytes in tumor microenvironment. Upon activation, lipid oxidation is downregulated in resting T cells, and glycolysis increases along with glutamine oxidation, in order to produce biosynthetic precursors required for rapid cell growth and proliferation. The checkpoint blockade antibodies against CTLA-4, PD-1, and PD-L1, which are used clinically, restore glucose in tumor microenvironment, permitting T-cell glycolysis and IFN-γ production. At the end of an immune response, the cells that survive to become memory T cells revert back to lipid oxidation with increased capacity for efficient energy generation. The mTOR pathway plays a key role in these metabolic shifts. Mechanistically, rapamycin treatment reduced mTORC1 activity and led to increased AMPK phosphorylation that correlated with an increased ability of $CD8^+$ T cells to perform lipid oxidation. *AMPK* AMP-activated protein kinase, *ATP* adenosine triphosphate, *CTLA-4* cytotoxic T-lymphocyte-associated protein 4, *GLUT1* glucose transporter 1, *G6P* glucose 6-phosphate, *mTOR* mammalian target of rapamycin, *PD-1* programmed cell death protein 1, *PDL-1* programmed death-ligand 1, *Pyr* pyruvate, *TCA* tricarboxylic acid cycle

The lipid requirements of membrane synthesis are essential for TE function as well. Inhibiting cholesterol esterification in $CD8^+$T cells led to potentiated effector function and enhanced proliferation [42, 43]. Inhibiting cholesterol esterification increases the plasma membrane cholesterol level of $CD8^+$T cells, which causes enhanced T-cell receptor clustering and signaling as well as more efficient formation of the immunological synapse [43, 44]. Sterol regulatory element-binding protein (SREBP) is another critical target for meeting the heightened lipid requirements of membrane synthesis during blastogenesis [45].

TE function has been regulated by transcriptional or posttranscriptional mechanisms. Aerobic glycolysis (the Warburg effect) is a metabolic hallmark of activated T cells and has been implicated in augmenting effector T-cell responses, including expression of the proinflammatory cytokine IFN-γ, via 3′ untranslated region (3′UTR)-mediated mechanisms [46]. Another study shows that lactate dehydrogenase A (LDHA) is induced in activated T cells to support aerobic glycolysis but promotes IFN-γ expression independently of its 3′UTR. Instead, LDHA maintains high concentrations of acetyl-coenzyme A to enhance histone acetylation and transcription of Ifng. Ablation of LDHA in T cells protects mice from immunopathology triggered by excessive IFN-γ expression or deficiency of regulatory T cells [47].

T-cell exhaustion is characterized by the stepwise and progressive loss of T-cell functions [1]. The exhausted T cells acquired a distinct metabolic profile from that of TE and TM cells. Two related studies by Bengsch et al. [48] and Scharping et al. [49] indicate dysfunctional mitochondria are identified as a key correlate of $CD8^+$ T-cell exhaustion. PD-1 pathway blockade resulted in transcriptional rewiring and reengagement of effector circuitry in the exhausted $CD8^+$ T cells' epigenetic landscape [50]. Manipulating glycolytic and mitochondrial metabolism might enhance checkpoint blockade outcomes [51].

7.3.2 Memory T Cell (TM)

A successful immune response relies not only on the ability of T cells to extensively proliferate and attain effector function but also to form long-lived memory T cells that can respond again to future antigen encounter. There is intense interest in understanding how long-lived cellular immunity is generated. $CD8^+$ cytotoxic T cells further differentiate into long-lived quiescent memory $CD8^+$ T cells (TM). TM cells require efficient energy generation to support basic cellular functions and prevent cell death [52, 53]. Pearce et al. reported that $CD8^+$TM cells possessed substantial mitochondrial spare respiratory capacity (SRC). SRC is the extra capacity available in cells to produce energy in response to increased stress or work and as such is associated with cellular survival. The mTOR pathway plays a key role in this metabolic remodeling. Tumor necrosis factor (TNF) receptor-associated factor 6 (TRAF6), an adaptor protein in the TNF-receptor and interleukin-1R/Toll-like receptor superfamily, is known to be required for the transition from effector to TM cells [52, 53]. TM cells express high levels of the mitochondrial lipid transporter-carnitine palmitoyltransferase 1a (CPT1a), and retroviral CPT1a expression enhanced $CD8^+$ TM generation in an adoptive transfer model.

Mitochondrial dynamics underlies TM fate. TE cells have punctate mitochondria, while TM cells possess fused networks. By altering cristae morphology, fusion protein Opa1 and fusion in TM cells are required for electron transport chain (ETC) complex associations favoring OXPHOS and FAO [54]. Enforcing fusion in TE cells imposes TM cell characteristics and enhances antitumor function. Thus, targeting mitochondrial dynamics may offer a novel adjuvant strategy to therapeutically influence cancer therapy.

Elevating L-arginine levels induced global metabolic changes including a shift from glycolysis to oxidative phosphorylation in activated T cells and promoted the generation of central memory-like cells endowed with higher survival capacity and, in a mouse model, antitumor activity [55]. During the process, three transcriptional regulators (BAZ1B, PSIP1, and TSN) sensed L-arginine levels and promoted T-cell survival [55].

7.3.3 Regulatory T Cells (Treg)

Treg cells respond to immune and inflammatory signals to mediate immunosuppression. The evidences show maintaining lineage and survival integrity of Treg cells require simultaneous response to both environmental signals and metabolic homeostasis [56, 57]. Treg cell-specific deletion of Atg7 or Atg5, two essential genes in autophagy, leads to loss of Treg cells and greater tumor resistance. Mechanistically, autophagy deficiency upregulates metabolic regulators mTORC1 and c-Myc and glycolysis, which contribute to defective Treg function. Furthermore, mTORC1 acts as a fundamental "rheostat" in Treg cells to link immunological signals from TCR and IL-2 to lipogenic pathways and functional fitness and highlight a central role of metabolic programming of Treg cell suppressive activity in immune homeostasis and tolerance [57].

Tissues with low-glucose and high-lactate concentrations frequently require immune responses to be more pro-tolerant, avoiding unwanted reactions against self-antigens or commensal bacteria. Angelin et al. report that the Treg transcription factor Foxp3 reprograms T-cell metabolism by suppressing Myc and glycolysis, enhancing OXPHS, and increasing nicotinamide adenine dinucleotide oxidation [58]. These adaptations allow Tregs a metabolic advantage in low-glucose, lactate-rich environments. This metabolic phenotype may explain how Tregs promote peripheral immune tolerance during tissue injury but also how cancer cells evade immune destruction in the tumor microenvironment.

Why do TE, TM, and Treg cells adopt markedly different types of metabolism upon activation? One possibility is that the shift to glycolysis in TE may be optimally suited to the rapid, short-term bursts of activation that are required at sites of infection or inflammation, whereas FAO in TM may be better able to energetically support cell survival, as T cells continue to fight virus infection or tumors over a long time period. Understanding T cells metabolism may therefore lead to novel approaches for selective immune modulation in cancer and autoimmune diseases.

7.4 Remaining Questions and Bottleneck

Intense ongoing investigation of immune cell metabolism is yielding an exponentially growing amount of information. Armed with new information and a comprehensive understanding of how metabolism dictates immune cell fate, researchers may discover novel therapeutic strategies for treatment of tumor. For example, as we discussed in Sect. 7.3, the long-term survival and antitumor immunity of adoptively transferred $CD8^{+}$T cells is dependent on their metabolic fitness. Scientists utilized a lipophilic cationic dye tetramethylrhodamine methyl ester (TMRM) to identify and isolate metabolically robust T cells based on their mitochondrial membrane potential ($\Delta\Psi m$) [59]. Transfer of these low-$\Delta\Psi m$ T cells was associated with superior long-term in vivo persistence and an enhanced capacity to eradicate established tumors compared with high-$\Delta\Psi m$ cells. The use of $\Delta\Psi m$-based sorting to enrich for cells with superior metabolic features was observed in $CD8^+$ and $CD4^+$ T-cell subsets.

At present, there are still obvious obstacles in reaching clinically effective conclusions about immunometabolism in tumor: the vast metabolic heterogeneity within a tumor and the lack of metabolic resemblance between tissue culture and in vivo conditions [60]. The in vitro studies performed cannot faithfully recapitulate the conditions of tumor local environment [61]. Therefore, the use of a reductionist approach to investigations, by focusing on specific nutrients, can produce misleading information which hampers reaching further conclusions. As a starting point, developing standardized techniques allowing assessment of the metabolism of human physiology is critical.

References

1. Jiang Y, Li Y, Zhu B (2015) T-cell exhaustion in the tumor microenvironment. Cell Death Dis 6:e1792
2. Hanahan D, Weinberg RA (2011) Hallmarks of cancer: the next generation. Cell 144(5):646–674
3. Son J et al (2013) Glutamine supports pancreatic cancer growth through a KRAS-regulated metabolic pathway. Nature 496(7443):101–105
4. Liberti MV, Locasale JW (2016) The Warburg effect: how does it benefit cancer cells? Trends Biochem Sci 41(3):211–218
5. DeBerardinis RJ et al (2008) The biology of cancer: metabolic reprogramming fuels cell growth and proliferation. Cell Metab 7(1):11–20
6. Sukumar M, Roychoudhuri R, Restifo NP (2015) Nutrient competition: a new axis of tumor immunosuppression. Cell 162(6):1206–1208
7. Chang CH et al (2015) Metabolic competition in the tumor microenvironment is a driver of cancer progression. Cell 162(6):1229–1241
8. Ho PC et al (2015) Phosphoenolpyruvate is a metabolic checkpoint of anti-tumor T cell responses. Cell 162(6):1217–1228
9. Mathis D, Shoelson SE (2011) Immunometabolism: an emerging frontier. Nat Rev Immunol 11(2):81
10. Murray PJ, Rathmell J, Pearce E (2015) SnapShot: immunometabolism. Cell Metab 22(1):190–190. e1
11. Mantovani A et al (2002) Macrophage polarization: tumor-associated macrophages as a paradigm for polarized M2 mononuclear phagocytes. Trends Immunol 23(11):549–555
12. Mantovani A et al (2004) Infiltration of tumours by macrophages and dendritic cells: tumour-associated macrophages as a paradigm for polarized M2 mononuclear phagocytes. Novartis Found Symp 256:137–145. discussion 146-8, 259-69
13. Mills EL, et al (2016) Succinate dehydrogenase supports metabolic repurposing of mitochondria to drive inflammatory macrophages. Cell
14. Mills EL, O'Neill LA (2016) Reprogramming mitochondrial metabolism in macrophages as an anti-inflammatory signal. Eur J Immunol 46(1):13–21
15. Rodriguez-Prados JC et al (2010) Substrate fate in activated macrophages: a comparison between innate, classic, and alternative activation. J Immunol 185(1):605–614
16. Tannahill GM et al (2013) Succinate is an inflammatory signal that induces IL-1beta through HIF-1alpha. Nature 496(7444):238–242

17. Lampropoulou V et al (2016) Itaconate links inhibition of succinate dehydrogenase with macrophage metabolic remodeling and regulation of inflammation. Cell Metab 24(1):158–166
18. Quatromoni JG, Eruslanov E (2012) Tumor-associated macrophages: function, phenotype, and link to prognosis in human lung cancer. Am J Transl Res 4(4):376–389
19. Rodriguez PC et al (2003) L-arginine consumption by macrophages modulates the expression of CD3 zeta chain in T lymphocytes. J Immunol 171(3):1232–1239
20. Nagaraj S et al (2007) Altered recognition of antigen is a mechanism of CD8+ T cell tolerance in cancer. Nat Med 13(7):828–835
21. Harlin H et al (2009) Chemokine expression in melanoma metastases associated with CD8+ T-cell recruitment. Cancer Res 69(7):3077–3085
22. Kelly B, O'Neill LA (2015) Metabolic reprogramming in macrophages and dendritic cells in innate immunity. Cell Res 25(7):771–784
23. Kelly B et al (2015) Metformin inhibits the production of reactive oxygen species from NADH:ubiquinone oxidoreductase to limit induction of interleukin-1beta (IL-1beta) and boosts interleukin-10 (IL-10) in lipopolysaccharide (LPS)-activated macrophages. J Biol Chem 290(33):20348–20359
24. Palsson-McDermott EM et al (2015) Pyruvate kinase M2 regulates Hif-1alpha activity and IL-1beta induction and is a critical determinant of the warburg effect in LPS-activated macrophages. Cell Metab 21(1):65–80
25. Huang SC et al (2016) Metabolic reprogramming mediated by the mTORC2-IRF4 signaling axis is essential for macrophage alternative activation. Immunity 45(4):817–830
26. Pearce EJ, Everts B (2015) Dendritic cell metabolism. Nat Rev Immunol 15(1):18–29
27. Krawczyk CM et al (2010) Toll-like receptor-induced changes in glycolytic metabolism regulate dendritic cell activation. Blood 115(23):4742–4749
28. Everts B et al (2012) Commitment to glycolysis sustains survival of NO-producing inflammatory dendritic cells. Blood 120(7):1422–1431
29. Everts B et al (2014) TLR-driven early glycolytic reprogramming via the kinases TBK1-IKKvarepsilon supports the anabolic demands of dendritic cell activation. Nat Immunol 15(4):323–332
30. Motz GT, Coukos G (2013) Deciphering and reversing tumor immune suppression. Immunity 39(1):61–73
31. Dong H, Bullock TN (2014) Metabolic influences that regulate dendritic cell function in tumors. Front Immunol 5:24
32. Sim WJ, Ahl PJ, Connolly JE (2016) Metabolism is central to Tolerogenic dendritic cell function. Mediat Inflamm 2016:2636701
33. Ferreira GB et al (2015) Vitamin D3 induces tolerance in human dendritic cells by activation of intracellular metabolic pathways. Cell Rep 10:711–725
34. Houtkooper RH, Pirinen E, Auwerx J (2012) Sirtuins as regulators of metabolism and healthspan. Nat Rev Mol Cell Biol 13(4):225–238
35. Lagouge M et al (2006) Resveratrol improves mitochondrial function and protects against metabolic disease by activating SIRT1 and PGC-1alpha. Cell 127(6):1109–1122
36. Zheng J et al (2012) Resveratrol improves insulin resistance of catch-up growth by increasing mitochondrial complexes and antioxidant function in skeletal muscle. Metabolism 61(7):954–965
37. Svajger U, Obermajer N, Jeras M (2010) Dendritic cells treated with resveratrol during differentiation from monocytes gain substantial tolerogenic properties upon activation. Immunology 129(4):525–535
38. Le Mercier I et al (2013) Tumor promotion by intratumoral plasmacytoid dendritic cells is reversed by TLR7 ligand treatment. Cancer Res 73(15):4629–4640
39. Wu D et al (2016) Type 1 interferons induce changes in core metabolism that are critical for immune function. Immunity 44(6):1325–1336
40. O'Neill LA, Pearce EJ (2016) Immunometabolism governs dendritic cell and macrophage function. J Exp Med 213(1):15–23

41. Gajewski TF, Schreiber H, Fu YX (2013) Innate and adaptive immune cells in the tumor microenvironment. Nat Immunol 14(10):1014–1022
42. MacIver NJ, Michalek RD, Rathmell JC (2013) Metabolic regulation of T lymphocytes. Annu Rev Immunol 31:259–283
43. Yang W et al (2016) Potentiating the antitumour response of CD8(+) T cells by modulating cholesterol metabolism. Nature 531(7596):651–655
44. Dustin ML (2016) Cancer immunotherapy: killers on sterols. Nature 531(7596):583–584
45. Kidani Y et al (2013) Sterol regulatory element-binding proteins are essential for the metabolic programming of effector T cells and adaptive immunity. Nat Immunol 14(5):489–499
46. Chang CH et al (2013) Posttranscriptional control of T cell effector function by aerobic glycolysis. Cell 153(6):1239–1251
47. Peng M et al (2016) Aerobic glycolysis promotes T helper 1 cell differentiation through an epigenetic mechanism. Science 354(6311):481–484
48. Bengsch B et al (2016) Bioenergetic insufficiencies due to metabolic alterations regulated by the inhibitory receptor PD-1 are an early driver of CD8(+) T cell exhaustion. Immunity 45(2):358–373
49. Scharping NE et al (2016) The tumor microenvironment represses T cell mitochondrial biogenesis to drive Intratumoral T cell metabolic insufficiency and dysfunction. Immunity 45(2):374–388
50. Pauken KE et al (2016) Epigenetic stability of exhausted T cells limits durability of reinvigoration by PD-1 blockade. Science 354(6316):1160–1165
51. Balmer ML, Hess C (2016) Feeling worn out? PGC1alpha to the rescue for dysfunctional mitochondria in T cell exhaustion. Immunity 45(2):233–235
52. Pearce EL et al (2009) Enhancing CD8 T-cell memory by modulating fatty acid metabolism. Nature 460(7251):103–107
53. Prlic M, Bevan MJ (2009) Immunology: a metabolic switch to memory. Nature 460(7251):41–42
54. Buck MD et al (2016) Mitochondrial dynamics controls T cell fate through metabolic programming. Cell 166(1):63–76
55. Geiger R et al (2016) L-arginine modulates T cell metabolism and enhances survival and antitumor activity. Cell 167(3):829–842. e13
56. Wei J et al (2016) Autophagy enforces functional integrity of regulatory T cells by coupling environmental cues and metabolic homeostasis. Nat Immunol 17(3):277–285
57. Zeng H et al (2013) mTORC1 couples immune signals and metabolic programming to establish T(reg)-cell function. Nature 499(7459):485–490
58. Angelin A, et al (2017) Foxp3 reprograms T cell metabolism to function in low-glucose, high-lactate environments. Cell Metab
59. Sukumar M et al (2016) Mitochondrial membrane potential identifies cells with enhanced Stemness for cellular therapy. Cell Metab 23(1):63–76
60. Schug ZT, Vande Voorde J, Gottlieb E (2016) The nurture of tumors can drive their metabolic phenotype. Cell Metab 23(3):391–392
61. Davidson SM et al (2016) Environment impacts the metabolic dependencies of Ras-driven non-small cell lung cancer. Cell Metab 23(3):517–528

MIX
Papier aus verantwortungsvollen Quellen
Paper from responsible sources
FSC® C105338

FSC
www.fsc.org

If you have any concerns about our products,
you can contact us on
ProductSafety@springernature.com

In case Publisher is established outside the EU,
the EU authorized representative is:
Springer Nature Customer Service Center GmbH
Europaplatz 3, 69115 Heidelberg, Germany

Printed by Libri Plureos GmbH
in Hamburg, Germany